Numerical Methods

A Programming-based Approach

Numerical Methods

A Programming-based Approach

Arun Kumar Jalan
MSc PhD
Professor in Mathematics
Dean (Students' Affairs)
MCKV Institute of Engineering,
Kolkata

Utpal Sarkar
MSc
Assistant Professor in Mathematics
MCKV Institute of Engineering,
Kolkata

Universities Press

UNIVERSITIES PRESS (INDIA) PRIVATE LIMITED
Registered Office
3-6-747/1/A & 3-6-754/1, Himayatnagar, Hyderabad 500 029, Telangana, India
e-mail: info@universitiespress.com: www.universitiespress.com

Distributed by
Orient Blackswan Private Limited

Registered Office
3-6-752 Himayatnagar, Hyderabad 500 029, Telangana, India

Other Offices
Bengaluru/ Bhopal/ Chennai/ Ernakulam/ Guwahati/ Hyderabad
Jaipur/ Kolkata/ Lucknow/ Mumbai/ New Delhi/ Noida/ Patna

ISBN: 978-81-7371-958-5

Typeset in Times New Roman PS 10/12 *by*
MacroTex Solutions, Madipakkam, Chennai 600 091.

Printed in India by
Yash Printographics, Noida

Published by
Universities Press (India) Private Limited
3-6-747/1/A & 3-6-754/1, Himayatnagar, Hyderabad 500 029, Telangana, India

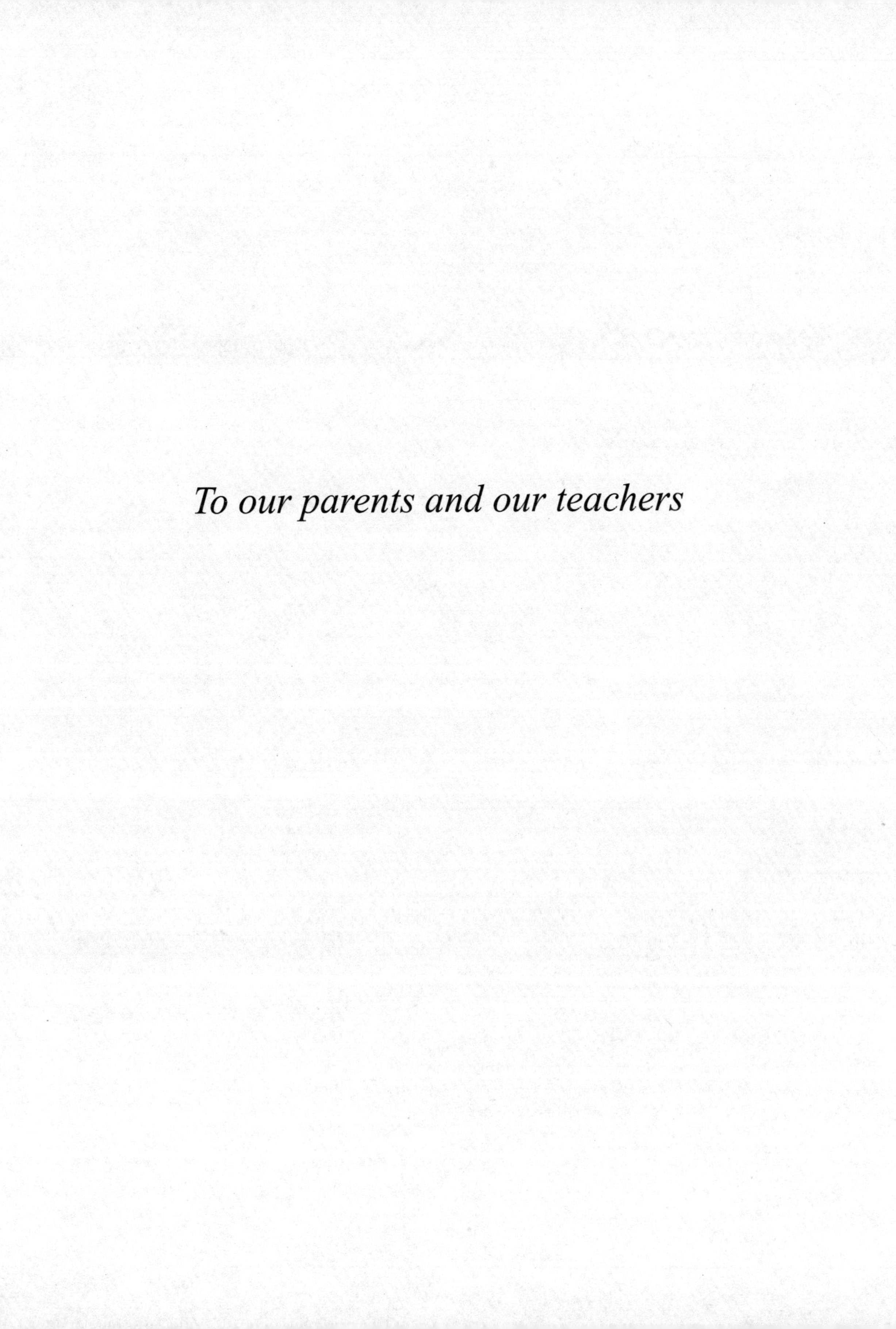

To our parents and our teachers

Preface and Acknowledgements

It is our proud privilege to offer this textbook *Numerical Methods: A Programming-based Approach* to the students of engineering in Indian universities. The book comprehensively covers all the topics prescribed in the syllabi of engineering courses all over India.

Every effort has been taken to present the concepts about the different tools of numerical methods for implementation in engineering problems in such a way that students can easily grasp these ideas. Numerous examples and diagrams have been provided to make the subject interesting. The theoretical as well as applied aspects of each topic have been presented in a systematic manner, with the emphasis on lucidity and simplicity of language. Many solved questions have been included at the end of each chapter to reinforce interest and to clarify concepts to the student.

We shall deem our labour amply rewarded if this book proves helpful to teachers and students.

We express our sincere thanks to Prof KS Chaudhuri, former Head, Department of Mathematics, Jadavpur University and Prof Bibhas Chandra Giri, Department of Mathematics, Jadavpur University. The legacy of scholarly and scientific enquiry and temperament that we inherited from them will remain with us as invaluable treasure to be invested in all that we do.

We wish to offer our sincere thanks to Prof Debapriya De, Head, Basic Science; Prof Premananda Jana, Head, Computer Science; Mr Supriyo Das, Computer Science; and all the faculty members of the Basic Science Department for their advice at various stages of this book.

Words are inadequate to express our gratitude to Shri SS Kejriwal (Chairman, MCKVIE) and Shri Kishan Kumar Kejriwal (MT, MCKVIE) for their wholehearted encouragement in developing the manuscript.

We also extend our thanks to Universities Press for their tireless efforts in publishing the book.

Our immeasurable gratitude and thanks are due to our parents, Shri Lachmi Narayan Jalan and Smt Gayatri Debi Jalan; late Biswanath Sarkar and Smt Sachi Rani Sarkar, who have been the driving force behind all our achievements. We would also like to thank our elder brothers, Shri Mahabir Prasad Jalan and Shri Asoke Kumar Jalan; Shri Nemai Sarkar, Shri Prabir Sarkar, Shri Prabal Sarkar, Shri Deepak Sarkar and Shri Biplab Sarkar. Their affection and blessings will forever be cherished.

Lastly, we express our sincere thanks to Mrs Seema Jalan and Mrs Krishna Sarkar for their wholehearted co-operation during the preparation of the book. We wish to thank Naman, Chaman, Tanush and Upasana for their pleasant disruption.

Any suggestions and comments for the improvement of the book are welcome and will be incorporated, if necessary.

Arun Kumar Jalan
Utpal Sarkar

Contents

1

Approximation in Numerical Computations

1.1 INTRODUCTION

Numerical methods or techniques are effective tools that can be used to solve certain problems of science and technology where analytical methods fail to yield a solution. Hence they have become essential to scientists, engineers and those who deal with mathematical problems. However, it should be remembered that numerical methods provide solutions to mathematical problems only to a desired level of accuracy, and so, in most cases, are not error-free. In this context, we shall study the different types of errors, their estimation and propagation while dealing with different numerical computations.

1.2 EXACT NUMBER, APPROXIMATE NUMBER AND SIGNIFICANT DIGITS/FIGURES

Exact Number A number is said to be exact if it represents any real number as a whole. For example, 5, 15 and 3/5.

Approximate Number A number that is used as an approximation to an exact number and which differs slightly from that exact number is called an approximate number. For example,

- $\sqrt{2}$ is an exact number while 1.4142 is an approximate number
- π is an exact number while 3.1416 is an approximate number

Significant Digits/Figures In the decimal system, all the non-zero digits – 1, 2, 3, 4, 5, 6, 7, 8 and 9 – are significant figures. The digit 0 is a significant figure for the following two cases:

- If 0 is placed between any **two non-zero** (i.e., significant) digits
- If 0 is at the terminal of an approximate number having an embedded decimal point

However, 0 is not significant when it is used **to fix the position of the decimal point** or to fill up the position of unknown digits, or if it precedes all non-zero digits. Moreover, all terminal 0s of an integral number (i.e., without a decimal point) are insignificant. For example,

- 84.725 and 8.4725 have **5** significant figures (8, 4, 7, 2, 5).
- 0.84725, 0.084725 and 0.00084725 all have **5** significant figures (8, 4, 7, 2, 5). Here, 0 (or 0s) has (have) been used to fix the position of the decimal point. So, 0 is not significant in these cases.
- 3075 has **4** significant digits (3, 0, 7, 5). In this case, 0 lies between **two non-zero** digits, 3 and 7. So, it is significant.

- 7800 has **2** significant digits (7, 8). The two 0s are terminal 0s of an integral number. So, these 0s are not significant.
- 0.004350×10^2 has **4** significant figures (4, 3, 5, 0). The first three 0s are not significant as they have been used to fix the position of the decimal point. However, the last **0** (10^2 is ignored) has been used in an approximate number having an **embedded decimal point**. So, this 0 is significant.

Procedure to calculate significant figures: First express the given real number as $0 \cdot a_1 a_2 \ldots a_m \times 10^n$, where $a_1 \neq 0$, n is an integer and m is a positive integer. Then the real number has m significant figures. For example,

- $600000 = 0.6 \times 10^6$ has 1 significant digit as $m = 1$.
- $0.03409 = 0.3409 \times 10^{-1}$ has 4 significant digits as $m = 4$.
- $1.00234 = 0.100234 \times 10^1$ has 6 significant digits as $m = 6$.
- $0.00303 = 0.303 \times 10^{-2}$ has 3 significant digits as $m = 3$.

Note: When a number is given with an embedded decimal point, it should not be expressed as shown above. In this case, we only observe the value of m and conclude. For example,

- 2.35×10^2 has 3 significant figures (2, 3, 5)
- 3.50 has 3 significant figures (3, 5, 0)
- 2.350×10^2 has 4 significant figures (2, 3, 5, 0)
- 2.3500×10^2 has 5 significant figures (2, 3, 5, 0, 0)

Note: In any number, the leftmost significant digit is the most significant digit and the rightmost significant digit is the least significant digit. For example,

- In the number 0.0432, the digits 4, 3 and 2 are significant. 4 is the most significant digit and 2 is the least significant digit.
- The number 7.004 has 4 significant digits, 7, 0, 0 and 4. Here the most significant digit is 7 and the least significant digit is 4.
- Every digit in 55.5444 is significant. The leftmost digit 5 is the most significant digit and the rightmost digit 4 is the least significant digit.
- 0.0003200 has four significant digits, 3, 2, 0 and 0. The digit 3 is the most significant digit and the rightmost 0 is the least significant digit.

1.3 ROUNDING NUMBERS

The process of dropping a certain number of digits of a real number from the right is called rounding a number. The rules are as follows:

Suppose we need to round a number to the nth digit from the left. Then,

- If the $(n + 1)$th place digit is less than 5, leave the nth place digit unaltered
- If the $(n + 1)$th place digit is greater than 5, add 1 to the nth place digit
- If the $(n + 1)$th place digit is exactly equal to 5, then
  - if there is any non-zero digit after the $(n + 1)$th digit, the nth digit is increased by 1
  - if all digits after the $(n + 1)$th digit are zero, then
    - the nth digit is increased by 1 if it is odd
    - the nth digit remains unaltered if it is even

Reference: For the second part of the third rule, students are advised to follow the well-known Banker's rule: *If the digit to be rounded is 5, make the preceding digit an even number.*

For example,

- $12.6\underline{7}5 \approx 12.68$ (to 2 decimal places)
- $12.6\underline{8}5 \approx 12.68$ (to 2 decimal places)
- $0.003\underline{2}\underline{56} \approx 0.0033$ (to 4 decimal places)
- $3.450\underline{6}\underline{531} \approx 3.4507$ (to 4 decimal places)

1.4 ERRORS, TRUNCATION AND ROUNDING ERRORS

1.4.1 Errors

The difference between the exact or true value of a quantity and its approximate value obtained by measurement is known as the **error**. Let x_A be the approximate value of the true value x_T. Then, we define

$$\begin{aligned}\text{Error } (E) &= \text{True value} - \text{Approximate value}\\ &= x_T - x_A\end{aligned}$$

$$\text{Absolute error } (E_a) = |E| = |x_T - x_A|$$

$$\begin{aligned}\text{Relative error } (E_r) &= \frac{|E|}{|\textit{True value}|} = \frac{E_a}{|\textit{True value}|}\\ &= \left|\frac{x_T - x_A}{x_T}\right|, \text{provided } x_T \neq 0.\end{aligned}$$

$$\text{Percentage error } (E_p) = E_r \times 100\%$$

For illustration, suppose 22/7 and 3.1429 are, respectively, the exact value and approximate value of π. Then

$$x_T = \frac{22}{7}$$

$$\text{and } x_A = 3.1429$$

Therefore, the error committed in this measurement is given by

$$\begin{aligned}E = x_T - x_A = \frac{22}{7} - 3.1429\\ = -0.0003\end{aligned}$$

Hence

$$\begin{aligned}\text{Absolute error } (E_a) &= |E| = |x_T - x_A|\\ &= 0.0003\end{aligned}$$

$$\begin{aligned}\text{Relative error } (E_r) &= \left|\frac{x_T - x_A}{x_T}\right|\\ &= \frac{0.0003}{22/7} \approx 0.0000955\end{aligned}$$

$$\begin{aligned}\text{Percentage error } (E_p) &= E_r \times 100\% \approx 0.0000955 \times 100\%\\ &= 0.00955\%\end{aligned}$$

1.4.2 Truncation Error

When the sum of an infinite series or an infinitesimal process (e.g., differentiation) is replaced by a finite one, **truncation errors** occur. For example,

- If we consider the sum up to n terms for the infinite series

$$e^x = 1 + \frac{x}{1!} + \frac{x^2}{2!} + \cdots,$$

we commit a truncation error.

- If $\frac{df}{dx}$ is approximated by $\frac{f(x+h) - f(x)}{h}$, a truncation error occurs.

1.4.3 Rounding Error

When an exact number, terminating or non-terminating, is replaced by a finite number of digits (i.e., rounded to a certain number of decimal places), we then commit a **rounding error**. For example,

- If we replace 2.53216 (true or exact value) with 2.53 (approximate value), the rounding error will be = 2.53216 – 2.53 = 0.00216.
- If 12.675 is to be rounded to two decimal places, then the approximate number will be 12.68. So, the rounding error will be = 12.675 – 12.68 = –0.005.

1.4.4 Chopping Error

When extra digits are simply cut off or discarded from an exact number, we commit an error known as the **chopping error**. For example,

- If the last three digits of the number 4.78549 are simply discarded or cut off, then it will be 4.78, and so the chopping error will be 4.78549 – 4.78 = 0.00549.
- If the number 12.456 is represented by 12.45, then the chopping error will be 12.456 – 12.45 = 0.006.

Rules to Remember

- If x_A is the approximate value of x_T, correctly **rounded to n decimal places**, then the absolute error,

$$E_a = |x_T - x_A| \le \frac{1}{2} \times 10^{-n}.$$

- If x_A is the approximate value of x_T, correctly **rounded to n significant digits**, then the relative error,

$$E_r = \left| \frac{x_T - x_A}{x_T} \right| < 10^{-n}.$$

- If a number is correct to n **significant figures** and the first significant digit of the number is s_1, then the relative error,

$$E_r < \frac{1}{s_1} \times 10^{-n+1}$$

1.5 FIXED POINT AND FLOATING POINT ARITHMETIC

Real numbers can be represented in a computer using two basic forms or systems:

(i) Fixed point form
(ii) Floating point form

Fixed Point Form In this type of representation, a fixed position for the decimal point is assumed by the user and all real numbers are stored after appropriate shifting, if necessary. For illustration, let us assume that each location, called a **word**, in computer memory can store 6 digits and has the provision to store one or more signs. Now, if a programmer assumes the position for the decimal point as

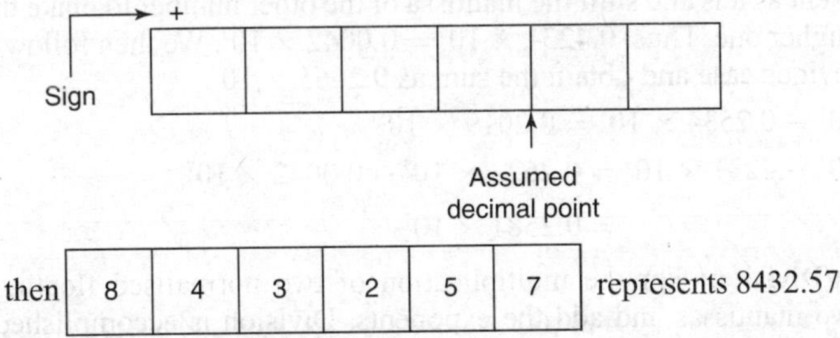

In this type of representation, the maximum and minimum possible numbers that can be stored are 9999.99 and 0000.01, respectively, in magnitude. However, this is quite insufficient in practice.

Floating Point Form To avoid the difficulties that arise in fixed point representation, this convention has been adopted. In this system, a real number is expressed as a combination of a **mantissa** and an **exponent**. For example, in the expression of the real number 13.5×10^4, 13.5 is the mantissa, 10 the base and 4 the exponent. However, this representation is not unique as the above number has the following equivalent expressions:

$$0.0135 \times 10^7 = 0.135 \times 10^6 = 1350 \times 10^2$$

$$= 13500 \times 10^1, \text{ etc.}$$

Clearly the only difference is in the position of the decimal point and the exponent of 10. Observe that the position of the decimal point is **floating** between the digits.

In computer memory, a convenient representation of real numbers is the **normalised floating point** mode. In this method, the mantissa is less than 1 and greater than or equal to 0.1 and the exponent is the power of 10 which multiplies the mantissa. For example, the number 45.78×10^7 is represented in this mode as 0.4578E9, where E9 stands for 10^9. The mantissa is 0.4578 and the exponent is 9. The number is stored in memory as

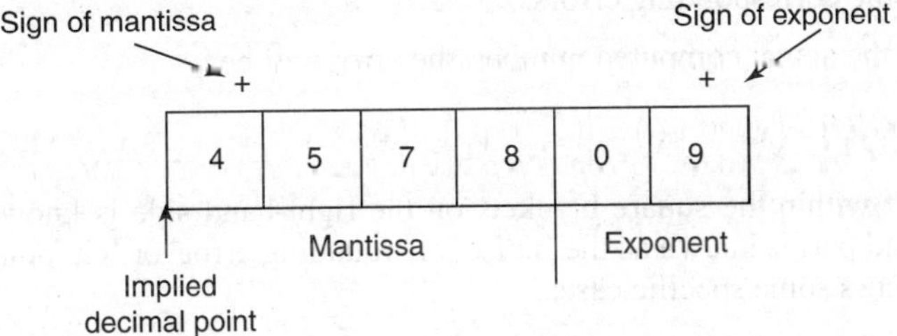

By adopting this mode, the range of numbers that may be stored is 9999×10^{99} to 0.1000×10^{-99} in magnitude, which is much larger.

Floating Point Arithmetic In this section, we shall discuss some arithmetic operations such as addition, subtraction, multiplication and division. The procedures are given below.

- **Addition/Subtraction:** For the addition of two normalised floating numbers, the exponents of both numbers must be made equal and then the mantissas must be added. The subtraction operation is nothing but the addition of a negative number to another number. For example,

- $0.4553 \times 10^5 + 0.2534 \times 10^5$. In this case, both exponents are the same. So, we just add the two mantissas and then multiply the result by the exponent, as $0.4553 + 0.2534 = 0.7087$, and so the sum is equal to 0.7087×10^5.
- $0.4231 \times 10^5 + 0.2623 \times 10^7$. Here, the exponents are not equal. We keep the operand with the larger exponent as it is and shift the mantissa of the other number to make the exponent equal to that of the higher one. Thus, $0.4231 \times 10^5 = 0.0042 \times 10^7$. We then follow the same procedure as in the previous case and obtain the sum as 0.2665×10^7.
- $0.4553 \times 10^5 - 0.2534 \times 10^5 = 0.2019 \times 10^5$
- $0.2623 \times 10^7 - .4231 \times 10^5 = 0.2623 \times 10^7 - 0.0042 \times 10^7$

$$= 0.2581 \times 10^7$$

- **Multiplication/Division:** For the multiplication of two normalised floating numbers, we just multiply the two mantissas and add the exponents. Division is accomplished by division of the mantissa and subtraction of the divisor exponent from the exponent of the dividend. For example,
  - $(0.3324 \times 10^3) \times (0.4432 \times 10^2) = (0.3324 \times 0.4432) \times 10^{3+2}$

$$= 0.14731968 \times 10^5$$
$$= 0.1473 \times 10^5$$

  - $(0.5324 \times 10^3) \div (0.4432 \times 10^2) = (0.5324 \div 0.4432) \times 10^{3-2}$

$$= 1.201263 \times 10^1$$
$$= 0.1201 \times 10^2$$

1.6 PROPAGATION OF ERRORS

This type of error arises due to errors in arithmetic computations with the numbers. Let us consider the arithmetic operations $+, -, \times$ and $\div$. Let * denote any of these operations and *' denote the same arithmetic operation as it is actually carried out by a computer. Let x_A and y_A be the approximations to the true values x_T and y_T, respectively. Then,

$$x_T = x_A + \epsilon_1$$
$$\text{and} \quad y_T = y_A + \epsilon_2$$

where ϵ_1 and ϵ_2 are the corresponding errors.

Now, if $x_A *' y_A$ is the actual computed number, the error will be

$$(x_T * y_T) - (x_A *' y_A) = [(x_T * y_T) - (x_A * y_A) + (x_A * y_A) - (x_A *' y_A)]$$

Then the first part within the square brackets on the right-hand side is known as the 'propagated error' while the second part is known as the 'noise' or 'rounding error' or 'chopping error'.

Now, we shall discuss some specific cases.

Propagation of Error Due to Addition or Subtraction For addition or subtraction, we have

$$(x_T \pm y_T) - (x_A \pm y_A) = (x_T - x_A) \pm (y_T - y_A) = \epsilon_1 \pm \epsilon_2$$

Thus, error in $(x_A \pm y_A) =$ error in $(x_A) \pm$ error in (y_A)

Propagation of Error Due to Multiplication In this case, we have

$$x_T y_T - x_A y_A = x_T y_T - (x_T - \epsilon_1)(y_T - \epsilon_2) = x_T \epsilon_2 + y_T \epsilon_1 - \epsilon_1 \epsilon_2$$

So, error in $(x_A y_A) = x_T \cdot$ error in $(y_A) + y_T \cdot$ error in $(x_A) -$ error in $(x_A) \cdot$ error in (y_A)
So, the relative error in $(x_A y_A)$ may be expressed as

$$E_r(x_A y_A) = \frac{x_T y_T - x_A y_A}{x_T y_T} = \frac{\epsilon_1}{x_T} + \frac{\epsilon_2}{y_T} - \frac{\epsilon_1}{x_T}\frac{\epsilon_2}{y_T}$$

$$= E_r(x_A) + E_r(y_A) - E_r(x_A)E_r(y_A)$$

Now, if $E_r(x_A), E_r(y_A) << 1$

Then $\boldsymbol{E_r(x_A y_A) \approx E_r(x_A) + E_r(y_A)}$

Propagation of Error Due to Division Proceeding as above, we can arrive at the relative error in division,

$$E_r(x_A/y_A) \approx E_r(x_A) - E_r(y_A)$$

Assuming $|E_r(y_A)| << 1$

1.7 GENERAL ERROR FORMULA

Let y be a function of independent variables $x_1, x_2, \ldots, x_n$ subject to the errors $\Delta x_1, \Delta x_2, \ldots, \Delta x_n$, respectively, which are very small in magnitude. Let Δy be the error in y. Then,

$$y = f(x_1, x_2, \ldots, x_n)$$

$$y + \Delta y = f(x_1 + \Delta x_1, x_2 + \Delta x_2, \ldots, x_n + \Delta x_n)$$

Now, assuming the function is differentiable, we expand the right-hand side in Taylor's series and obtain

$$y + \Delta y = f(x_1, x_2 \ldots, x_n) + \Delta x_1 \frac{\partial f}{\partial x_1} + \Delta x_2 \frac{\partial f}{\partial x_2} + \cdots + \Delta x_n \frac{\partial f}{\partial x_n}$$

$$+ \text{ terms involving higher powers of } \Delta x_n,\ n = 1, 2, \ldots, n.$$

As $\Delta x_1, \Delta x_2, \ldots, \Delta x_n$ are very small, neglecting the higher powers of Δx_n, we have

$$\Delta y \simeq \Delta x_1 \frac{\partial f}{\partial x_1} + \Delta x_2 \frac{\partial f}{\partial x_2} + \cdots + \Delta x_n \frac{\partial f}{\partial x_n} \tag{1.1}$$

Equation (1.1) is known as the **general error formula** for computing the error of a function

$$y = f(x_1, x_2, \ldots, x_n).$$

Now, dividing (1.1) by y, we get

$$\frac{\Delta y}{y} \approx \frac{1}{y}\left[\Delta x_1 \frac{\partial f}{\partial x_1} + \Delta x_2 \frac{\partial f}{\partial x_2} + \cdots + \Delta x_n \frac{\partial f}{\partial x_n}\right]$$

$$= \frac{\Delta x_1}{y}\frac{\partial f}{\partial x_1} + \frac{\Delta x_2}{y}\frac{\partial f}{\partial x_2} + \cdots + \frac{\Delta x_n}{y}\frac{\partial f}{\partial x_n}$$

Thus, the maximum relative error in y is given by

$$E_r(y) = \left|\frac{\Delta y}{y}\right| \le \left|\frac{\Delta x_1}{y}\frac{\partial f}{\partial x_1}\right| + \left|\frac{\Delta x_2}{y}\frac{\partial f}{\partial x_2}\right| + \cdots + \left|\frac{\Delta x_n}{y}\frac{\partial f}{\partial x_n}\right| \tag{1.2}$$

1.7.1 Error Formula for Addition of Numbers

Let $y = x_1 + x_2 + \cdots x_n$. Then using equation (1.2), we get the maximum relative error in y as

$$E_r(y) \le \left|\frac{\Delta x_1}{y}\cdot 1\right| + \left|\frac{\Delta x_2}{y}\cdot 1\right| + \cdots + \left|\frac{\Delta x_n}{y}\cdot 1\right|$$

$$= \left|\frac{\Delta x_1}{y}\right| + \left|\frac{\Delta x_2}{y}\right| + \cdots + \left|\frac{\Delta x_n}{y}\right|$$

For example, suppose $x = 0.1429$ and $y = 0.0909$ are the approximations, correct to 4 decimal places, and $z = x + y = 0.1429 + 0.0909 = 0.2338$. As the numbers are correct to 4 decimal places, the maximum absolute error in each case is

$$\frac{1}{2}\times 10^{-4} = 0.00005$$

i.e.,

$$\Delta x = \Delta y = 0.00005$$

Hence, the maximum relative error in z, i.e., in 0.2338, is given by

$$E_r(z)\big|_{max.} = \left|\frac{\Delta x}{z}\right| + \left|\frac{\Delta y}{z}\right| = \frac{0.00005}{0.2338} + \frac{0.00005}{0.2338}$$
$$= 0.00043$$

1.7.2 Error Formula for Subtraction of Numbers

If $y = x_1 - x_2$, then using equation (1.2), we get the maximum relative error in y as

$$E_r(y) \le \left|\frac{\Delta x_1}{y}\cdot 1\right| + \left|\frac{\Delta x_2}{y}\cdot(-1)\right|$$

$$= \left|\frac{\Delta x_1}{y}\right| + \left|\frac{\Delta x_2}{y}\right|$$

To illustrate this, we consider the previous example again.
In this case, $z = x - y = 0.1429 - 0.0909 = 0.052$.
Therefore, the maximum relative error in z, i.e., in 0.052, will be

$$E_r(z)\big|_{max.} = \left|\frac{\Delta x}{z}\right| + \left|\frac{\Delta y}{z}\right| = \frac{0.00005}{0.052} + \frac{0.00005}{0.052}$$
$$= 0.0019$$

1.7.3 Error Formula for Multiplication of Numbers

Let us consider the product of n terms $y = x_1 x_2 x_3 \ldots x_n$. From equation (1.2), we have

$$E_r(y) \le \left|\frac{x_2 x_3 \ldots x_n}{y}\Delta x_1\right| + \left|\frac{x_1 x_3 \ldots x_n}{y}\Delta x_2\right| + \cdots + \left|\frac{x_1 x_2 x_3 \ldots x_{n-1}}{y}\Delta x_n\right|$$

$$= \left|\frac{x_2 x_3 \ldots x_n}{x_1 x_2 x_3 \ldots x_n}\Delta x_1\right| + \left|\frac{x_1 x_3 \ldots x_n}{x_1 x_2 x_3 \ldots x_n}\Delta x_2\right| + \cdots + \left|\frac{x_1 x_2 x_3 \ldots x_{n-1}}{x_1 x_2 x_3 \ldots x_n}\Delta x_n\right|$$

$$= \left|\frac{\Delta x_1}{x_1}\right| + \left|\frac{\Delta x_2}{x_2}\right| + \cdots + \left|\frac{\Delta x_n}{x_n}\right|$$

$$= E_r(x_1) + E_r(x_2) + \cdots + E_r(x_n)$$

Thus, the **maximum relative error** in the **product of *n* terms** is approximately equal to the **algebraic sum of their relative errors.**

For example, consider the approximate numbers $x = 7.342$ and $y = 4.312$ which are correct to 3 decimal places, and $p = xy = 7.342 \times 4.312$.

As the numbers are correct to 3 decimal places, the maximum absolute error in each case is

$$\frac{1}{2} \times 10^{-3} = 0.0005$$

i.e.,
$$\Delta x = \Delta y = 0.0005$$

So, the maximum relative error in p is given by

$$E_r(p)\Big|_{max.} = \left|\frac{\Delta x}{x}\right| + \left|\frac{\Delta y}{y}\right| = \frac{0.0005}{7.342} + \frac{0.0005}{4.312}$$

$$= 0.0002501$$

1.7.4 Error Formula for Division of Numbers

If $y = x_1/x_2$, then using equation (1.2), we get the maximum relative error in y as

$$E_r(y) \le \left|\frac{\Delta x_1}{y} \cdot \frac{1}{x_2}\right| + \left|\frac{\Delta x_2}{y} \cdot \frac{-x_1}{x_2^2}\right| = \left|\frac{\Delta x_1}{x_1/x_2} \cdot \frac{1}{x_2}\right| + \left|\frac{\Delta x_2}{x_1/x_2} \cdot \frac{-x_1}{x_2^2}\right|$$

$$= \left|\frac{\Delta x_1}{x_1}\right| + \left|\frac{\Delta x_2}{x_2}\right| = E_r(x_1) + E_r(x_2)$$

Therefore, the **maximum relative error** in the **division of two terms** is approximately equal to the **algebraic sum of their relative errors.**

Let $x = 7.342$ and $y = 0.241$ be two approximate numbers, correct to 3 decimal places and $= x/y = 7.342/0.241$.

As the numbers are correct to 3 decimal places, the maximum absolute error in each case is

$$\frac{1}{2} \times 10^{-3} = 0.0005$$

i.e.,
$$\Delta x = \Delta y = 0.0005$$

So, the maximum relative error in d is given by

$$E_r(d)\Big|_{max.} = \left|\frac{\Delta x}{x}\right| + \left|\frac{\Delta y}{y}\right| = \frac{0.0005}{7.342} + \frac{0.0005}{0.241}$$

$$= 0.0021$$

1.7.5 Error Formula for Product of Exponent Terms

Suppose $y = ax_1^{m_1} x_2^{m_2} \ldots x_n^{m_n}$, where a is a constant. Then from equation (1.2), we get the maximum relative error in y as

$$E_r(y) \le \left|\frac{am_1x_1^{m_1-1}x_2^{m_2}\ldots x_n^{m_n}}{y}\Delta x_1\right| + \left|\frac{am_2x_1^{m_1}x_2^{m_2-1}\ldots x_n^{m_n}}{y}\Delta x_2\right| + \cdots + \left|\frac{am_nx_1^{m_1}x_2^{m_2}\ldots x_n^{m_n-1}}{y}\Delta x_n\right|.$$

$$= m_1\left|\frac{\Delta x_1}{x_1}\right| + m_2\left|\frac{\Delta x_2}{x_2}\right| + \cdots + m_n\left|\frac{\Delta x_n}{x_n}\right|$$

$$= m_1E_r(x_1) + m_2E_r(x_2) + \cdots + m_nE_r(x_n)$$

An example is given to illustrate this. Suppose $u = 10x^3y^2z^4$ and errors in x, y, z are 0.03, 0.01, 0.02 at $x = 3, y = 2, z = 4$. Then we have the maximum relative error in u, using the above formula, as

$$E_r(u)\Big|_{max.} = 3\left|\frac{\Delta x}{x}\right| + 2\left|\frac{\Delta y}{y}\right| + 4\left|\frac{\Delta z}{z}\right| = 3\times\frac{0.03}{3} + 2\times\frac{0.01}{2} + 4\times\frac{0.02}{4}$$

$$= 0.06$$

SOLVED EXAMPLES

EXAMPLE 1.1 Find the number of significant figures of the following numbers:

1.0086, 1200, 1200.0, 0.01005, 0.001, 0.005006, 0.050060.

Solution: The number of significant figures is given in the following list:

Number(s)	Significant figures	No. of significant figures
1.0086	1,0,0,8,6	5
1200	1,2	2
1200.0	1,2,0,0,0	5
0.01005	1,0,0,5	4
0.001	1	1
0.005006	5,0,0,6	4
0.050060	5,0,0,6,0	5

EXAMPLE 1.2 Find the number of significant figures in the following numbers:

(a) 0.0340986
(b) 0.0340980
(c) 1.00234
(d) 0.0128742
(e) 6,00,000
(f) 0.0001234
(g) 0.00303
(h) 1.00234

Solution: (a) 6, (b) 6, (c) 6, (d) 6, (e) 1, (f) 4, (g) 3, (h) 6

EXAMPLE 1.3 Find the number of significant figures in $va = 0.3941$ when its absolute error is given as 0.25×10^{-2}.

Solution: We know that if x_A is the approximate value of x_T, correctly rounded to n decimal places, the absolute error

$$E_a = |x_T - x_A| \leq \frac{1}{2} \times 10^{-n}$$

Here the absolute error is

$$0.25\times10^{-2} = \frac{1}{4}\times10^{-2} = \frac{1}{2}\times\left(\frac{1}{2}\times10^{-2}\right) < \frac{1}{2}\times10^{-2}$$

This shows that the value of n is 2. Therefore, there are 2 significant figures.

EXAMPLE 1.4 Determine the most and least significant digits in the following numbers:

12.433, 0.001100, 40.004, 0.0123.

Solution:

Number	Most significant digit	Least significant digit
12.433	1	rightmost digit 3
0.001100	leftmost digit 1	rightmost digit 0
40.004	leftmost digit 4	rightmost digit 4
0.0123	1	3

EXAMPLE 1.5 Round the following numbers to 4 significant figures:

45.453897, 25.487566, 43.45532, 77.435326.

Solution: 45.453897 ≈ 45.45, 25.487566 ≈ 25.49, 43.45532 ≈ 43.46, 77.435326 ≈ 77.44

EXAMPLE 1.6 Round the following numbers to 3 significant digits:

15.253000, 0.087555, 21.45522, 70.33826.

Solution: 15.253000 ≃ 15.3, 0.087555 ≃ 0.0876, 21.45522 ≃ 21.5, 70.33826 ≃ 70.3

EXAMPLE 1.7 Round the following numbers correct to 5 significant digits:

1.5002836, 0.251065, 0.2510655, 0.251055.

Solution: 1.5002836 ≈ 1.5003, 0.251065 ≈ 0.25106, 0.2510655 ≈ 0.25107, 0.251055 ≈ 0.25106

EXAMPLE 1.8 Round the following numbers correct to 3 significant digits:

1234567, 0.046064, 0.2510601, 0.10155.

Solution: 1234567 ≈ 123×10^4, 0.046064 ≈ 0.0461, 0.2510601 ≈ 0.251, 0.10155 ≈ 0.102

EXAMPLE 1.9 Round the following numbers:

(a) 123456 to 3 significant figures
(b) 31.520457 to 1 significant digit
(c) 0.0063945 to 4 significant figures

(d) 0.006059991 to 5 and 4 significant figures
(e) 0.0036945 to 4 significant figures
(f) 633040 to 4 significant figures
(g) 0.00326945 to 5 significant figures
(h) 31.5218 to 1 significant figure

Solution: (a) 123×10^3, (b) 0.3×10^2, (c) 0.006394, (d) 0.0060600 and 0.006060, (e) 0.003694, (f) 0.6330×10^4, (g) 0.0032694, (h) 0.3×10^2

EXAMPLE 1.10 Round the following numbers to 3 decimal places:

42.463877, 35.287523, 23.55559, 65.735336.

Solution: $42.463877 \approx 42.464$, $35.287523 \approx 35.288$, $23.55559 \approx 23.556$, $65.735336 \approx 65.735$

EXAMPLE 1.11 Compute the value of the figure 47,616 approximated to the 100th place.

Solution: 47,600

EXAMPLE 1.12 Round the following numbers to 4 decimal places:

5.254018, 8.76475, 44.00795, 0.00327.

Solution: $5.254018 \approx 5.2540$, $8.76475 \approx 8.7648$, $44.00795 \approx 44.0080$, $0.00327 \approx 0.0033$

EXAMPLE 1.13 Round the following numbers:

(a) 3.4506531to 4 decimal places
(b) 86.238527 to 3 decimal places
(c) 0.003256 to 4 decimal places
(d) 9.6506531 to 4 decimal places

Solution: (a) 3.4507, (b) 86.239, (c) 0.0033, (d) 9.6507

EXAMPLE 1.14 Find the percentage error if x is rounded off to 3 decimal places, given $x = 0.007898$.

Solution: Here, true value of x, $x_T = 0.007898$ and its approximate value, $x_A = 0.008$

$\therefore$ The error

$$E = x_T - x_A = 0.007898 - 0.008$$
$$= -1.02 \times 10^{-4}$$

$\therefore$ The absolute error

$$E_a = |E| = 1.02 \times 10^{-4}$$

Relative error

$$E_r = \frac{E_a}{v_T} = \frac{1.02 \times 10^{-4}}{0.007898}$$
$$= 0.0129$$

So, the percentage error

$$E_p = E_r \times 100\% = 1.29\%$$

EXAMPLE 1.15 Find the sum of the following approximate numbers, if the numbers are correct to the last digit: 2.56, 4.56273, 1.253, 1.05342.

Solution: As the number with the least number of digits (only two) after the decimal point is 2.56, the result of the sum must have two digits after the decimal point.

Hence $$2.56 + 4.56273 + 1.253 + 1.05342 \approx 9.43$$

EXAMPLE 1.16 Give the approximate representation of 2/3 correct to 4 significant figures and then find (a) absolute error, (b) relative error and (c) percentage error.

Solution: Here, true value, $v_T = 2/3$ and approximate value, $v_A = 0.6667$

∴ The error

$$E = v_T - v_A = \frac{2}{3} - 0.6667$$

$$= \frac{-0.0001}{3}$$

∴ (a) Absolute error

$$E_a = |v_T - v_A| = \frac{0.0001}{3}$$

$$\approx 0.000033$$

(b) Relative error

$$E_r = \frac{|v_T - v_A|}{v_T} = \frac{\frac{0.0001}{3}}{\frac{2}{3}}$$

$$= 0.00005$$

(c) Percentage error

$$E_p = E_r \times 100\% = 0.005\%$$

EXAMPLE 1.17 What will be the percentage error in approximating 4/3 to 1.3333?

Solution: Here, true value, $v_T = 4/3$ and approximate value, $v_A = 1.3333$

∴ The error

$$E = v_T - v_A = \frac{4}{3} - 1.3333$$

$$= \frac{0.0001}{3}$$

∴ The absolute error

$$E_a = |v_T - v_A| = \frac{0.0001}{3}$$

Relative error

$$E_r = \frac{|v_T - v_A|}{v_T} = \frac{\frac{0.0001}{3}}{\frac{4}{3}}$$

$$= 0.000025$$

So, the percentage error

$$E_p = E_r \times 100\% = 0.0025\%$$

EXAMPLE 1.18 If $\pi = 3.14$ is used in place of 3.14156, find the absolute error, relative error and percentage error.

Solution: Here we are given true value, $v_T = 3.14156$ and approximate value, $v_A = 3.14$

∴ The error

$$E = v_T - v_A = 3.14156 - 3.14$$

$$= 0.00156$$

∴ The absolute error

$$E_a = |v_T - v_A| = 0.00156$$

Relative error

$$E_r = \frac{|v_T - v_A|}{v_T} = \frac{0.00156}{3.14156}$$

$$= 4.966 \times 10^{-4}$$

Hence, the percentage error

$$E_p = E_r \times 100\% = 4.966 \times 10^{-2}\%$$

EXAMPLE 1.19 If 5/3 is approximated to 1.6667, find the absolute error.

Solution: Here, true value, $v_T = 5/3$ and approximate value, $v_A = 1.6667$

∴ The error

$$E = v_T - v_A = \frac{5}{3} - 1.6667$$

$$= \frac{-0.0001}{3}$$

∴ The absolute error

$$E_a = |v_T - v_A| = \frac{0.0001}{3}$$

$$= 0.000033$$

EXAMPLE 1.20 If 1.414 is used as an approximation of $\sqrt{2}$, determine the absolute and relative errors, provided the true value of $\sqrt{2}$ is 1.41421356.

Solution: Here, the true value of $\sqrt{2}$ is 1.41421356 and the approximate value is 1.414

∴ The absolute error $= |1.41421356 - 1.414| = 0.00021356$

$$= 0.21356 \times 10^{-3}$$

Relative error $= \dfrac{0.21356 \times 10^{-3}}{1.41421356} \approx 0.151 \times 10^{-3}$

EXAMPLE 1.21 Let 0.01 and 0.02 be the errors in measuring x and y, respectively. Find the percentage error in $z = x/y$ when $x = 1$ and $y = 1$.

Solution: Here, the true value of z, $z_T = \dfrac{1}{1} = 1$ and its approximate value,

$$z_A = \frac{1-0.01}{1-0.02} = \frac{99}{98}$$

∴ The error

$$E = z_T - z_A = 1 - \frac{99}{98}$$

$$= \frac{-1}{98}$$

∴ The absolute error $E_a = |E| = \dfrac{1}{98}$

Relative error

$$E_r = \frac{E_a}{z_T} = \frac{\frac{1}{98}}{1}$$

$$= 0.0102$$

So, the percentage error

$$E_p = E_r \times 100\% = 1.02\%$$

EXAMPLE 1.22 The approximate values of the number 1/3 are given as 0.30, 0.33 and 0.34. Which of these is the best approximation?

Solution: Here, true value, $v_T = \dfrac{1}{3}$

(i) If the approximate value, v_A is 0.30, then

Relative error $= \dfrac{|v_T - v_A|}{v_T} = \dfrac{|1/3 - 0.30|}{1/3}$

$$= 0.1$$

(ii) If the approximate value, v_A is 0.33, then

Relative error $= \dfrac{|v_T - v_A|}{v_T} = \dfrac{|1/3 - 0.33|}{1/3}$

$$= 0.01$$

(iii) If the approximate value, v_A is 0.34, then

$$\text{Relative error} = \frac{|v_T - v_A|}{v_T} = \frac{|1/3 - 0.34|}{1/3}$$
$$= 0.02$$

We see from the above that the relative error is least when the approximate value is 0.33. Hence, 0.33 is the best approximation for the number 1/3.

EXAMPLE 1.23 Calculate the sum of the positive square roots of 3, 5 and 7 to 4 significant figures and find its relative error.

Solution: Here, $\sqrt{3} = 1.732, \sqrt{5} = 2.236, \sqrt{7} = 2.646$

Hence, sum $= 1.732 + 2.236 + 2.646 = 6.614$

The total absolute error $= 0.0005 + 0.0005 + 0.0005 = 0.0015$, which shows that the sum is correct to 3 significant figures. So, sum $= 6.61$.

Therefore, the relative error $= \dfrac{0.0015}{6.61} = 0.0002$

EXAMPLE 1.24 If the number 0.347 is correct to the last digit, find the number of trustworthy figures in $(0.347)^3$.

Solution: We know that

Relative error (E_r) in $(0.347)^3 \leq 3 \times$ Relative error in 0.347

It is given that the number 0.347 is correct to the last digit. Therefore, the absolute error in 0.347 is

$$\frac{1}{2} \times 10^{-3} = 0.0005$$

Hence

$$E_r \leq 3 \times \frac{0.0005}{0.347}$$

Therefore, absolute error in $(0.347)^3$ is

$$\leq 3 \times \frac{0.0005}{0.347} \times (0.347)^3 \approx 0.00018$$

As the error affects the fourth decimal place, $(0.347)^3 \approx 0.042$ is correct to 3 decimal places.

EXAMPLE 1.25 Suppose 0.1429 and 0.0909 are the approximate values of 1/7 and 1/11, respectively. If they are correct to 4 decimal places, determine the maximum relative error in the result of

(i) $0.1429 + 0.0909$
(ii) $0.1429 - 0.0909$
(iii) 0.1429×0.0909
(iv) $0.1429 / 0.0909$

Solution: Let $x = 0.1429$ and $y = 0.0909$. We know that if x_A is the approximate value of x_T, correctly rounded to n decimal places, the absolute error

$$E_a = |x_T - x_A| \le \frac{1}{2} \times 10^{-n}$$

As the numbers are correct to 4 decimal places, the absolute error in each approximation is

$$= \frac{1}{2} \times 10^{-4} = 0.00005$$

i.e.,
$$|\Delta x| = 0.00005 = |\Delta y|$$

(i) Let $z = x + y = 0.1429 + 0.0909 = 0.2338$. Therefore,
Maximum relative error in

$$z = \left|\frac{\Delta x}{z}\right| + \left|\frac{\Delta y}{z}\right| = \frac{0.00005}{0.2338} + \frac{0.00005}{0.2338}$$
$$\approx 0.00043$$

(ii) Let $z = x - y = 0.1429 - 0.0909 = 0.052$. Therefore,
Maximum relative error in

$$z = \left|\frac{\Delta x}{z}\right| + \left|\frac{\Delta y}{z}\right| = \frac{0.00005}{0.052} + \frac{0.00005}{0.052}$$
$$\approx 0.0019$$

(iii) Let $z = xy$. Therefore,
Maximum relative error in

$$z = \left|\frac{\Delta x}{x}\right| + \left|\frac{\Delta y}{y}\right| = \frac{0.00005}{0.1429} + \frac{0.00005}{0.0909}$$
$$\approx 0.0009$$

(iv) Let $z = x/y$. Therefore,
Maximum relative error in

$$z = \left|\frac{\Delta x}{x}\right| + \left|\frac{\Delta y}{y}\right| = \frac{0.00005}{0.1429} + \frac{0.00005}{0.0909}$$
$$\approx 0.0009$$

EXAMPLE 1.26 Find the maximum relative error in the calculation of 343.4×856.3 and determine the interval in which the true result lies (assume that each number is correct to 4 significant figures).

Solution: Let $p = xy$, where $x = 343.4$ and $y = 856.3$.
As the given numbers are correct to 4 significant figures, $\Delta x = 0.05 = \Delta y$.
Therefore, the maximum relative error in p is given by

$$E_r(p)\Big|_{max.} = \left|\frac{\Delta p}{p}\right| = \left|\frac{\Delta x}{x}\right| + \left|\frac{\Delta y}{y}\right| = \frac{0.05}{343.4} + \frac{0.05}{856.3}$$
$$= 0.000204$$

The absolute error in p will be

$$|\Delta p| \leq |p|\left[\left|\frac{\Delta x}{x}\right| + \left|\frac{\Delta y}{y}\right|\right] = |y\Delta x| + |x\Delta y| = 0.05(856.3 + 343.4)$$

$$= 59.985$$

Now
$$p = xy = 343.4 \times 856.3$$

$$= 294053.42$$

Therefore, the true value of p lies between 294053.42 − 59.985 and 294053.42 + 59.985, i.e., 293993.435 and 294113.405 or 2940×10^2 and 2941×10^2.

EXAMPLE 1.27 Calculate the product 5.734×6.315, given that the numbers are correct to the last digit and state how many figures of the result are trustworthy.

Solution: Let $u = pq$, where $p = 5.734$ and $q = 6.315$.

As the given numbers are correct to their last digit, $\Delta p = 0.0005 = \Delta q$.

Therefore, the absolute error in u will be

$$|\Delta u| \leq |u|\left[\left|\frac{\Delta p}{p}\right| + \left|\frac{\Delta q}{q}\right|\right] = |q\Delta p| + |p\Delta q| = 0.0005(6.315 + 5.734)$$

$$= 0.0060245$$

Now, $u = pq = 5.734 \times 6.315 = 36.210210$

Therefore, the true value of u lies between 36.210210 − 0.0060245 and 36.210210 − 0.0060245 i.e., 36.2041855 and 36.2162345.

Hence, only the first three figures (3, 6 and 2) are trustworthy.

EXAMPLE 1.28 Compute the relative error in the time period $T = 2\pi.\sqrt{\dfrac{l}{g}}$ for l – 1m if the error in the measurement of l is 0.01.

Solution: Here we have

$$T = 2\pi.\sqrt{\frac{l}{g}}$$

Taking log on both sides, we get

$$\log T = \log 2\pi + \frac{1}{2}\log l - \frac{1}{2}\log g$$

$$\Rightarrow \frac{1}{T}\Delta T = \frac{1}{2}\frac{\Delta l}{l}$$

Therefore, the relative error in T is

$$\frac{1}{2}\frac{\Delta l}{l} = \frac{0.01}{2\times 1} = 0.005$$

EXAMPLE 1.29 How accurately should the length and time of vibration of a pendulum be measured in order that the computed value of g is correct to 0.01%?

Solution: The period of vibration, say T, of a pendulum of length l is given by

$$T = 2\pi.\sqrt{\frac{l}{g}}$$

Therefore
$$g = \frac{4\pi^2 l}{T^2} \Rightarrow \frac{\partial g}{\partial l} = \frac{4\pi^2}{T^2}$$

and
$$\frac{\partial g}{\partial T} = \frac{-8\pi^2 l}{T^3}$$

Again
$$\Delta l = \frac{\Delta g}{2\left(\dfrac{\partial g}{\partial l}\right)}$$

and
$$\Delta T = \frac{\Delta g}{2\left(\dfrac{\partial g}{\partial T}\right)}$$

Now, given that the percentage error in g is 0.01%.

$$\frac{\Delta g}{g} \times 100 = 0.01$$

$$\Rightarrow \frac{\Delta g}{\left(\dfrac{4\pi^2 l}{T^2}\right)} \times 100 = 0.01$$

Therefore,

(i) Percentage error in
$$l = \frac{\Delta l}{l} \times 100$$

$$= \frac{\Delta g}{2l\left(\dfrac{\partial g}{\partial l}\right)} \times 100 = \frac{1}{2l} \frac{\Delta g}{\left(4\pi^2 / T^2\right)} \times 100$$

$$= \frac{1}{2} \frac{\Delta g}{\left(4\pi^2 l / T^2\right)} \times 100 = \frac{1}{2} \times 0.01$$

$$= 0.005$$

(ii) Percentage error in
$$T = \frac{\Delta T}{T} \times 100$$

$$= \frac{\Delta g}{2T\left(\dfrac{\partial g}{\partial T}\right)} \times 100 = \frac{1}{2T} \frac{\Delta g}{\left(8\pi^2 l / T^3\right)} \times 100$$

$$= \frac{1}{4} \frac{\Delta g}{\left(4\pi^2 l / T^2\right)} \times 100 = \frac{1}{4} \times 0.01$$

$$= 0.0025$$

EXAMPLE 1.30 Round the following numbers to 1 significant digit:
11111, 55555, 71111, 75555, 76666, 77777, 88888, 89999, 99999.

Solution: 0.1×10^5, 0.6×10^5, 0.7×10^5, 0.8×10^5, 0.8×10^5, 0.8×10^5, 0.9×10^5, 0.9×10^5, 0.1×10^6

EXAMPLE 1.31 Find the truncation error in the expansion of e^x at $x = 1/2$, if

(i) The first three terms are retained in expansion
(ii) The first four terms are retained in expansion

Solution: We know

$$e^x = 1+\frac{x}{1!}+\frac{x^2}{2!}+\frac{x^3}{3!}+\frac{x^4}{4!}+\cdots\infty$$

which gives the true value of e^x at

$$x = \frac{1}{2}.$$

(i) In this case, the truncation error (E_T) is given by

$$E_T = \text{True value} - \text{Approximate value}$$

$$= \left[1+\frac{x}{1!}+\frac{x^2}{2!}+\frac{x^3}{3!}+\frac{x^4}{4!}+\cdots\right]-\left[1+\frac{x}{1!}+\frac{x^2}{2!}\right]$$

$$= \frac{x^3}{3!}+\frac{x^4}{4!}+\frac{x^5}{5!}+\cdots$$

Therefore, at

$$x = \frac{1}{2} = 0.5$$

$$E_T = \frac{(0.5)^3}{3!}+\frac{(0.5)^4}{4!}+\frac{(0.5)^5}{5!}+\cdots$$

$$\approx 0.0208333+0.0026042+0.0002604+\cdots$$

$$\approx 0.0236979$$

(ii) Here we consider only four terms. Hence the truncated error

$$E_T = \left[1+\frac{x}{1!}+\frac{x^2}{2!}+\frac{x^3}{3!}+\frac{x^4}{4!}+\cdots\right]-\left[1+\frac{x}{1!}+\frac{x^2}{2!}+\frac{x^3}{3!}\right]$$

$$= \frac{x^4}{4!}+\frac{x^5}{5!}+\frac{x^6}{6!}+\cdots$$

Therefore, at

$$x = \frac{1}{2} = 0.5,$$

$$E_T = \frac{(0.5)^4}{4!}+\frac{(0.5)^5}{5!}+\frac{(0.5)^6}{6!}+\cdots$$

$$\approx 0.0026042+0.0002604+0.0000217+\cdots$$

$$\approx 0.0028863$$

EXAMPLE 1.32 If $X = x + e$, show that $\sqrt{X} - \sqrt{x} \approx \frac{e}{2\sqrt{X}}$.

Solution:

$$\sqrt{X} - \sqrt{x} = \sqrt{X} - \sqrt{X - e} = \sqrt{X} - \sqrt{X}\left(1 - \frac{e}{\sqrt{X}}\right)^{1/2}$$

$$\approx \sqrt{X} - \sqrt{X}\left(1 - \frac{e}{2\sqrt{X}}\right) = \frac{e}{2\sqrt{X}}$$

EXAMPLE 1.33 Calculate absolute and relative errors in computing $y = 3\sin x - 5x^2 + 9$ for $x = 0$ when the error in x is 0.003.

Solution: Here we have $y = 3\sin x - 5x^2 + 9$ and so at $x = 0$, $y = 3\sin 0 - 5 \times 0^2 + 9 = 9$

$\therefore$ From the general formula, we have

$$\Delta y \approx \frac{dy}{dx}.\Delta x = (3\cos x - 10x).\Delta x$$

Now, at $x = 0$, $\Delta x = 0.003$

$\therefore$ The error in y at $x = 0$ is given by

$$\Delta y \approx (3\cos 0 - 10 \times 0) \times 0.003 = 0.009$$

So, the absolute error is 0.009 and hence the relative error is

$$\frac{|\Delta y|}{y} = \frac{0.009}{9} = 0.001$$

MULTIPLE CHOICE QUESTIONS

1. The error inherent in a numerical method is called computational error
   (a) true (b) false
2. The number 3.4506531 when rounded to 4 decimal places will give
   (a) 3.4506 (b) 3.4507 (c) 3.451 (d) none
3. The number of significant figures in 0.0340986 is
   (a) 5 (b) 6 (c) 7 (d) 4
4. If a is the actual value and e its estimated value, the formula for relative error is
   (a) $\frac{a}{e}$ (b) $\frac{|a-e|}{e}$ (c) $\frac{a-e}{a}$ (d) $\frac{|a-e|}{a}$
5. Rounding error is a form of
   (a) truncation error (b) numerical error (c) inherent error (d) none
6. The number of significant digits in 1.00234 is
   (a) 4 (b) 6 (c) 3 (d) 5
7. When rounded after 4 decimal places, 0.003256 becomes
   (a) 0.0032 (b) 0.0033 (c) 0.00326 (d) none
8. The number of significant digits in 6,00,000 is
   (a) 1 (b) 7 (c) 0 (d) 6

9. The ratio of absolute error and true value is called
(a) relative error (b) absolute error (c) truncation error (d) inherent error
10. The number of significant digits in 0.0001234 is
(a) 7 (b) 4 (c) 8 (d) 6
11. The percentage error in approximating 4/3 to 1.3333 is
(a) 0.0025% (b) 25% (c) 0.00025% (d) 0.25%
12. The number 0.0063945 when rounded to 4 significant figures, is
(a) 0.0064 (b) 0.0063 (c) 0.006395 (d) 0.006394
13. The figure 47,616 approximated to the 100th place, is
(a) 47,630 (b) 47,620 (c) 47,700 (d) 47,600
14. The number of significant figures in 0.0340980 is
(a) 5 (b) 6 (c) 7 (d) 3
15. The number 86.238527 rounded to 3 decimal places is
(a) 86.2 (b) 86.238 (c) 86.239 (d) 86.23
16. The number of significant digits in 0.00303 is
(a) 6 (b) 5 (c) 3 (d) 2
17. The number of significant figures in 0.0128742 is
(a) 5 (b) 6 (c) 7 (d) 3
18. If 5/3 is approximated to 1.6667, then absolute error is
(a) 0.000033 (b) 0.000043 (c) 0.000045 (d) 0.000051
19. Relative error is measured by which of the following expressions?
(a) $\text{mod}\left[\dfrac{\text{absolute error}}{\text{approximate error}}\right]$ (b) $\text{mod}\left[\dfrac{\text{absolute error}}{\text{exact error}}\right]$
(c) mod[exact value − approximate value] (d) none
20. If E_a is the absolute error in a quantity whose true and approximate values are given by x_t and x_a, then the relative error is given by
(a) $\left|\dfrac{E_a}{x_a}\right|$ (b) $\left|\dfrac{E_a}{x_t}\right|$ (c) $\left|\dfrac{E_a}{x_t - x_a}\right|$ (d) $|E_a|$
21. After rounding to 3 decimal places, the number 68.2094 becomes
(a) 68.209 (b) 68.21 (c) 68.3 (d) 0.209
22. When rounded to 3 decimal places, the number 0.299574 will be
(a) 0.299 (b) 0.300 (c) 0.290 (d) 0.3
23. If 123456 is rounded to 3 significant figures, then it will be
(a) 123 (b) 1230 (c) 12300 (d) 123000
24. After rounding to 1 significant digit, the number 999999 will be
(a) 1 (b) 10 (c) 0 (d) 1000000
25. When rounded to 3 decimal places, the number 9.999827 becomes
(a) 9.999 (b) 10 (c) 10.000 (d) 9.10
26. When rounded to 1 significant figure, the number 31.520457 becomes
(a) 3 (b) 30 (c) 31 (d) 32
27. The number of significant digits in 1.00000 is
(a) 1 (b) 2 (c) 6 (d) none
28. The number of significant digits in 7.0048 is
(a) 3 (b) 5 (c) 1 (d) 2

29. Which digit is not significant in the number 0.00087?
(a) 7 (b) 8 (c) 0 (d) none
30. When 2/3 is approximated by 0.6667, the relative error will be
(a) 0.005 (b) 0.0005 (c) 0.00005 (d) 0.000005
31. The number 9.6506531 when rounded to 4 decimal places will give
(a) 9.6506 (b) 9.6507 (c) 9.6505 (d) none
32. The number of significant figures in 0.03409 is
(a) 5 (b) 6 (c) 7 (d) 4
33. In the problem "Find the area of a circle having radius 2; given $\pi = 3.14$", the type of error of the approximation 3.14 for π is
(a) inherent error (b) truncation error (c) rounding error (d) percentage error

Answers

1. (a)	2. (b)	3. (b)	4. (d)	5. (b)	6. (b)	7. (b)	8. (a)	9. (a)	10. (b)	11. (a)
12. (d)	13. (d)	14. (b)	15. (c)	16. (c)	17. (b)	18. (a)	19. (d)	20. (b)	21. (a)	22. (b)
23. (d)	24. (d)	25. (c)	26. (b)	27. (c)	28. (b)	29. (c)	30. (c)	31. (b)	32. (d)	33. (c)

SHORT ANSWER QUESTIONS

1. Write short note(s) on the following:
(a) absolute error (b) rounding error (c) relative error (d) inherent error
2. Find the number of significant figures:
(i) 3900 (ii) 0.3699 (iii) 0.00033 (iv) 0.002501
(v) 0.000009 (vi) 0.000090 (vii) 12345 (viii) 0.20510
(ix) 130.00 (x) 1.021×10^5 (xi) 1.00000×10^{-5}
3. Round the following numbers to 3 decimal places:
(i) 3.5075 (ii) 0.3685 (iii) 0.00833 (iv) 0.00257
(v) 0.00409 (vi) 0.0549
4. Round the following numbers to 3 significant figures:
(i) 350.75 (ii) 30685 (iii) 0.0833 (iv) 0.0267
(v) 40.19 (vi) 256425
5. Find the percentage error(s) in the approximation of
(i) 7/6 by 1.16 (ii) 5/6 by 0.8333 (iii) 4/3 by 1.33
6. Find the absolute error(s) in the approximation of
(i) 5/3 by 1.6667 (ii) 8.03567 to 4 significant figures (iii) 5.0214 to 3 significant figures
7. Round the number 75162 to 4 significant digits and calculate its percentage error.
8. Calculate the error in evaluating $f(x) = \cos x e^{\log 2}$ around $x = 2$ if the absolute error in x is 10^{-6}.
9. Calculate the absolute error in the approximation of 2/3 to 4 significant figures.
10. What will be the percentage error if $e = 2.71828$ is approximated by 2.71937?
11. If $y = (0.31x + 2.73)/(x + 0.35)$, find the relative error in y when $x = 0.5 \pm 0.1$.
12. Find the number of significant figures in $v_a = 0.3941$ when its relative error is given as 0.1×10^{-2}

13. N is a function of different measurable quantities u,v,w,x,y and is given by $N = \dfrac{u^p v^q w^r}{x^s y^t}$ find an upper limit to the relative error in the measure of N.

Answers

2. (i) 2 (ii) 4 (iii) 2 (iv) 4 (v) 1 (vi) 2 (vii) 5 (viii) 5 (ix) 5 (x) 4 (xi) 6
3. (i) 3.508 (ii) 0.368 (iii) 0.008 (iv) 0.002 (v) 0.004 (vi) 0.055
4. (i) 351 (ii) 307 (iii) 0.833×10^{-1} (iv) 0.267×10^{-1} (v) 2564
5. (i) 0.571% (ii) 0.004% (iii) 0.25% 6. (i) 0.000033 (ii) 0.00033 (iii) 0.0014 7. 0.00265%
8. 0.0053×10^{-3} 9. 3.3×10^{-5} 10. 4.02×10^{-2} 11. 0.9464, 1.225 12. 3

LONG ANSWER QUESTIONS

1. Find the maximum absolute error in computing $u = \dfrac{x^3 y^2}{z}$ when $x = y = z = 0.1$, $\Delta x = \Delta y = \Delta z = 0.002$.
2. Find the relative error in computing $x - y$ for $x = 9.05$ and $y = 6.56$, having absolute errors $\Delta x = 0.001$ and $\Delta y = 0.003$, respectively.
3. If $u = 4x^2y^3/z^3$ and errors in x,y,z are 0.001, 0.001, 0.002 then compute the relative error in u at $x = y = z = 1$.
4. If $u = 10x^3y^2z^2$ and errors in x,y,z are 0.03, 0.01, 0.02 at $x = 3$, $y = 1$, $z = 2$, respectively, calculate the absolute and relative errors in u.
5. If $\Delta r = \Delta h = 0.1$, find the absolute and relative errors up to 3 significant figures in $v = \frac{1}{3}\pi r^2 h$ when $r = 2$ and $h = 3$.
6. Find the absolute, relative and percentage errors when
   (i) 1/3 is approximated to 0.333
   (ii) 5/6 is approximated to 0.8333
   (iii) 3.45234 is approximated to 3.45678
   (iv) 5/3 is approximated to 1.6667
7. What is the percentage error in approximating 4/3 to 1.3333?
8. Find the percentage error in the approximate representation of 2/3 correct to 4 significant digits.
9. Find the absolute error in calculating the area of a circle of radius 5 if the error in radius is 0.1.
10. Find the relative error in the calculation of $\dfrac{7.342}{0.241}$, where the numbers are correct to 3 decimal places and determine the smallest interval in which the true result lies. Also state how many digits of the result are trustworthy.
11. The length and breadth of a room measured correct to 1 cm are 5.43 m and 3.82 m, respectively. Determine the percentage error in the measurement of the area of the room 20.7426 sq. m.

Answers

1. 0.012 2. 0.00034 3. 0.004 4. 140.4, 0.13 5. 0.168, 0.0135
6. (i) 3.333×10^{-4}, 0.001, 0.1 (ii) 3.333×10^{-5}, 4.0×10^{-5}, 4.0×10^{-3}
(iii) 0.00444, 0.128443×10^{-2}, 0.128443 (iv) 0.000033 7. 0.0025% 8. 0.005% 9. 55
10. relative error is 0.0021, interval is [30.4008, 30.5286], first two digits are trustworthy 11. 45%

2

Calculus of Finite Differences

2.1 INTRODUCTION

The calculus of finite differences deals with changes or variations in the dependent variable that occur with changes in the independent variable. In many practical fields, we may come across tabular values of two variables, one dependent on the other. In such cases, we may face difficulties in finding their relation using analytical methods. However, calculus of finite differences can provide the solution to these types of problems. Moreover, the study of this topic helps us to solve problems like numerical differentiation and integration. In this chapter, we shall discuss the types of difference operators, their relations and applications.

2.2 FINITE DIFFERENCES

Let a real-valued function $y = f(x)$ be known only for a set of $(n + 1)$ distinct equispaced points $x_0, x_1, \ldots, x_n$ such that $x_i = x_0 + ih$, $(i = 1, 2, \ldots, n)$, where h is the length of each subinterval $[x_i, x_{i+1}]$, known as the **space length**. The corresponding values of the function are $y_0 = f(x_0)$, $y_1 = f(x_1)$, …, $y_n = f(x_n)$. The x values are called **arguments** or **nodes** while the y values are known as **entries**.

Now, we shall discuss the various types of difference operators and how they work with these tabular values. The purpose of our study is to determine the values of $f(x)$ for some intermediate values of x.

2.3 FORWARD DIFFERENCES

The differences $y_1 - y_0, y_2 - y_1, y_3 - y_2, \ldots, y_n - y_{n-1}$ are called **1st order forward differences** for the entries $y_0, y_1, y_2, y_3, \ldots, y_{n-1}, y_n$ and are denoted by $\Delta y_0, \Delta y_1, \Delta y_2, \ldots, \Delta y_{n-1}$, respectively.

In general, $\Delta y_i = y_{i+1} - y_i$, $i = 0, 1, 2, \ldots, n - 1$. This Δ is known as the **forward difference operator**.

The 2nd order differences are the forward differences of these 1st order differences and are denoted by $\Delta^2 y_0, \Delta^2 y_1, \Delta^2 y_2, \ldots$ etc.

Thus

$$\Delta^2 y_0 = \Delta y_1 - \Delta y_0 = (y_2 - y_1) - (y_1 - y_0)$$
$$= y_2 - 2y_1 + y_0$$
$$\Delta^2 y_1 = \Delta y_2 - \Delta y_1 = (y_3 - y_2) - (y_2 - y_1)$$
$$= y_3 - 2y_2 + y_1$$
$$\Delta^2 y_2 = \Delta y_3 - \Delta y_2 = (y_4 - y_3) - (y_3 - y_2)$$
$$= y_4 - 2y_3 + y_2$$

Similarly, the 3rd order forward differences are given as

$$\Delta^3 y_0 = \Delta^2 y_1 - \Delta^2 y_0 = (y_3 - 2y_2 + y_1) - (y_2 - 2y_1 + y_0)$$
$$= y_3 - 3y_2 + 3y_1 - y_0$$
$$\Delta^3 y_1 = \Delta^2 y_2 - \Delta^2 y_1 = (y_4 - 2y_3 + y_2) - (y_3 - 2y_2 + y_1)$$
$$= y_4 - 3y_3 + 3y_2 - y_1$$
$$\Delta^3 y_2 = \Delta^2 y_3 - \Delta^2 y_2 = (y_5 - 2y_4 + y_3) - (y_4 - 2y_3 + y_2)$$
$$= y_5 - 3y_4 + 3y_3 - y_2$$

In general, the *n*th order forward differences are defined as

$$\Delta^n y_k = y_{n+k} - n_{C_1} y_{n-1+k} + \cdots + (-1)^n y_k, k = 0,1,2,\ldots$$

In functional form, the 1st order forward difference is given by $\Delta f(x) = f(x + h) - f(x)$ and it is shown in Fig. 2.1.

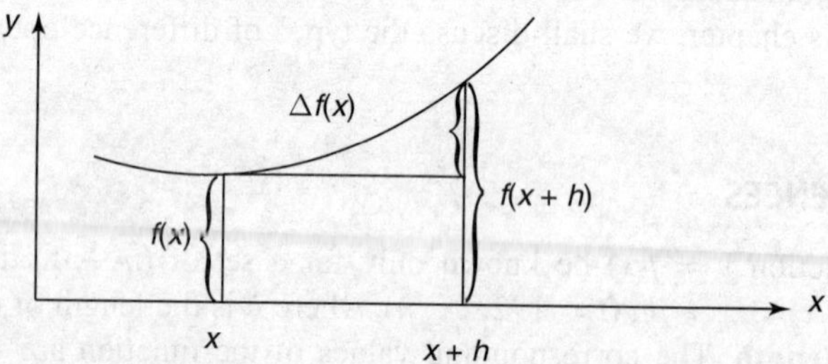

Fig. 2.1 $\Delta f(x) = f(x+4) - f(x)$

Similarly

$$\Delta^2 f(x) = \Delta\{\Delta f(x)\}$$
$$= \Delta\{f(x+h) - f(x)\}$$
$$= \Delta f(x+h) - \Delta f(x)$$
$$= \{f(x+2h) - f(x+h)\} - \{f(x+h) - f(x)\}$$
$$= f(x+2h) - 2f(x+h) + f(x)$$

This is known as the 2nd order forward difference of $f(x)$, and so on.

Construction of Forward Difference Table Let us consider the following tabular values:

x:	x_0	x_1	x_2	x_3	x_4	x_5
y:	y_0	y_1	y_2	y_3	y_4	y_5

where all the arguments $x_0, x_1, x_2, x_3, x_4, x_5$ are equally spaced and $h = x_1 - x_0 = x_2 - x_1 = \ldots = x_5 - x_4$. Then

$$\Delta y_0 = y_1 - y_0, \Delta y_1 = y_2 - y_1, \Delta y_2 = y_3 - y_2, \Delta y_3 = y_4 - y_3, \Delta y_4 = y_5 - y_4$$

$$\Delta^2 y_0 = \Delta y_1 - \Delta y_0,\ \Delta^2 y_1 = \Delta y_2 - \Delta y_1,\ \Delta^2 y_2 = \Delta y_3 - \Delta y_2,\ \Delta^2 y_3 = \Delta y_4 - \Delta y_3$$

$$\Delta^3 y_0 = \Delta^2 y_1 - \Delta^2 y_0,\ \Delta^3 y_1 = \Delta^2 y_2 - \Delta^2 y_1,\ \Delta^3 y_2 = \Delta^2 y_3 - \Delta^2 y_2$$

$$\Delta^4 y_0 = \Delta^3 y_1 - \Delta^3 y_0,\ \Delta^4 y_1 = \Delta^3 y_2 - \Delta^3 y_1$$

$$\Delta^5 y_0 = \Delta^4 y_1 - \Delta^4 y_0$$

Now, we can show these forward differences in two different tables.

TABLE 2.1 Diagonal forward differences

x	y	Δy	$\Delta^2 y$	$\Delta^3 y$	$\Delta^4 y$	$\Delta^5 y$
x_0	y_0					
		Δy_0				
x_1	y_1		$\Delta^2 y_0$			
		Δy_1		$\Delta^3 y_0$		
x_2	y_2		$\Delta^2 y_1$		$\Delta^4 y_0$	
		Δy_2		$\Delta^3 y_1$		$\Delta^5 y_0$
x_3	y_3		$\Delta^2 y_2$		$\Delta^4 y_1$	
		Δy_3		$\Delta^3 y_2$		
x_4	y_4		$\Delta^2 y_3$			
		Δy_4				
x_5	y_5					

TABLE 2.2 Horizontal forward differences

x	y	Δy	$\Delta^2 y$	$\Delta^3 y$	$\Delta^4 y$	$\Delta^5 y$
x_0	y_0	Δy_0	$\Delta^2 y_0$	$\Delta^3 y_0$	$\Delta^4 y_0$	$\Delta^5 y_0$
x_1	y_1	Δy_1	$\Delta^2 y_1$	$\Delta^3 y_1$	$\Delta^4 y_1$	
x_2	y_2	Δy_2	$\Delta^2 y_2$	$\Delta^3 y_2$		
x_3	y_3	Δy_3	$\Delta^2 y_3$			
x_4	y_4	Δy_4				
x_5	y_5					

For example, let us consider the following table:

x	0	5	10	15	20	25
$f(x)$	7	11	14	18	24	32

The diagonal forward difference table is given below:

x	$f(x)$	$\Delta f(x)$	$\Delta^2 f(x)$	$\Delta^3 f(x)$	$\Delta^4 f(x)$	$\Delta^5 f(x)$
0	7					
		4				
5	11		−1			
		3		2		
10	14		1		−1	
		4		1		0
15	18		2		−1	
		6		0		
20	24		2			
		8				
25	32					

And the horizontal forward difference table is given below:

x	$f(x)$	$\Delta f(x)$	$\Delta^2 f(x)$	$\Delta^3 f(x)$	$\Delta^4 f(x)$	$\Delta^5 f(x)$
0	7	4	−1	2	−1	0
5	11	3	1	1	−1	
10	14	4	2	0		
15	18	6	2			
20	24	8				
25	32					

Here

$$x_0 = 0,\ f(x_0) = 7,\ \Delta f(x_0) = 4,\ \Delta^2 f(x_0) = -1,\ \Delta^3 f(x_0) = 2,\ \Delta^4 f(x_0) = -1,\ \Delta^5 f(x_0) = 0$$

$$x_1 = 5,\ f(x_1) = 11,\ \Delta f(x_1) = 3,\ \Delta^2 f(x_1) = 1,\ \Delta^3 f(x_1) = 1,\ \Delta^4 f(x_1) = -1$$

$$x_2 = 10,\ f(x_2) = 14,\ \Delta f(x_2) = 4,\ \Delta^2 f(x_2) = 2,\ \Delta^3 f(x_2) = 0$$

$$x_3 = 15,\ f(x_3) = 18,\ \Delta f(x_3) = 6,\ \Delta^2 f(x_3) = 2$$

$$x_4 = 20,\ f(x_4) = 24,\ \Delta f(x_0) = 8$$

Properties of Forward Difference Operator The forward difference operator has the following properties:

(i) $\Delta c = 0$, where c is a constant

(ii) $\Delta\{cf(x)\} = c\Delta f(x)$, where c is a constant

(iii) $\Delta\{f(x).g(x)\} = f(x+h)\Delta g(x) + g(x)\Delta f(x)$

(iv) $\Delta\left\{\dfrac{f(x)}{g(x)}\right\} = \dfrac{g(x)\Delta f(x) - f(x)\Delta g(x)}{g(x)g(x+h)}$

Proof: (i) Let $f(x) = c$.

Then $\Delta f(x) = f(x+h) - f(x) = c - c = 0$

(ii) $\Delta\{cf(x)\} = cf(x+h) - cf(x) = c\{f(x+h) - f(x)\}$

$= c\Delta f(x)$

(iii) $\Delta\{f(x).g(x)\} = f(x+h)g(x+h) - f(x)g(x)$

$= f(x+h)g(x+h) - f(x+h)g(x) + f(x+h)g(x) - f(x)g(x)$

$= f(x+h)\{g(x+h) - g(x)\} + g(x)\{f(x+h) - f(x)\}$

$= f(x+h)\Delta g(x) + g(x)\Delta f(x)$

(iv) $$\Delta\left[\frac{f(x)}{g(x)}\right] = \frac{f(x+h)}{g(x+h)} - \frac{f(x)}{g(x)}$$

$$= \frac{f(x+h)g(x) - g(x+h)f(x)}{g(x+h)g(x)}$$

$$= \frac{f(x+h)g(x) - f(x)g(x) + f(x)g(x) - g(x+h)f(x)}{g(x+h)g(x)}$$

$$= \frac{g(x)\{f(x+h) - f(x)\} - f(x)\{g(x+h) - g(x)\}}{g(x+h)g(x)}$$

$$= \frac{g(x)\Delta f(x) - f(x)\Delta g(x)}{g(x)g(x+h)}$$

EXAMPLE 2.1 Show that $\Delta \log f(x) = \log\left[1 + \frac{\Delta f(x)}{f(x)}\right]$.

Solution: $$\Delta \log f(x) = \log f(x+h) - \log f(x) = \log\left[\frac{f(x+h)}{f(x)}\right]$$

$$= \log\left[\frac{f(x+h) - f(x) + f(x)}{f(x)}\right] = \log\left[\frac{\Delta f(x) + f(x)}{f(x)}\right]$$

$$= \log\left[1 + \frac{\Delta f(x)}{f(x)}\right]$$

EXAMPLE 2.2 Prove that $\Delta^2\left(\frac{5x+12}{x^2+5x+6}\right) = \frac{2(5x+16)}{(x+2)(x+3)(x+4)(x+5)}$ when $h = 1$.

Solution: $$\Delta^2\left(\frac{5x+12}{x^2+5x+6}\right) = \Delta\left[\Delta\left(\frac{5x+12}{(x+2)(x+3)}\right)\right]$$

$$= \Delta\left[\frac{5(x+1)+12}{(x+3)(x+4)} - \frac{5x+12}{(x+2)(x+3)}\right]$$

$$= \Delta\left[\frac{-(5x+14)}{(x+2)(x+3)(x+4)}\right]$$

$$= -\left[\frac{5(x+1)+14}{(x+3)(x+4)(x+5)} - \frac{5x+14}{(x+2)(x+3)(x+4)}\right]$$

$$= \frac{10x+32}{(x+2)(x+3)(x+4)(x+5)}$$

$$= \frac{2(5x+16)}{(x+2)(x+3)(x+4)(x+5)}$$

EXAMPLE 2.3 Prove that $f(4) = f(3) + \Delta f(2) + \Delta^2 f(1) + \Delta^3 f(1)$.

Solution: We have $\Delta f(3) = f(4) - f(3)$

Therefore

$$\begin{aligned} f(4) &= f(3) + \Delta f(3) \\ &= f(3) + \Delta[f(2) + \Delta f(2)] \\ &= f(3) + \Delta f(2) + \Delta^2 f(2) \\ &= f(3) + \Delta f(2) + \Delta^2[f(1) + \Delta f(1)] \\ &= f(3) + \Delta f(2) + \Delta^2 f(1) + \Delta^3 f(1) \end{aligned}$$

EXAMPLE 2.4 Evaluate $\Delta^2(\cos 2x)$.

Solution:

$$\begin{aligned} \Delta^2(\cos 2x) &= \Delta[\Delta \cos 2x] \\ &= \Delta[\cos 2(x+h) - \cos 2x] \\ &= \Delta[2\sin(2x+h)\sin(-h)] \\ &= -2\sin h\Delta[\sin(2x+h)] \\ &= -2\sin h[\sin(2x+3h) - \sin(2x+h)] \\ &= -2\sin h[2\cos(2x+2h).\sin h] \\ &= -4\sin^2 h\cos(2x+2h) \end{aligned}$$

Fundamental Theorem of Difference Calculus If $f(x)$ is a polynomial of degree n, then the nth order and the $(n+1)$th forward differences of the polynomial are, respectively, a **constant** and **zero**.

Proof: Let $f(x) = a_0 x^n + a_1 x^{n-1} + \cdots + a_{n-1}x + a_n$ be the polynomial of degree n, where $a_0, a_1, \ldots, a_n$ are constant and $a_0 \neq 0$.

Then

$$\Delta f(x) = f(x+h) - f(x)$$

$$= \left\{a_0\left(x+h\right)^n + a_1\left(x+h\right)^{n-1} + \cdots + a_{n-1}\left(x+h\right) + a_n\right\} - \left\{a_0x^n + a_1x^{n-1} + \cdots + a_{n-1}x + a_n\right\}$$

$$= a_0\left\{\left(x+h\right)^n - x^n\right\} + a_1\left\{\left(x+h\right)^{n-1} - x^{n-1}\right\} + \cdots + a_{n-1}\left\{x+h-x\right\}$$

$$= a_0nhx^{n-1} + b_1x^{n-2} + \cdots + b_{n-2}x + ha_{n-1}, \text{ where } b_1, b_2, \ldots, b_{n-2} \text{ are constants.}$$

Clearly $\Delta f(x)$ is a polynomial of degree $(n - 1)$.

Again, $\Delta^2 f(x) = \Delta f(x + h) - \Delta f(x)$

$$= a_0nh\left\{\left(x+h\right)^{n-1} - x^{n-1}\right\} + b_1\left\{\left(x+h\right)^{n-2} - x^{n-2}\right\} + \cdots + b_{n-2}\left\{x+h-x\right\}$$

$$= a_0n\left(n-1\right)h^2x^{n-2} + c_1x^{n-3} + \cdots + c_{n-2}, \text{ where } c_1, c_2, \ldots, c_{n-2} \text{ are constants.}$$

We see that $\Delta^2 f(x)$ is a polynomial of degree $(n - 2)$.

Continuing this process for the nth difference, we get a polynomial of degree $(n - n) = 0$, i.e.,

$$\Delta^n f\left(x\right) = a_0n\left(n-1\right)\left(n-2\right)\ldots 3\cdot 2\cdot 1\cdot h^n = a_0n!h^n, \text{ a constant and hence,}$$

$$\Delta^{n+1} f\left(x\right) = 0$$

Note: It is clear from this theorem that if there are n known values of y for n values of x, then the polynomial that suits these values will be of degree less than or equal to $(n - 1)$. In that case, the nth order forward difference of this polynomial will vanish. We use this concept to solve the problem of missing terms.

EXAMPLE 2.5 Find the missing term in the following table:

x	0	1	2	3	4
$f(x)$	4	8	15	-	72

Solution: There are four known values of $f(x)$. So $f(x)$ is a polynomial of degree less than 4. Therefore, $\Delta^4 f(x) = 0$. Now, we construct the forward difference table, with the missing term being denoted as z:

$f(x)$	$\Delta f(x)$	$\Delta^2 f(x)$	$\Delta^3 f(x)$	$\Delta^4 f(x)$
4				
	4			
8		3		
	7		$z - 25$	
15		$z - 22$		$134 - 4z$
	$z - 15$		$109 - 3z$	
z		$87 - 2z$		
	$72 - z$			
72				

Therefore $\Delta^4 f(x) = 0$

$\Rightarrow$ $134 - 4z = 0$

$\Rightarrow$ $z = 33.5$

EXAMPLE 2.6 Find the missing terms in the following table:

x	0	5	10	15	20	25
$f(x)$	1	-	38	82	-	260

Solution: There are four known values of $f(x)$. So $f(x)$ is a polynomial of degree less than 4. Therefore, $\Delta^4 f(x) = 0$. Now, we construct the forward difference table, with the missing terms denoted as p and q:

$f(x)$	$\Delta f(x)$	$\Delta^2 f(x)$	$\Delta^3 f(x)$	$\Delta^4 f(x)$
1				
	$p - 1$			
p		$39 - 2p$		
	$38 - p$		$3p - 33$	
38		$6 + p$		$q - 4p - 99$
	44		$q - p - 132$	
82		$q - 126$		$600 - 4q + p$
	$q - 82$		$468 - 3q$	
q		$342 - 2q$		
	$260 - q$			
260				

Therefore
$$\Delta^4 f(x) = 0 \Rightarrow \begin{cases} q - 4p - 99 = 0 \\ 600 - 4q + p = 0 \end{cases}$$

Solving these, we get $p = 13.6$ and $q = 153.4$.

EXAMPLE 2.7 If the interval of differencing is unity, evaluate

(i) $\Delta^3(4x^3 - 2x^2 + 7x + 11)$
(ii) $\Delta^{10}(1 - ax)(1 - bx^2)(1 - cx^3)(1 - dx^4)$

Solution: (i) As $\Delta^n x^m = 0$ for $n > m$, we have

$$\Delta^3\left(4x^3 - 2x^2 + 7x + 11\right)$$
$$= 4\Delta^3\left(x^3\right) - 2\Delta^3\left(x^2\right) + 7\Delta^3(x) + \Delta^3(11)$$
$$= 4 \times 3! - 2 \times 0 + 7 \times 0 + 0, \text{ since } h = 1.$$
$$= 24$$

(ii) The polynomial in this case is of degree 10 and the coefficient of x^{10} is $abcd$. So, we get

$$\Delta^{10}(1 - ax)\left(1 - bx^2\right)\left(1 - cx^3\right)\left(1 - dx^4\right) = abcd\Delta^{10}\left(x^{10}\right)$$
$$= abcd \times 10!, \text{ as } h = 1.$$

EXAMPLE 2.8 Evaluate $\Delta^4 f(x)$ where $f(x) = 5x^4 + 2x^3 - 8x^2 + 3x - 13$, taking $h = 2$.

Solution: We know $\Delta^n x^m = 0$ for $n > m$.

Therefore $$\Delta^4 f(x) = 5\Delta^4(x^4) + 2\Delta^4(x^3) - 8\Delta^4(x^2) + 3\Delta^4(x) - \Delta^4(13)$$

$$= 5\times 4!\times 2^4 + 2\times 0 - 8\times 0 + 3\times 0 - 0 = 1920, \text{ as } h = 2.$$

EXAMPLE 2.9 Find the polynomial whose first forward difference is $3x^2 + 7x + 6$, assuming that the interval of differencing is unity.

Solution: As the given polynomial is of degree 2, the required solution is of degree 3. So, let the required polynomial be

$$f(x) = ax^3 + bx^2 + cx + d$$

Using the given condition, we get

$$\Delta f(x) = 3x^2 + 7x + 6$$

$$\Rightarrow \quad \left\{a(x+1)^3 + b(x+1)^2 + c(x+1) + d\right\} - \left\{ax^3 + bx^2 + cx + d\right\} = 3x^2 + 7x + 6$$

$$\Rightarrow \quad 3ax^2 + (3a + 2b)x + (a + b + c) = 3x^2 + 7x + 6$$

Then, equating the like powers of x, we get

$$3a = 3,\ 3a + 2b = 7 \text{ and } a + b + c = 6$$

Solving the above equations, we get $a = 1$, $b = 2$ and $c = 3$.

Therefore, the required polynomial is

$$f(x) = x^3 + 2x^2 + 3x + d, \text{ where } d \text{ is a constant.}$$

EXAMPLE 2.10 Find $f(x)$ where $\Delta f(x) = x^2 + 11x + 5$, interval of differencing is unity and Δ is the forward difference operator.

Solution: As the given polynomial is of degree 2, let the required polynomial be

$$f(x) = ax^3 + bx^2 + cx + d$$

Using the given condition, we get

$$\Delta f(x) = x^2 + 11x + 5$$

$$\Rightarrow \quad \left\{a(x+1)^3 + b(x+1)^2 + c(x+1) + d\right\} - \left\{ax^3 + bx^2 + cx + d\right\} = x^2 + 11x + 5$$

$$\Rightarrow \quad 3ax^2 + (3a + 2b)x + (a + b + c) = x^2 + 11x + 5$$

Then, equating the like powers of x, we get

$$3a = 1,\ 3a + 2b = 11 \text{ and } a + b + c = 5$$

Solving the above equations, we get $a = 1/3$, $b = 5$ and $c = -\,1/3$.

Therefore, the required polynomial is

$$f(x)=\frac{1}{3}x^3+5x^2-\frac{1}{3}x+d, \text{ where } d \text{ is a constant.}$$

2.4 BACKWARD DIFFERENCES

The differences $y_1 - y_0, y_2 - y_1, y_3 - y_2, \ldots, y_n - y_{n-1}$ are called **1st order backward differences** for the entries $y_0, y_1, y_2, y_3, \ldots, y_{n-1}, y_n$ and are denoted by $\nabla y_1, \nabla y_2, \nabla y_3, \ldots, \nabla y_n$, respectively.

In general, $\nabla y_i = y_i - y_{i-1}$, $i = 1, 2, \ldots, n$. This ∇ is known as the **backward difference operator**.

The 2nd order differences are the backward differences of the 1st order differences and are denoted by $\nabla^2 y_2, \nabla^2 y_3, \ldots, \nabla^2 y_n$, etc.

Thus

$$\nabla^2 y_2 = \nabla y_2 - \nabla y_1 = (y_2 - y_1) - (y_1 - y_0)$$
$$= y_2 - 2y_1 + y_0$$
$$\nabla^2 y_3 = \nabla y_3 - \nabla y_2 = (y_3 - y_2) - (y_2 - y_1)$$
$$= y_3 - 2y_2 + y_1$$
$$\nabla^2 y_4 = \nabla y_4 - \nabla y_3 = (y_4 - y_3) - (y_3 - y_2)$$
$$= y_4 - 2y_3 + y_2$$

Similarly, the 3rd order backward differences are given as

$$\nabla^3 y_3 = \nabla^2 y_3 - \nabla^2 y_2 = (y_3 - 2y_2 + y_1) - (y_2 - 2y_1 + y_0)$$
$$= y_3 - 3y_2 + 3y_1 - y_0$$
$$\nabla^3 y_4 = \nabla^2 y_4 - \nabla^2 y_3 = (y_4 - 2y_3 + y_2) - (y_3 - 2y_2 + y_1)$$
$$= y_4 - 3y_3 + 3y_2 - y_1$$

In general, the nth order backward differences are defined as

$$\nabla^n y_k = y_k - {}^nC_1\, y_{k-1} + \cdots + (-1)^n\, y_{k-n}, n = 1, 2, \ldots \text{ and } n \le k.$$

In functional form, the 1st order backward difference is given by $\nabla f(x) = f(x) - f(x - h)$ and it is shown in Fig. 2.2.

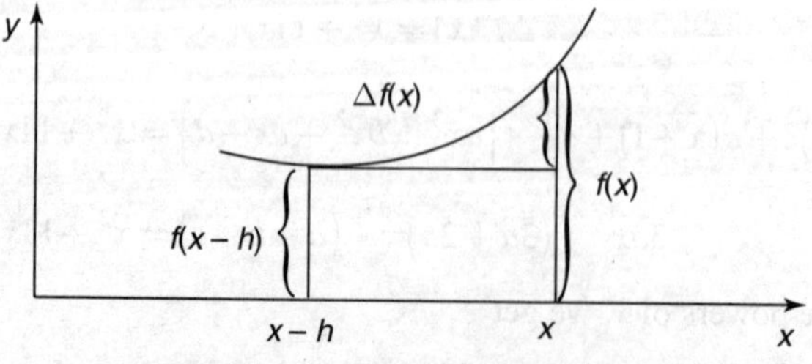

Fig. 2.2 $\nabla f(x) = f(x) - f(x - h)$

Similarly

$$\begin{aligned}\nabla^2 f(x) &= \nabla\{\nabla f(x)\}\\ &= \nabla\{f(x)-f(x-h)\}\\ &= \nabla f(x)-\nabla f(x-h)\\ &= \{f(x)-f(x-h)\}-\{f(x-h)-f(x-2h)\}\\ &= f(x)-2f(x-h)+f(x-2h)\end{aligned}$$

This is known as the 2nd order backward difference of $f(x)$, and so on.

Construction of Backward Difference Table Let us consider the following tabular values:

$$\begin{matrix} x: & x_0 & x_1 & x_2 & x_3 & x_4 & x_5 \\ y: & y_0 & y_1 & y_2 & y_3 & y_4 & y_5 \end{matrix}$$

where all the arguments $x_0, x_1, x_2, x_3, x_4, x_5$ are equally spaced and $h = x_1 - x_0 = x_2 - x_1 = \ldots = x_5 - x_4$.

Then

$$\nabla y_1 = y_1 - y_0,\ \nabla y_2 = y_2 - y_1,\ \nabla y_3 = y_3 - y_2,\ \nabla y_4 = y_4 - y_3,\ \nabla y_5 = y_5 - y_4$$

$$\nabla^2 y_2 = \nabla y_2 - \nabla y_1,\ \nabla^2 y_3 = \nabla y_3 - \nabla y_2,\ \nabla^2 y_4 = \nabla y_4 - \nabla y_3,\ \nabla^2 y_5 = \nabla y_5 - \nabla y_4$$

$$\nabla^3 y_3 = \nabla^2 y_3 - \nabla^2 y_2,\ \nabla^3 y_4 = \nabla^2 y_4 - \nabla^2 y_3,\ \nabla^3 y_5 = \nabla^2 y_5 - \nabla^2 y_4$$

$$\nabla^4 y_4 = \nabla^3 y_4 - \nabla^3 y_3,\ \nabla^4 y_5 = \nabla^3 y_5 - \nabla^3 y_4$$

$$\nabla^5 y_5 = \nabla^4 y_5 - \nabla^4 y_4$$

Now, we can show these backward differences in two different tables:

Table 2.3 Diagonal backward differences

x	y	∇y	$\nabla^2 y$	$\nabla^3 y$	$\nabla^4 y$	$\nabla^5 y$
x_0	y_0					
		∇y_1				
x_1	y_1		$\nabla^2 y_2$			
		∇y_2		$\nabla^3 y_3$		
x_2	y_2		$\nabla^2 y_3$		$\nabla^4 y_4$	
		∇y_3		$\nabla^3 y_4$		$\nabla^5 y_5$
x_3	y_3		$\nabla^2 y_4$		$\nabla^4 y_5$	
		∇y_4		$\nabla^3 y_5$		
x_4	y_4		$\nabla^2 y_5$			
		∇y_5				
x_5	y_5					

Table 2.4 Horizontal backward differences

x	y	∇y	$\nabla^2 y$	$\nabla^3 y$	$\nabla^4 y$	$\nabla^5 y$
x_0	y_0					
x_1	y_1	∇y_1				
x_2	y_2	∇y_2	$\nabla^2 y_2$			
x_3	y_3	∇y_3	$\nabla^2 y_3$	$\nabla^3 y_3$		
x_4	y_4	∇y_4	$\nabla^2 y_4$	$\nabla^3 y_4$	$\nabla^4 y_4$	
x_5	y_5	∇y_5	$\nabla^2 y_5$	$\nabla^3 y_5$	$\nabla^4 y_4$	$\nabla^5 y_5$

For example, let us consider the following table:

x	0	2	4	6	8	10
$f(x)$	5	8	10	13	14	16

The diagonal backward difference table is given below:

x	$f(x)$	$\nabla f(x)$	$\nabla^2 f(x)$	$\nabla^3 f(x)$	$\nabla^4 f(x)$	$\nabla^5 f(x)$
0	5					
		3				
2	8		− 1			
		2		2		
4	10		1		− 5	
		3		− 3		11
6	13		− 2		6	
		1		3		
8	14		1			
		2				
10	16					

And the horizontal forward difference table is given below:

x	$f(x)$	$\nabla f(x)$	$\nabla^2 f(x)$	$\nabla^3 f(x)$	$\nabla^4 f(x)$	$\nabla^5 f(x)$
0	5					
2	8	3				
4	10	2	− 1			
6	13	3	1	2		
8	14	1	− 2	− 3	− 5	
10	16	2	1	3	6	11

Here, $x_1 = 2,\ f(x_1) = 8,\ \nabla f(x_1) = 3$

$$x_2 = 4,\ f(x_2) = 10,\ \nabla f(x_2) = 2,\ \nabla^2 f(x_2) = -1$$

$$x_3 = 6,\ f(x_3) = 13,\ \nabla f(x_3) = 3,\ \nabla^2 f(x_3) = 1,\ \nabla^3 f(x_3) = 2$$

$$x_4 = 8, f(x_4) = 14, \nabla f(x_4) = 1, \nabla^2 f(x_4) = -2, \nabla^3 f(x_4) = -3, \nabla^4 f(x_4) = -5$$

$$x_5 = 10, f(x_5) = 16, \nabla f(x_5) = 2, \nabla^2 f(x_5) = 1, \nabla^3 f(x_5) = 3, \nabla^4 f(x_5) = 6, \nabla^5 f(x_5) = 11$$

2.5 SHIFT OPERATOR

The shift operator is denoted by E and is defined by $Ef(x) = f(x + h)$, h being the length of differencing. This is depicted in Fig. 2.3

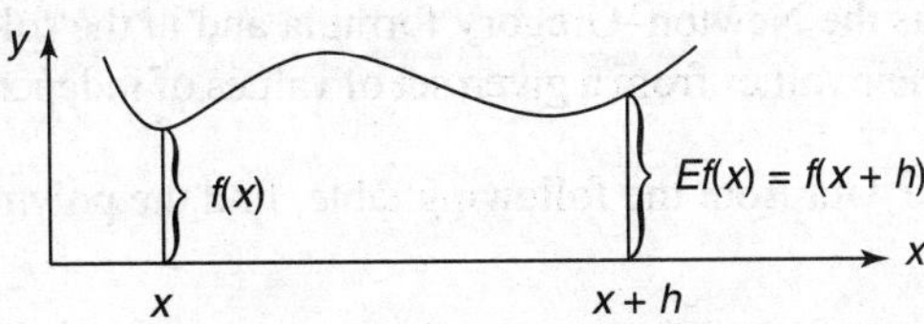

Fig. 2.3 $Ef(x) = f(x + h)$

Now $E^2f(x) = Ef(x + h) = f(x + 2h)$, $E^3f(x) = Ef(x + 2h) = f(x + 3h)$ and so on.

In general $E^nf(x) = f(x + nh)$, $n \epsilon R^+$

Similarly, the inverse shift operator is defined as

$$E^{-1}f(x) = f(x - h),\ E^{-2}f(x) = f(x - 2h),\ E^{-3}f(x) = f(x - 3h) \text{ and so on.}$$

In general $E^{-n}f(x) = f(x - nh)$, $n \epsilon R^+$

Some Identities

(i) We see that

$$\Delta f(x) = f(x + h) - f(x)$$

and $Ef(x) = f(x + h)$.

So
$$Ef(x) = f(x + h) = \Delta f(x) + f(x)$$
$$= (1 + \Delta)f(x).$$

Therefore $E \equiv 1 + \Delta$

(ii) For the backward difference, we have

$$\nabla f(x) = f(x) - f(x - h)$$

and from the shift operator $E^{-1}f(x) = f(x - h)$.

Hence
$$E^{-1}f(x) = f(x - h) = f(x) - \nabla f(x)$$
$$= (1 - \nabla)f(x)$$

which gives $E^{-1} \equiv 1 - \nabla$

Using the identity $E \equiv 1 + \Delta$, we shall derive one important formula which is very useful for obtaining the polynomial for a given set of data.

Newton–Gregory Formula

We have $f(x + nh) = E^nf(x) = (1 + \Delta)^nf(x)$

$$= \left[1 + \binom{n}{1}\Delta + \binom{n}{2}\Delta^2 + \cdots + \binom{n}{n}\Delta^n\right] f(x)$$

$$= f(x) + \binom{n}{1}\Delta f(x) + \binom{n}{2}\Delta^2 f(x) + \cdots + \binom{n}{n}\Delta^n f(x)$$

$$= \sum_{i=0}^{n} \binom{n}{i}\Delta^i f(x)$$

This formula is known as the Newton–Gregory formula and in the following examples it is used for deriving polynomials and their values from a given set of values of independent and dependent variables.

EXAMPLE 2.11 Using the data from the following table, find the polynomial $f(x)$ and hence calculate $f(0.5)$ and $f(3.5)$.

x	0	1	2	3	4
$f(x)$	3	6	11	18	27

Solution: The forward difference table is given below:

x	$f(x)$	$\Delta f(x)$	$\Delta^2 f(x)$	$\Delta^3 f(x)$	$\Delta^4 f(x)$
0	3				
		3			
1	6		2		
		5		0	
2	11		2		0
		7		0	
3	18		2		
		9			
4	27				

As $\Delta^2 f(x)$ is constant and $\Delta^3 f(x)$ is zero, $y = f(x)$ is a polynomial of degree 2.

Now $\quad f(x_0 + nh) = E^n f(x_0) = (1 + \Delta)^n f(x_0)$ and $x_0 = 0$, $h = 1$.

Let $\quad x = x_0 + nh$ so that $n = x$.

Therefore $\quad f(x) = (1 + \Delta)^n f(x_0)$

$$= \left[1 + n\Delta + \frac{n(n-1)}{2!}\Delta^2 + \frac{n(n-1)(n-2)}{3!}\Delta^3 + \cdots\right] f(x_0)$$

$$= f(x_0) + n\Delta f(x_0) + \frac{n(n-1)}{2!}\Delta^2 f(x_0) + \frac{n(n-1)(n-2)}{3!}\Delta^3 f(x_0) + \cdots$$

$$= 3 + x \times 3 + \frac{x(x-1)}{2!} \times 2 + \frac{x(x-1)(x-2)}{3!} \times 0$$

$= x^2 + 2x + 3$, which is the required polynomial.

Now $$f(0.5) = 0.25 + 2 \times 0.5 + 3 = 4.25$$

and $$f(3.5) = 12.25 + 2 \times 3.5 + 3 = 22.25$$

EXAMPLE 2.12 Find the polynomial $y = f(x)$ suitable for the following data, and hence calculate $f(1.5)$.

x	0	1	2	3
$f(x)$	1	2	1	10

Solution: The forward difference table is given below:

x	$f(x)$	$\Delta f(x)$	$\Delta^2 f(x)$	$\Delta^3 f(x)$
0	1			
		1		
1	2		−2	
		−1		12
2	1		10	
		9		
3	10			

From the difference table, it is clear that $y = f(x)$ is a polynomial of degree 3.

Now $$f(x_0 + nh) = E^n f(x_0) = (1 + \Delta)^n f(x_0) \text{ and } x_0 = 0,\ h = 1$$

Let $$x = x_0 + nh \text{ so that } n = x.$$

Therefore
$$f(x) = (1 + \Delta)^n f(x_0)$$
$$= \left[1 + n\Delta + \frac{n(n-1)}{2!}\Delta^2 + \frac{n(n-1)(n-2)}{3!}\Delta^3 + \cdots\right] f(x_0)$$
$$= f(x_0) + n\Delta f(x_0) + \frac{n(n-1)}{2!}\Delta^2 f(x_0) + \frac{n(n-1)(n-2)}{3!}\Delta^3 f(x_0) + \cdots$$
$$= 1 + x \times 1 + \frac{x(x-1)}{2!} \times (-2) + \frac{x(x-1)(x-2)}{3!} \times 12$$
$$= 2x^3 - 7x^2 + 6x + 1, \text{ which is the required polynomial.}$$

So $$f(1.5) = 2 \times (1.5)^3 - 7 \times (1.5)^2 + 6 \times 1.5 + 1 = 1.2$$

EXAMPLE 2.13 Find the polynomial $f(x)$ for the following data, and hence calculate $f(5)$.

x	0	2	4	6	8
$f(x)$	−1	13	43	89	151

Solution: The forward difference table is given below:

x	$f(x)$	$\Delta f(x)$	$\Delta^2 f(x)$	$\Delta^3 f(x)$	$\Delta^4 f(x)$
0	−1				
		14			
2	13		16		
		30		0	
4	43		16		0
		46		0	
6	89		16		
		62			
8	151				

As $\Delta^2 f(x)$ is constant and $\Delta^3 f(x)$ is zero, $y = f(x)$ is a polynomial of degree 2.

Now $\quad f(x_0 + nh) = E^n f(x_0) = (1 + \Delta)^n f(x_0)$ and $x_0 = 0$, $h = 2$.

Let $\quad x = x_0 + nh$ so that $n = x/2$.

Therefore

$$f(x) = (1 + \Delta)^n f(x_0)$$

$$= \left[1 + n\Delta + \frac{n(n-1)}{2!}\Delta^2 + \frac{n(n-1)(n-2)}{3!}\Delta^3 + \cdots\right] f(x_0)$$

$$= f(x_0) + n\Delta f(x_0) + \frac{n(n-1)}{2!}\Delta^2 f(x_0) + \frac{n(n-1)(n-2)}{3!}\Delta^3 f(x_0) + \cdots$$

$$= -1 + \left(\frac{x}{2}\right) \times 14 + \frac{\frac{x}{2}\left(\frac{x}{2} - 1\right)}{2!} \times 16 + \frac{\frac{x}{2}\left(\frac{x}{2} - 1\right)\left(\frac{x}{2} - 2\right)}{3!} \times 0$$

$= 2x^2 + 3x - 1$, which is the required polynomial.

Hence $\quad f(5) = 2 \times 25 + 3 \times 5 - 1 = 64$

EXAMPLE 2.14 Find the polynomial $f(x)$ for the following data, and hence calculate $f(0.5)$.

x	0	1	2	3	4
$f(x)$	3	7	13	24	37

Solution: The forward difference table is given below:

x	$f(x)$	$\Delta f(x)$	$\Delta^2 f(x)$
0	3		
		4	
1	7		2
		6	
2	13		2
		8	
3	21		2
		10	
4	31		

As $\Delta^2 f(x)$ is constant and $\Delta^3 f(x)$ is zero, $y = f(x)$ is a polynomial of degree 2.

Now $\qquad f(x_0 + nh) = E^n f(x_0) = (1 + \Delta)^n f(x_0)$ and $x_0 = 0$, $h = 1$.

Let $\qquad x = x_0 + nh$ so that $n = x$.

Therefore $\qquad f(x) = (1 + \Delta)^n f(x_0)$

$$= \left[1 + n\Delta + \frac{n(n-1)}{2!}\Delta^2 + \frac{n(n-1)(n-2)}{3!}\Delta^3 + \cdots\right] f(x_0)$$

$$= f(x_0) + n\Delta f(x_0) + \frac{n(n-1)}{2!}\Delta^2 f(x_0) + \frac{n(n-1)(n-2)}{3!}\Delta^3 f(x_0) + \cdots$$

$$= 3 + x \times 4 + \frac{x(x-1)}{2!} \times 2 + 0$$

$= x^2 + 2x + 3$, which is the required polynomial.

Hence $\qquad f(0.5) = 0.25 + 2 \times 0.5 + 3 = 4.25$

2.6 CENTRAL DIFFERENCE OPERATOR AND AVERAGING OPERATOR

The central difference operator is denoted by δ and is defined as

$$\delta f(x) = f\left(x + \frac{h}{2}\right) - f\left(x - \frac{h}{2}\right), \ h \text{ being the step length.}$$

Clearly $\qquad \delta f(x) = E^{\frac{1}{2}} f(x) - E^{\frac{-1}{2}} f(x)$,

which implies that $\qquad \delta \equiv E^{\frac{1}{2}} - E^{\frac{-1}{2}}$

The averaging operator μ is defined as

$$\mu f(x) = \frac{1}{2}\left\{f\left(x + \frac{h}{2}\right) + f\left(x - \frac{h}{2}\right)\right\}, \ h \text{ being the step length.}$$

So $$\mu f(x) = \frac{1}{2}\left\{E^{\frac{1}{2}} f(x) + E^{\frac{-1}{2}} f(x)\right\} \Rightarrow \mu = \frac{1}{2}\left\{E^{\frac{1}{2}} + E^{\frac{-1}{2}}\right\}$$

2.7 DIVIDED DIFFERENCES

Let a real-valued function $y = f(x)$ be known only for a set of $(n + 1)$ distinct points $x_0, x_1, \ldots, x_n$. The corresponding values of the function are $y_0 = f(x_0)$, $y_1 = f(x_1)$, …, $y_n = f(x_n)$. In this case, the arguments are not necessary equispaced. Then, we define divided differences of $f(x)$ as follows:

The 1st order divided difference

- for the arguments x_0, x_1 is denoted by $[x_0, x_1]$ or $[x_1, x_0]$ and defined as

$$[x_0, x_1] = \frac{y_1 - y_0}{x_1 - x_0} = \frac{y_0 - y_1}{x_0 - x_1} = [x_1, x_0]$$

- for the arguments x_1, x_2 is denoted by $[x_1, x_2]$ or $[x_2, x_1]$ and defined as

$$[x_1,x_2]=\frac{y_2-y_1}{x_2-x_1}=\frac{y_1-y_2}{x_1-x_2}=[x_2,x_1]$$

and so on.

Similarly, the 2nd order divided difference

- for the arguments x_0, x_1, x_2 is denoted by $[x_0, x_1, x_2]$ or $[x_2, x_1, x_0]$ and is defined by

$$[x_0,x_1,x_2]=\frac{[x_1,x_2]-[x_0,x_1]}{x_2-x_0}=\frac{[x_0,x_1]-[x_1,x_2]}{x_0-x_2}$$
$$=[x_2,x_1,x_0]$$

- for the arguments x_1, x_2, x_3 is denoted by $[x_1, x_2, x_3]$ or $[x_3, x_2, x_1]$ and is defined by

$$[x_1,x_2,x_3]=\frac{[x_2,x_3]-[x_1,x_2]}{x_3-x_1}=\frac{[x_1,x_2]-[x_2,x_3]}{x_1-x_3}$$
$$=[x_3,x_2,x_1]$$

and so on.

Construction of Divided Difference Table Let us consider the data given in the following table:

x	2	4	5	7	8
$f(x)$	3	43	138	778	1515

The divided difference table is shown below:

x	$f(x)$	1st order	2nd order	3rd order	4th order
2	3				
		20			
4	43		25		
		95		10	
5	138		75		1
		320		16	
7	778		139		
		737			
8	1515				

Here

$$x_0 = 2, x_1 = 4, x_2 = 5, x_3 = 7 \text{ and } x_4 = 8$$
$$y_0 = 3, y_1 = 43, y_2 = 138, y_3 = 778 \text{ and } y_4 = 1515$$

- **Calculation of 1st order divided differences:**

$$[x_0,x_1]=\frac{y_1-y_0}{x_1-x_0}=\frac{43-3}{4-2}$$
$$=20$$

$$[x_1,x_2]=\frac{y_2-y_1}{x_2-x_1}=\frac{138-43}{5-4}$$
$$=95$$

$$[x_2, x_3] = \frac{y_3 - y_2}{x_3 - x_2} = \frac{778 - 138}{7 - 5}$$
$$= 320$$

$$[x_3, x_4] = \frac{y_4 - y_3}{x_4 - x_3} = \frac{1515 - 778}{8 - 7}$$
$$= 737$$

- **Calculation of 2nd order divided differences:**

$$[x_0, x_1, x_2] = \frac{[x_1, x_2] - [x_0, x_1]}{x_2 - x_0} = \frac{95 - 20}{5 - 2}$$
$$= 25$$

$$[x_1, x_2, x_3] = \frac{[x_2, x_3] - [x_1, x_2]}{x_3 - x_1} = \frac{320 - 95}{7 - 4}$$
$$= 75$$

$$[x_2, x_3, x_4] = \frac{[x_3, x_4] - [x_2, x_3]}{x_4 - x_2} = \frac{737 - 320}{8 - 5}$$
$$= 139$$

- **Calculation of 3rd order divided differences:**

$$[x_0, x_1, x_2, x_3] = \frac{[x_1, x_2, x_3] - [x_0, x_1, x_2]}{x_3 - x_0}$$
$$= \frac{75 - 25}{7 - 2} = 10$$

$$[x_1, x_2, x_3, x_4] = \frac{[x_2, x_3, x_4] - [x_1, x_2, x_3]}{x_4 - x_1} = \frac{139 - 75}{8 - 4}$$
$$= 16$$

- **Calculation of 4th order divided differences:**

$$[x_0, x_1, x_2, x_3, x_4] = \frac{[x_1, x_2, x_3, x_4] - [x_0, x_1, x_2, x_3]}{x_4 - x_0}$$
$$= \frac{16 - 10}{8 - 2} = 1$$

2.8 FACTORIAL NOTATION

The **factorial function** of x is defined as

$$x^{(n)} = \begin{cases} x(x-h)(x-2h)\ldots(x-\overline{n-1}h) \\ x(x-1)(x-2)\ldots(x-\overline{n-1}),\ h=1 \end{cases}$$

with the assumption $x^{(0)} = 1$

Now $\Delta x^{(n)} = (x + h)^{(n)} - x^{(n)}$

$$= \left\{(x+h)x(x-h)(x-2h)\ldots\left(x-\overline{n-2h}\right)\right\} - \left\{x(x-h)(x-2h)\ldots\left(x-\overline{n-1h}\right)\right\}$$

$$= nh.x(x-h)(x-2h)\ldots\left(x-\overline{n-2h}\right) = nhx^{(n-1)}$$

Again $\Delta^2 x^{(n)} = \Delta\left(\Delta x^{(n)}\right) = \Delta\left(nhx^{(n-1)}\right) = nh.\Delta x^{(n-1)}$

$$= nh\left[\left\{(x+h)x(x-h)(x-2h)\ldots\left(x-\overline{n-3h}\right)\right\} - \left\{x(x-h)(x-2h)\ldots\left(x-\overline{n-2h}\right)\right\}\right]$$

$$= n(n-1)h^2.x(x-h)(x-2h)\ldots\left(x-\overline{n-3h}\right)$$

$$= n(n-1)h^2 x^{(n-2)}$$

Therefore, proceeding in this way, we get

$$\Delta^n x^{(n)} = \begin{cases} n!h^n \\ n!\,, h=1 \end{cases}$$

Clearly $\Delta^{n+1} x^{(n)} = 0$

2.9 PROPAGATION OF ERROR IN THE DIFFERENCE TABLE

Let $y = f(x)$ be a real-valued function that takes the values $y_0, y_1, \ldots, y_6$ at the nodes $x_0, x_1, \ldots, x_6$, respectively. Let the functional value y_3 be affected by an error ϵ, such that the erroneous value of y_3 is $y_3 + \epsilon$. Then the following difference table shows the effect of this error in the differences.

Table 2.5 Propagation of error in differences

x	y	Δy	$\Delta^2 y$	$\Delta^3 y$	$\Delta^4 y$
x_0	Y_0				
		Δy_0			
x_1	y_1		$\Delta^2 y_0$		
		Δy_1		$\Delta^3 y_0 + \epsilon$	
x_2	y_2		$\Delta^2 y_1 + \epsilon$		$\Delta^4 y_0 - 4\epsilon$
		$\Delta y_2 + \epsilon$		$\Delta^3 y_1 - 3\epsilon$	
x_3	$y_3 + \epsilon$		$\Delta^2 y_2 - 2\epsilon$		$\Delta^4 y_1 + 6\epsilon$
		$\Delta y_3 - \epsilon$		$\Delta^3 y_2 + 3\epsilon$	
x_4	y_4		$\Delta^2 y_3 + \epsilon$		$\Delta^4 y_2 - 4\epsilon$
		Δy_4		$\Delta^3 y_3 - \epsilon$	
X_5	Y_5		$\Delta^2 y_4$		
		Δy_5			
x_6	y_6				

From the above table, we conclude:

- *As the order of differences increases, the effect of error increases.*
- *The algebraic sum of the errors in any difference column vanishes.*

- *The coefficients of ϵ in a column, say $\Delta^n y$, are the binomial coefficients of $(1 - \epsilon)^n$, e.g., the coefficients of ϵ in $\Delta^3 y$ are 1, −3, 3 and −1.*
- *The maximum error in each column exists opposite to the erroneous tabular value y_3.*

This concept helps us detect errors in a difference table. It is illustrated in the following examples.

EXAMPLE 2.15 $y = f(x)$ is a polynomial of degree 5 with $y_0 = f(0) = 0$, $y_1 = f(1) = 3$, $y_2 = f(2) = 14$, $y_3 = f(3) = 45$, $y_4 = f(4) = 84$, $y_5 = f(5) = 155$ and $y_6 = f(6) = 258$. It is found that there is an error in the value of y_3. Find the correct value of y_3.

Solution: Let the erroneous value of y_3 be $y_3 + \epsilon$, ϵ being the error in y_3. Now, let us construct the forward difference table.

x	y	Δy	$\Delta^2 y$	$\Delta^3 y$	$\Delta^4 y$	$\Delta^5 y$	$\Delta^6 y$
0	0						
		3					
1	3		8				
		11		$12 + \epsilon$			
2	14		$20 + \epsilon$		$-24 - 4\epsilon$		
		$31 + \epsilon$		$-12 - 3\epsilon$		$60 + 10\epsilon$	
3	$45 + \epsilon$		$8 - 2\epsilon$		$36 + 6\epsilon$		$-120 - 20\epsilon$
		$39 - \epsilon$		$24 + 3\epsilon$		$-60 - 10\epsilon$	
4	84		$32 + \epsilon$		$-24 - 4\epsilon$		
		71		$-\epsilon$			
5	155		32				
		103					
6	258						

As the given polynomial is of degree 5, $\Delta^6 y = 0$. Using this, we get $\epsilon = -6$.
Therefore, the correct value of y_3 is 45 − 6, i.e., 39.

SOLVED EXAMPLES

EXAMPLE 2.16 Prove the following identities:

(i) $\Delta.\nabla \equiv \Delta - \nabla$

(ii) $\dfrac{\Delta}{\nabla} - \dfrac{\nabla}{\Delta} \equiv \Delta + \nabla$

(iii) $E \equiv \Delta + 1$

(iv) $E^{-1} \equiv 1 - \nabla$

(v) $\nabla E \equiv \Delta$

(vi) $(1 + \Delta)(1 - \nabla) \equiv 1$

(vii) $e^{hD} \equiv E \equiv \Delta + 1, D = \dfrac{d}{dx}$

(viii) $e^{-hD} \equiv E^{-1} \equiv 1-\nabla, D = \dfrac{d}{dx}$

Proof of (i):

$$\begin{aligned}\Delta.\nabla f(x) &= \Delta\{\nabla f(x)\}\\ &= \Delta\{f(x) - f(x-h)\}\\ &= \Delta f(x) - \Delta f(x-h)\\ &= \Delta f(x) - \{f(x-h+h) - f(x-h)\}\\ &= \Delta f(x)\{f(x) - f(x-h)\}\\ &= \Delta f(x) - \nabla f(x) = (\Delta - \nabla)f(x).\end{aligned}$$

Proof of (ii):

$$\begin{aligned}\frac{\Delta}{\nabla} - \frac{\nabla}{\Delta} &\equiv \frac{\Delta^2 - \nabla^2}{\Delta.\nabla} \equiv \frac{(\Delta+\nabla)(\Delta-\nabla)}{\Delta-\nabla} \qquad \text{using (i).}\\ &\equiv \Delta + \nabla,\end{aligned}$$

Proof of (iii):

$$\begin{aligned}Ef(x) &= f(x+h)\\ &= f(x) + \{f(x+h) - f(x)\}\\ &= f(x) + \Delta f(x)\\ &= (1+\Delta)f(x)\end{aligned}$$

Proof of (iv):

$$\begin{aligned}\nabla Ef(x) &= \nabla f(x+h) = f(x+h) - f(x)\\ &= \Delta f(x)\end{aligned}$$

Proof of (v):

$$\begin{aligned}E^{-1}f(x) &= f(x-h)\\ &= f(x) - \{f(x) - f(x-h)\}\\ &= f(x) - \nabla f(x)\\ &= (1-\nabla)f(x)\end{aligned}$$

Proof of (vi):

$$(1+\Delta)(1-\nabla) = E.E^{-1} = 1 \quad \text{using (iii) and (iv).}$$

Proof of (vii):

$$\begin{aligned}Ef(x) &= f(x+h)\\ &= f(x) + hf'(x) + \frac{h^2}{2!}f''(x) + \cdots \quad \text{[by Taylor's series]}\\ &= f(x) + hDf(x) + \frac{h^2}{2!}D^2 f(x) + \cdots\end{aligned}$$

$$= \left[1 + hD + \frac{h^2}{2!}D^2 + \cdots\right] f(x)$$

$$= e^{hD} f(x)$$

$\therefore$ $$e^{hD} = E = 1 + \Delta, \qquad \text{using (iii).}$$

Proof of (viii):

$$E^{-1} f(x) = f(x-h)$$

$$= f(x) - hf'(x) + \frac{h^2}{2!} f''(x) - \cdots \qquad \text{[by Taylor's series]}$$

$$= f(x) - hDf(x) + \frac{h^2}{2!} D^2 f(x) - \cdots$$

$$= \left[1 - hD + \frac{h^2}{2!} D^2 - \cdots\right] f(x)$$

$$= e^{-hD} f(x)$$

$\therefore$ $$e^{-hD} = E - 1 = 1 - \nabla, \qquad \text{using (iv).}$$

EXAMPLE 2.17 Prove the following results:

(i) $\delta E^{1/2} = \Delta$ (ii) $E^{1/2} = \mu + \frac{1}{2}\delta$ (iii) $E^{-1/2} = \mu - \frac{1}{2}\delta$

Solution:

(i) $\delta E^{\frac{1}{2}} f(x) = \left(E^{\frac{1}{2}} - E^{\frac{-1}{2}}\right) E^{\frac{1}{2}} f(x) = (E-1) f(x) = \Delta f(x)$

(ii) $\mu + \frac{1}{2}\delta = \frac{1}{2}\left\{E^{\frac{1}{2}} + E^{\frac{-1}{2}}\right\} + \frac{1}{2}\left\{E^{\frac{1}{2}} - E^{\frac{-1}{2}}\right\} = E^{\frac{1}{2}}$

(iii) $\mu - \frac{1}{2}\delta = \frac{1}{2}\left\{E^{\frac{1}{2}} + E^{\frac{-1}{2}}\right\} - \frac{1}{2}\left\{E^{\frac{1}{2}} - E^{\frac{-1}{2}}\right\} = E^{-\frac{1}{2}}$

EXAMPLE 2.18 Prove that $D = \frac{1}{h}\left\{\Delta - \frac{\Delta^2}{2} + \frac{\Delta^3}{3} - \cdots\right\} = \frac{1}{h}\left\{\nabla + \frac{\nabla^2}{2} + \frac{\nabla^3}{3} + \cdots\right\}.$

Solution: We know $e^{hD} = \Delta + 1$

$\therefore$ $$hD = \log_e(1+\Delta) = \Delta - \frac{\Delta^2}{2} + \frac{\Delta^3}{3} - \cdots$$

$\Rightarrow$ $$D = \frac{1}{h}\left\{\Delta - \frac{\Delta^2}{2} + \frac{\Delta^3}{3} - \cdots\right\}$$

Again, we have $$e^{-hD} = 1 - \nabla$$

$\therefore$ $$-hD = \log_e(1-\nabla) = -\left\{\nabla + \frac{\nabla^2}{2} + \frac{\nabla^3}{3} + \cdots\right\}$$

$\Rightarrow$ $$D = \frac{1}{h}\left\{\nabla + \frac{\nabla^2}{2} + \frac{\nabla^3}{3} + \cdots\right\}$$

EXAMPLE 2.19 Find the value of

(i) $\left(\dfrac{\Delta^2}{E}\right)x^3$ where $h = 1$

(ii) $\Delta \cos x$

(iii) $(\Delta - \nabla)\, x^2$

Solution:

(i) $$\left(\frac{\Delta^2}{E}\right)x^3 = \left(\frac{(E-1)^2}{E}\right)x^3 = \left(\frac{E^2+1-2E}{E}\right)x^3$$

$$= \left(E + E^{-1} - 2\right)x^3$$

$$= E\left(x^3\right) + E^{-1}\left(x^3\right) - 2x^3$$

$$= (x+1)^3 + (x+1)^3 - 2x^3, \text{ as } h = 1.$$

$$= 2\left(x^3 + 3h^2\right) - 2x^3 = 6h^2$$

(ii) $$\Delta \cos x = \cos(x+h) - \cos x$$

$$= -2\sin\left(x + \frac{h}{2}\right)\cdot \sin\frac{h}{2}$$

(iii) $$(\Delta - \nabla)x^2 = \Delta x^2 - \nabla x^2$$

$$= \left\{(x+h)^2 - x^2\right\} - \left\{x^2 - (x-h)^2\right\}$$

$$= 2h^2$$

EXAMPLE 2.20 Prove that $[a,b,c,d] = -\dfrac{1}{abcd}$ for the function $f(x) = 1/x$.

Solution: We have $f(x) = 1/x$

So

$$[a,\, b] = \frac{\frac{1}{b} - \frac{1}{a}}{b-a} = -\frac{1}{ab}$$

$$[b,c]=\frac{\frac{1}{c}-\frac{1}{b}}{c-b}=-\frac{1}{bc}$$

$$[c,d]=\frac{\frac{1}{d}-\frac{1}{c}}{d-c}=-\frac{1}{cd}$$

$$[a,b,c]=\frac{[b,c]-[a,b]}{c-a}=\frac{-\frac{1}{bc}+\frac{1}{ab}}{c-b}$$

$$=\frac{1}{abc}$$

$$[b,c,d]=\frac{[c,d]-[b,c]}{d-b}=\frac{-\frac{1}{cd}+\frac{1}{bc}}{d-b}$$

$$=\frac{1}{bcd}$$

$$[a,b,c,d]=\frac{[b,c,d]-[a,b,c]}{d-a}=\frac{\frac{1}{bcd}-\frac{1}{abc}}{d-a}$$

$$=-\frac{1}{abcd}$$

EXAMPLE 2.21 Find the divided difference [a, b] when the function is $f(x) = 1/x^2$.

Solution: As $f(x) = 1/x^2$

we have

$$[a,b]=\frac{1/b^2-1/a^2}{b-a}=\frac{a^2-b^2}{b^2a^2(b-a)}$$

$$=\frac{-(a+b)}{a^2b^2}$$

EXAMPLE 2.22 If $f(a)=\frac{1}{a^2+2}$, find $f[x, y]$ and $f[x, y, z]$.

Solution: Given that $f(a)=\frac{1}{a^2+2}$

Therefore

$$f[x,y]=\frac{\frac{1}{x^2+2}-\frac{1}{y^2+2}}{x-y}=\frac{y^2-x^2}{(x^2+2)(y^2+2)(x-y)}$$

$$=-\frac{(x+y)}{(x^2+2)(y^2+2)}$$

Similarly $$f[y,z] = -\frac{(y+z)}{(y^2+2)(z^2+2)}$$

$$f[x,y,z] = \frac{-\frac{(x+y)}{(x^2+2)(y^2+2)} + \frac{(y+z)}{(y^2+2)(z^2+2)}}{x-z} = \frac{(y+z)(x^2+2)-(x+y)(z^2+2)}{(x^2+2)(y^2+2)(z^2+2)(x-z)}$$

$$= \frac{xy+yz+zx-2}{(x^2+2)(y^2+2)(z^2+2)}$$

EXAMPLE 2.23 If $f(x) = e^{ax}$, show that $\Delta^n f(x) = (e^{ah} - 1)^n e^{ax}$.

Solution: We shall prove the result by the method of induction.

$$\Delta f(x) = f(x+h) - f(x) = e^{a(x+h)} - e^{ax}$$
$$= (e^{ah} - 1)^1 e^{ax}$$

Hence, the result is true for $n = 1$.

Let us assume that the result is true for $n = m$. Then, we must have

$$\Delta^m f(x) = (e^{ah} - 1)^m e^{ax}$$

Therefore

$$\Delta^{m+1} f(x) = \Delta^m[\Delta f(x)] = \Delta^m[f(x+h) - f(x)]$$
$$= \Delta^m f(x+h) - \Delta^m f(x)$$
$$= (e^{ah} - 1)^m e^{a(x+h)} - (e^{ah} - 1)^m e^{ax}$$
$$= (e^{ah} - 1)^{m+1} e^{ax}$$

Thus, we see that the result holds for $n = m + 1$ when it holds for $n = m$. Therefore, by the principle of mathematical induction, the result holds for all values of n.

EXAMPLE 2.24 Estimate the missing values in the following table:

X	1	3	5	7	9	11
Y	2	?	27	64	?	216

Solution: There are four known values of $f(x)$. So $f(x)$ is a polynomial of degree less than 4. Therefore, $\Delta^4 f(x) = 0$. Now, we construct the forward difference table, with the missing terms denoted by p and q:

$f(x)$	$\Delta f(x)$	$\Delta^2 f(x)$	$\Delta^3 f(x)$	$\Delta^4 f(x)$
2				
	$p-2$			
p		$29-2p$		
	$27-p$		$3p-19$	
27		$10+p$		$q-4p-92$
	37		$q-p-111$	
64		$q-101$		$492-4q+p$
	$q-64$		$381-3q$	
q		$280-2q$		
	$216-q$			
216				

Therefore $$\Delta^4 f(x) = 0$$

$$\Rightarrow \quad \begin{cases} q - 4p - 92 = 0 \\ 492 - 4q + p = 0 \end{cases}$$

Solving these, we get $p = 8.27 \approx 8$ and $q = 125.08 \approx 125$

EXAMPLE 2.25 Find the missing term:

x	1	2	3	4	5	6	7
$f(x)$	2	4	8	-	32	64	128

Solution: There are six known values of $f(x)$. So $f(x)$ is a polynomial of degree $6 - 1 = 5$. Therefore, $\Delta^6 f(x) = 0$. Now, we construct the forward difference table assuming that the missing term is denoted by a:

$f(x)$	$\Delta f(x)$	$\Delta^2 f(x)$	$\Delta^3 f(x)$	$\Delta^4 f(x)$	$\Delta^5 f(x)$	$\Delta^6 f(x)$
2						
	2					
4		2				
	4		$a - 14$			
8		$a - 12$		$66 - 4a$		
	$a - 8$		$52 - 3a$		$10a - 158$	
a		$40 - 2a$		$6a - 92$		$322 - 20a$
	$32 - a$		$3a - 40$		$164 - 10a$	
32		a		$72 - 4a$		
	32		$32 - a$			
64		32				
	64					
128						

Therefore $$\Delta^4 f(x) = 0$$

$$\Rightarrow \quad 322 - 20u - 0$$

$$\Rightarrow \quad a = 16.1 \approx 16$$

EXAMPLE 2.26 Show that $\Delta^r y_k = \nabla^r y_{k+r}$, where notations have their usual meaning.

Solution: We have $\Delta^r y_k = (E - 1)^r y_k$, as $E = 1 + \Delta$

$$= (E - 1)^r E^{-r} E^r y_k$$

$$= \left(\frac{E - 1}{E}\right)^r E^r y_k$$

$$= (1 - E^{-1})^r y_{k+r}$$

$$= \nabla^r y_{k+r}, \text{ since } E^{-1}$$

$$= 1 - \nabla$$

EXAMPLE 2.27 Show that $U_x = U_{x-1} + \Delta U_{x-2} + \Delta^2 U_{x-3} + \cdots + \Delta^{n-1} U_{x-n} + \Delta^n U_{x-n}$, where notations have their usual meaning.

Solution: We have

$$\begin{aligned} U_x - \Delta^n U_{x-n} &= U_x - \Delta^n E^{-n} U_x \\ &= (1 - \Delta^n E^{-n}) U_x \\ &= E^{-n}(E^n - \Delta^n) U_x \\ &= E^{-n}\left(\frac{E^n - \Delta^n}{E - \Delta}\right) U_x \end{aligned}$$

As

$$\begin{aligned} E &= 1 + \Delta E - \Delta = 1. \\ &= E^{-n}(E^{n-1} + \Delta E^{n-2} + \Delta^2 E^{n-3} + \cdots + \Delta^{n-1}) U_x \\ &= (E^{-1} + \Delta E^{-2} + \Delta 2 E^{-3} + \cdots + \Delta^{n-1} E^{-n}) U_x \\ &= U_{x-1} + \Delta U_{x-2} + \Delta^2 U_{x-3} + \cdots + \Delta^{n-1} U_{x-n} \end{aligned}$$

Therefore

$$U_x = U_{x-1} + \Delta U_{x-2} + \Delta^2 U_{x-3} + \cdots + \Delta^{n-1} U_{x-n} + \Delta^n U_{x-n}.$$

EXAMPLE 2.28 If f_i is the value of $f(x)$ at $x = x_i$ where $x_i = x_0 + ih$, $(i = 1, 2, 3, \ldots)$ and $h > 0$, then prove that $f_i - E^i f_0 - \sum_{j=0}^{i} \binom{i}{j} \Delta^j f_0$ for $j \leq i$.

Solution: We have

$$\begin{aligned} f_i &= f(x_i) = f(x_0 + ih) \\ &= E^i f_0 = (1 + \Delta)^i f_0 \\ &= \left[1 + \binom{i}{1}\Delta + \binom{i}{2}\Delta^2 + \cdots + \binom{i}{i}\Delta^i\right] f_0 \\ &= f_0 + \binom{i}{1}\Delta f_0 + \binom{i}{2}\Delta^2 f_0 + \cdots + \binom{i}{i}\Delta^i f_0 \\ &= \sum_{j=0}^{i} \binom{i}{j} \Delta^j f_0 \end{aligned}$$

EXAMPLE 2.29 If h is very small, show that $\Delta^{n+1} f(x_0) \approx h^{n+1} f^{n+1}(x_0)$.

Solution: We use the method of induction to show the result.

$$\lim_{h \to 0} \frac{f(x_0 + h) - f(x_0)}{h} = f'(x_0), \text{ provided the limit exists.}$$

$$\Rightarrow \qquad \lim_{h \to 0} \frac{\Delta f(x_0)}{h} = f'(x_0)$$

$$\Rightarrow \qquad \Delta f(x_0) \approx h f'(x_0)$$

Again $$\lim_{h\to 0}\frac{f'(x_0+h)-f'(x_0)}{h}=f''(x_0)$$

$$\Rightarrow \quad \Delta f'(x_0)\approx hf''(x_0)$$

$$\Rightarrow \quad \Delta\{hf'(x_0)\}\approx h^2 f''(x_0)$$

Hence $$\Delta^2 f(x_0)\approx h^2 f''(x_0)$$

Continuing the process, we have $$\Delta^{n+1} f(x_0)\approx h^{n+1} f^{n+1}(x_0).$$

MULTIPLE CHOICE QUESTIONS

1. If $f(x)$ be a polynomial of degree n, then … order difference is a constant.
   (a) n^{th} (b) $(n-1)^{th}$ (c) $(n+1)^{th}$ (d) none
2. Which of the following relations is true?
   (a) $E = 1 + \Delta$ (b) $E = 1 - \Delta$ (c) $E = 1/\Delta$ (d) none
3. The relation between the shift operator E and the central difference operator δ is
   (a) $\delta = E^{1/2} - E^{-1/2}$ (b) $\delta = E^{1/2} + E^{-1/2}$
   (c) $\delta = \frac{1}{2}\left(E^{1/2} - E^{-1/2}\right)$ (d) $\delta = \frac{1}{2}\left(E^{1/2} + E^{-1/2}\right)$
4. Which of the following relations is/are true?
   (a) $\Delta\nabla = \Delta - \nabla$ (b) $\Delta\nabla = \Delta + \nabla$ (c) $\Delta\nabla = \Delta/\nabla$ (d) none
5. The value of $\left(\frac{\Delta^2}{E}\right)x^3$, when length of differencing $h = 1$ is
   (a) $3x^2$ (b) $6x$ (c) $6x^2$ (d) 6
6. If $f(x) = 1/x$, the divided difference $[a, b, c]$ is
   (a) $\frac{1}{a+b+c}$ (b) $\frac{1}{abc}$ (c) $\frac{1}{a^2+b^2}$ (d) $\frac{1}{a+b-c}$
7. Which of the following results is correct($h = 1$)?
   (a) $\Delta x^n = nx^{n-1}$ (b) $\Delta x^{(n)} = nx^{(n-1)}$ (c) $\Delta^n e^x = e^x$ (d) $\Delta \cos x = -\sin x$
8. $(\Delta - \nabla)x^2$ is equal to (the notations have their usual meanings)
   (a) h^2 (b) $-2h^2$ (c) $2h^2$ (d) none
9. If the interval of differencing is unity and $f(x) = ax^2$ ('a' is a constant), which of the following choices is wrong?
   (a) $\Delta f(x) = a(2x + 1)$ (b) $\Delta^2 f(x) = 2a$
   (c) $\Delta^3 f(x) = 2$ (d) $\Delta^4 f(x) = 0$
10. If the interval of differencing is unity, which of the following is true?
    (a) $\Delta^n x^n = (n + 1)!$ (b) $\Delta^n x^n = n!$ (c) $\Delta^n x^n = 0$ (d) $\Delta^n x^n = n$

11. The nth order forward difference of the nth degree polynomial is
(a) $n!$ (b) $(n+1)!$ (c) 0 (d) none
12. If $f(x) = 1/x^2$, then the divided difference $f(a, b)$ is
(a) $\frac{a+b}{(ab)^2}$ (b) $-\frac{a+b}{(ab)^2}$ (c) $\frac{1}{a^2-b^2}$ (d) $\frac{1}{a^2}-\frac{1}{b^2}$
13. $\delta E^{1/2}$ is equal to
(a) Δ (b) ∇ (c) E (d) none
14. The expression for $\Delta^3 y_0$ (the symbol Δ has its usual meaning) is
(a) $y_3 - 3y_2 + 3y_1 - y_0$ (b) $y_2 - 2y_1 + y_0$
(c) $y_3 + 3y_2 + 3y_1 + y_0$ (d) none
15. If $f(x)$ be a polynomial of degree n, the nth order difference is a constant.
(a) true (b) false
16. The relation $(1 + \Delta)(1 - \nabla) = 1$, where the symbols have their usual meaning, is
(a) true (b) false
17. The relation between E and Δ is
(a) $E = 1 + \Delta$ (b) $E = 1 - \Delta$ (c) $E = \Delta - 1$ (d) $E = \frac{1}{1+\Delta}$
18. The relation between shift operator 'E' and forward difference operator 'Δ' is given by
(a) $\Delta = 1 + E$ (b) $E = 1 + \Delta$ (c) $E = \Delta$ (d) $E = \Delta + 2$
19. If the n values of $f(x)$ are given, then $f(x)$ can be approximated by a polynomial of degree
(a) n (b) $n - 1$ (c) $n + 1$ (d) none
20. If $f(x) = 1/x$, the third divided difference with arguments a, b, c and d is equal to
(a) $\frac{1}{abcd}$ (b) $\frac{-1}{abcd}$ (c) $\frac{ad}{bc}$ (d) none
21. The $(n + 1)^{th}$ order forward difference of the nth degree polynomial is
(a) $n!$ (b) $(n + 1)!$ (c) 0 (d) none
22. Which of the following is not true (the notations have their usual meaning)?
(a) $\Delta = E - 1$ (b) $\Delta.\nabla = \Delta - \nabla$ (c) $\frac{\Delta}{\nabla} = \Delta + \nabla$ (d) $\nabla = 1 - E^{-1}$
23. The shift operator E is equal to
(a) e^{-hD} (b) e^h (c) e^{hD} (d) e
24. If $f(x) = be^{ax}$, then $\Delta f(x)$ is
(a) $be^{ax}(e^{ah} - 1)$ (b) $be^{ax(a-1)}$ (c) $be^{ax}(1 - e^{ax})$ (d) none
25. The value of $\Delta^2(ax^2 + bx + c)$ is
(a) $2ah + b$ (b) $2ah$ (c) $2ah^2$ (d) $2a$
26. First order forward difference of a constant is
(a) 0 (b) 1 (c) 2 (d) 3
27. If $f(x) = 2x^3 - 3x^2 + 4x + 5$, then $\Delta^3 f(x)$, where $h = 1$, is
(a) 8 (b) 12 (c) 200 (d) none
28. If the interval of differencing is unity, the value of $\Delta^n x^{(n)}$ is
(a) $n!$ (b) 0 (c) n (d) none

29. The value of $\Delta^{n+1}x^{(n)}$ is
(a) $n!$ (b) 0 (c) n (d) none
30. If the interval of differencing is unity, the value of $\Delta^{-1}c$, c being a constant, is
(a) c (b) 0 (c) cx (d) none
31. The divided difference of a constant function is
(a) 0 (b) constant (c) ∞ (d) none
32. Which of the following is not true (the notations have their usual meaning)?
(a) $2\mu\delta = \Delta - \nabla$ (b) $E = e^{hD}$ (c) $\Delta = \Delta E$ (d) $(1 + \Delta)(1 - \nabla) = 1$
33. ΔE is equal to (the notations have their usual meaning)
(a) E (b) ∇^2 (c) Δ (d) E^2
34. If the interval of differencing is unity, the value of $\Delta^3[(1 - x)(1 - 3x)(1 - 5x)]$ is
(a) -60 (b) -90 (c) -80 (d) -70
35. The divided difference is symmetrical in all their arguments.
(a) true (b) false
36. The value of $(1 + \Delta)(1 - \nabla)$ is
(a) 1 (b) 0 (c) -1 (d) 2
37. $\Delta(1/x)$ is equal to
(a) $\frac{1}{x(x+h)}$ (b) $\frac{1}{x+h}$ (c) $\frac{-h}{x+h}$ (d) $\frac{-h}{x(x+h)}$

Answers

1. (a) 2. (a) 3. (a) 4. (a) 5. (b) 6. (b) 7. (b) 8. (c) 9. (c) 10. (b) 11. (a)
12. (b) 13. (a) 14. (a) 15. (a) 16. (a) 17. (a) 18. (b) 19. (b) 20. (b) 21. (c) 22. (c)
23. (c) 24. (a) 25. (c) 26. (a) 27. (b) 28. (a) 29. (b) 30. (c) 31. (a) 32. (a) 33. (c)
34. (b) 35. (a) 36. (a) 37. (d)

SHORT ANSWER QUESTIONS

1. Prove the following relations:
(i) $\Delta.\nabla \equiv \Delta - \nabla$
(ii) $E \equiv 1 + \Delta$ and $E^{-1} \equiv 1 - \nabla$
(iii) $\nabla E \equiv \Delta$
(iv) $E\Delta \equiv \Delta E$
(v) $\Delta + \nabla = \frac{\Delta}{\nabla} - \frac{\nabla}{\Delta}$
(vi) $(1 + \Delta)(1 - \nabla) \equiv 1$
(vii) $\delta \equiv \mathrm{E}^{1/2} - \mathrm{E}^{-1/2}$ and $\mu \equiv \frac{1}{2}\left[E^{1/2} + E^{-1/2}\right]$
(viii) $\Delta.\nabla \equiv \Delta - \nabla \equiv \delta^2$
(ix) $\delta E^{-1/2} \equiv \nabla$, $\delta E^{1/2} \equiv \Delta$ and $E \equiv \left(\frac{\delta}{\nabla}\right)^2 \equiv \left(\frac{\Delta}{\delta}\right)^2$
(x) $2\mu\delta = \Delta + \nabla$

(xi) $E^{1/2} \equiv \mu + \frac{1}{2}\delta$ and $E^{-1/2} \equiv \mu - \frac{1}{2}\delta$

2. If $D \equiv \frac{d}{dx}$, prove that

(i) $E \equiv e^{hD}$

(ii) $\Delta \equiv e^{hD} - 1$ or $D \equiv \frac{1}{h}\log(1+\Delta)$

(iii) $\nabla \equiv 1 - e^{-hD}$ or $D \equiv -\frac{1}{h}\log(1-\nabla)$

3. Show that

(i) $\left(\frac{\Delta^2}{E}\right)x^3 = 6xh^2$

(ii) $\left(\frac{\Delta^2}{E}e^x\right)\frac{Ee^x}{\Delta^2 e^x} = e^x$

(iii) $(\Delta - \nabla)x^2 = 2h^2$

(iv) $\Delta^2(ax^2 + bx + c) = 2ah^2$

(v) $\Delta^2(e^{3x+2}) = (e^{3h} - 1)2e^{3x+2}$

4. If the interval of differencing be unity, show that

(i) $\left(\frac{\Delta^2}{E}\right)x^3 = 6x$

(ii) $\Delta^2(x^2 + 2x + 7) = 2$

(iii) $\Delta\left(\frac{1}{x+7}\right) = -\frac{1}{(x+7)(x+8)}$

(iv) $\Delta e^{ax+b} = (e^a - 1)e^{ax+b}$

(v) $\Delta\left(\frac{2^x}{x!}\right) = \frac{2^x(1-x)}{(x+1)!}$

(vi) $f(4) = f(3) + \Delta f(2) + \Delta^2 f(1) + \Delta^3 f(1)$

5. Taking $h = 2$, prove that

(i) $\left(\frac{\Delta^2}{E}\right)x^2 = 8$

(ii) $\Delta f(x) = 240$, where $f(x) = 5x^3 - 9x^2 + 2x - 13$

6. Prove that

(i) $\Delta \log f(x) = \log\left[1 + \frac{\Delta f(x)}{f(x)}\right]$

(ii) $\Delta\left(\frac{f_n}{g_n}\right) = \frac{\Delta f_n . g_n - f_n . \Delta g_n}{g_n . g_{n+1}}$

(iii) $f(x_0,x_1) = u(x_0)v(x_0,x_1) + u(x_0,x_1)v(x_0)$, where $f(x) = u(x)v(x)$

7. Evaluate the following, taking $h = 1$:
   (i) $(\Delta + \nabla)^2(x^2 + x)$
   (ii) $\Delta\left(\dfrac{x}{x^2-1}\right)$
   (iii) $\Delta\left(\dfrac{1}{x^2+5x+6}\right)$
   (iv) $\Delta(e^{2x}\log 3x)$
   (v) $\Delta^2(e^x)$
8. Evaluate by taking $h = 2$.
   (i) $\Delta^3\{(1 + x)(1 + 2x)(1 + 3x)\}$
   (ii) $\Delta^{10}\{(1 - x)(1 - 2x^2)(1 - 3x^3)(1 - 4x^4)\}$
9. Prove that $\Delta^r y_k = \nabla^r y_{k+r}$.
10. If $f(x)$ is a polynomial of degree 2 and $f(1) = 3, f(2) = 7, f(3) = 13, f(4) = 21$, find $f(5)$ using the difference table.
11. Prove that $hD = \sinh^{-1}(\mu\delta)$, where notations have their usual meaning.

Answers

7. (i) 8 (ii) $-\dfrac{x^2+x+1}{(x-1)x(x+1)(x+2)}$ (iii) $-\dfrac{2}{(x+2)(x+3)(x+4)}$ (iv) $e^{2x}[e^2\log 3(x + 1) - \log 3x]$
(v) $(e - 1)^2e^x$ 8. (i) 288 (ii) $24\times 10! \times 2^{10}$ 10. 31

LONG ANSWER QUESTIONS

1. Evaluate
   (i) $\Delta^2\left(\dfrac{5x+12}{x^2+5x+6}\right), h = 1$
   (ii) $\left(\dfrac{\Delta^2}{E}\right)x^4,\ h = 1$
2. Show that $\Delta^n(e^x) = (e^h - 1)^n e^x$
3. If D stands for $\dfrac{d}{dx}$, then prove that
   (i) $D \equiv \dfrac{1}{h}\left\{\Delta - \dfrac{\Delta^2}{2} + \dfrac{\Delta^3}{3} - \cdots\right\}$
   (ii) $D \equiv \dfrac{1}{h}\left\{\nabla + \dfrac{\nabla^2}{2} + \dfrac{\nabla^3}{3} + \cdots\right\}$

4. Show that $\nabla y_{n+1} = h\left\{1+\frac{1}{2}\nabla+\frac{5}{12}\nabla^2+\cdots\right\}Dy_n$, where $D=\frac{d}{dx}$.

5. Prove that the nth order forward difference of an nth degree polynomial is constant.

6. If $f(x) = x^2$, prove that $f[x_0, x_1, x_2] = 1$.

7. Prove that $f[a,b,c,d] = -\frac{abc+bcd+acd+abd}{a^2b^2c^2d^2}$, where $f(x) = 1/x^2$.

8. If $f(x)$ is a polynomial of degree 3 and $f(0) = -1, f(1) = 5, f(2) = 13, f(3) = 36, f(4) = 69$ where $f(2)$ is not correct, find the error in $f(2)$.

9. Prove the following results:

(i) $u_0 + \frac{x}{1!}u_1 + \frac{x^2}{2!}u_2 + \cdots = e^x\left[1+\frac{x}{1!}\Delta+\frac{x^2}{2!}\Delta^2+\cdots\right]u_0$

(ii) $u_0 - u_1 + u_2 - \cdots = \frac{1}{2}\left[1-\frac{1}{2}\Delta+\frac{1}{2^2}\Delta^2-\cdots\right]u_0$

10. Show that

(i) $U_x = U_{x-1} + \Delta U_{x-2} + \Delta^2 U_{x-3} + \cdots + \Delta^{n-1}U_{x-n} + \Delta^n U_{x-n}$.

(ii) $u_{n+x} = u_x + \binom{x}{1}\Delta u_{n-1} + \binom{x+1}{2}\Delta^2 u_{n-2} + \binom{x+2}{3}\Delta^3 u_{n-3} + \cdots$

(iii) $\Delta^n u_{x-n} = u_x - \binom{n}{1}u_{x-1} + \binom{n}{2}u_{x-2} - \binom{n}{3}u_{x-3} + \cdots$

11. Following is a table of values of a polynomial of degree 5. It is given that $f(3)$ is in error. Correct the error.

x	0	1	2	3	4	5	6
y	1	2	33	254	1025	3126	7777

12. Given the following data, find f(x) assuming it to be a polynomial of degree 3 in x.

x	0	1	2	3
f(*x*)	1	2	11	34

13. Find the equation of the cubic curve that passes through the points (0, – 5), (1, – 10), (2, – 9), (3,4) and (4,35).

14. If the third order differences are constant, find y_4 if $y_0 = 9, y_1 = 18, y_2 = 20, y_3 = 24$.

15. Evaluate $\Delta^n\left(\frac{1}{x}\right)$, taking $h = 1$.

16. Prove that $\Delta^m\left(\frac{1}{x}\right) = \frac{(-1)^m m!h^m}{x(x+h)(x+2h)\ldots(x+mh)}$.

17. Given $u_{50} = 92345$, $u_{51} = 91556$, $u_{52} = 90748$ and $u_{55} = 88204$, find u_{53} and u_{54}.
18. Find the seventh term of the sequence 1, 4, 10, 20, 35, 56, …
19. Find the sixth term of the series 8, 12, 19, 29, 42, …
20. Evaluate $\left(\frac{\Delta^2}{E}\right)\left[\sin(x+h)\right]+\frac{\Delta^2\left[\sin(x+h)\right]}{E\left[\sin(x+h)\right]}$.
21. Find the missing term(s) in the following tables:

(a)

x	0	1	2	3	4
f(x)	1	3	9	-	81

(b)

x	0	1	2	3	4	5
f(x)	0	-	6	13	-	25

(c)

x	5	10	15	20	25	30	35
f(x)	7	-	12	17	-	28	34

(d)

x	0	5	10	15	20	25
f(x)	1	-	38	82	-	260

(e)

x	0	5	10	15	20	25
f(x)	6	10	-	17	-	31

(f)

x	1	2	3	4	5	6	7
f(x)	2	4	8	-	32	64	128

Answers

1. $\frac{2(5x+16)}{(x+2)(x+3)(x+4)(x+5)}$ 2. $12x^2 + 2$ 8. The correct value of $f(2)$ is 16

11. Error is 10 and $f(3) = 244$ 12. $x^3 + x^2 - x + 1$ 13. $x^3 - 6x - 5$ 14. 30

15. $\frac{(-1)^n n!}{x(x+1)(x+2)\ldots(x+n)}$ 17. $u_{53} = 89920$ and $u_{54} = 89073$ 18. 84

20. $2(\cos h - 1)\{\sin(x + h) - 1\}$ 21. (a). 31 (b). 1, 20 (c). 8.22, 22.42 (d). 13.6, 153.4 (e). 13.25, 22.5 (f). 17

3 Interpolation

3.1 INTRODUCTION

According to Theile, "Interpolation is the art of reading between the lines of the table". It is a technique used to establish a relation between the dependent and independent variables from a given set of values of the dependent variable corresponding to the number of values of the argument. From this relation, the value of the function for any intermediate value of the argument is then obtained. Interpolation also means insertion of intermediate terms of a series.

3.2 INTERPOLATION AND EXTRAPOLATION

Suppose $y = f(x)$ is a real-valued function whose analytic form is not known but is known for a given set of $(n + 1)$ values of x, say $x = x_0, x_1, \ldots, x_n$, where $x_0 < x_1 < \ldots < x_{n-1} < x_n$. Then,

- the process of estimating the value of $f(x)$ for a given $\boldsymbol{x} \in [\boldsymbol{x}_0, \boldsymbol{x}_n]$, $x \neq x_i$, $i = 0, 1, \ldots, n$ is called **interpolation** (Fig. 3.1)
- the process of estimating the value of $f(x)$for a given $\boldsymbol{x} \notin [\boldsymbol{x}_0, \boldsymbol{x}_n]$ is called **extrapolation** (Fig. 3.1)

For example, let us consider the following tabulated values of the independent variable x and the dependent variable y.

x	1	2	3	4
y	7	9	12	17

Clearly, $x_0 = 1$ and $x_3 = 4$, i.e., the interval is $[x_0, x_3] = [1, 4]$. Now, if we want to calculate y at $\boldsymbol{x} = \mathbf{2.2}$, then it is interpolation, while the process of estimating y at $\boldsymbol{x} = \mathbf{0}$ is extrapolation.

Interpolating Polynomial and Error As the analytical form, i.e., the explicit nature of $f(x)$ is not known, we find, in general, a polynomial in x, say $p(x)$, so that

$$p(x_i) = f(x_i)$$

for all $i = 0, 1, 2, \ldots, n$

The polynomial $p(x)$ is called the interpolating polynomial for $f(x)$ and we write

$$f(x) \approx p(x)$$

It is to be noted that although $p(x)$ gives the exact value of $f(x)$ at any given node, it may not be true for any other points (x values) which are not given in the tabular values of x and $f(x)$. In that case, an error is committed and it is given by

$$R(x) = f(x) - p(x)$$

This error $R(x)$ is called the **remainder term** or the **error term** of the interpolation formula. If $(n + 1)$ values of x are given and the interpolating point is x, then

$$R(x) = (x - x_0)(x - x_1)\ldots(x - x_n)\frac{f^{n+1}(\xi)}{(n+1)!}$$

where $\xi \epsilon [x_0, x_n]$

3.3 INTERPOLATION WITH EQUAL INTERVALS

For interpolation, if $x_0, x_1, \ldots, x_n$ are equidistant and

- if x is at the top of the table, then we apply the forward method
- if x is at the bottom of the table, then we apply the backward method
- if x lies near the centre of (x_0, x_n), we apply the central method

Fig. 3.1 illustrates these concepts.

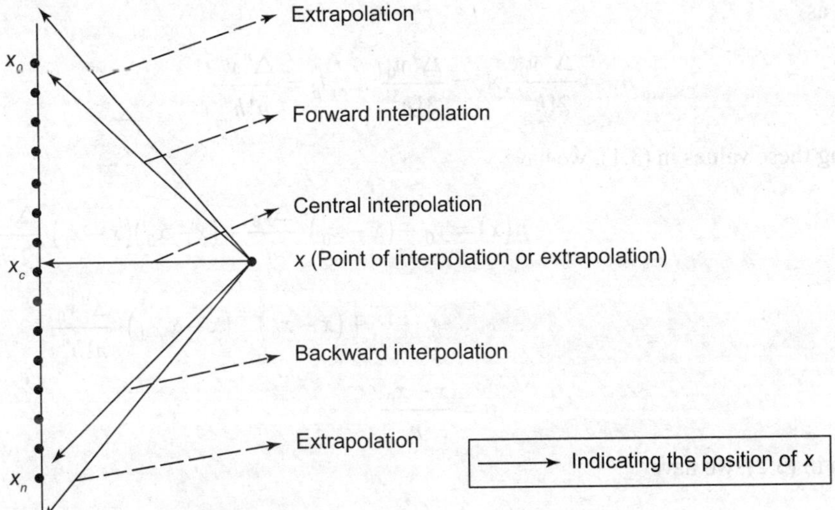

Fig. 3.1 *Interpolation and extrapolation*

3.3.1 Newton's Forward Interpolation

When the arguments $x_0, x_1, \ldots, x_n$ are **equally spaced** and the point, say x, is very close to x_0, i.e., at the **top of the table,** then **Newton's forward interpolation** is used.

Interpolating Polynomial Let a function $y = f(x)$ be known only for a set of $(n + 1)$ distinct equispaced points $x_0, x_1, \ldots, x_n$ such that $x_i = x_0 + ih$, $(i = 1, 2, \ldots, n)$, where h is the length of each subinterval $[x_i, x_{i+1}]$ and the corresponding values of the function are $y_0 = f(x_0)$, $y_1 = f(x_1)$, $\ldots$, $y_n = f(x_n)$. As there are $(n + 1)$ values of $y = f(x)$, we can assume a polynomial $p(x)$ of degree not greater than n, such that

$$p(x_i) = f(x_i) = y_i \quad (i = 1, 2, \ldots, n)$$

With this assumption, the interpolating polynomial $p(x)$ can be written as

$$p(x) = A_0 + A_1(x - x_0) + A_2(x - x_0)(x - x_1) + \cdots + A_n(x - x_0)\ldots(x - x_{n-1}) \tag{3.1}$$

where $A_0, A_1, \ldots, A_n$ are constants to be determined.
We have $h = x_1 - x_0 = x_2 - x_1 = \ldots = x_n - x_{n-1}$.
Putting $x = x_0$ in (3.1), we get

$$p(x_0) = f(x_0) = A_0$$

$$\Rightarrow \qquad A_0 = y_0$$

where $\qquad y_0 = f(x_0)$

Putting $x = x_1$ in (3.1), we get

$$p(x_1) = f(x_1) = A_0 + A_1(x_1 - x_0)$$

$$\Rightarrow \qquad y_1 = y_0 + hA_1$$

$$\Rightarrow \qquad A_1 = \frac{\Delta y_0}{h}$$

Similarly, putting $x = x_2, x_3, \ldots, x_n$ successively in (3.1), we determine the values of $A_2, A_3, \ldots, A_n$, respectively, as

$$A_2 = \frac{\Delta^2 y_0}{2!h^2}, A_3 = \frac{\Delta^3 y_0}{3!h^3}, \ldots, A_n = \frac{\Delta^n y_0}{n!h^n}$$

Now, using these values in (3.1), we have

$$p(x) = y_0 + (x - x_0)\cdot\frac{\Delta y_0}{h} + (x - x_0)(x - x_1)\cdot\frac{\Delta^2 y_0}{2!h^2}$$

$$+ \cdots + (x - x_0)\ldots(x - x_{n-1})\cdot\frac{\Delta^n y_0}{n!h^n} \tag{3.2}$$

Let $\qquad u = \frac{x - x_0}{h}.$

Using this in (3.2), we have

$$f(x) \approx p(uh + x_0)$$

$$= y_0 + \frac{u\Delta y_0}{1!} + u(u-1)\frac{\Delta^2 y_0}{2!}$$

$$+ \cdots + u(u-1)\ldots(u - \overline{n-1})\frac{\Delta^n y_0}{n!} \tag{3.3}$$

Equation (3.3) is known as **Newton's forward interpolating polynomial.**
Now, adding the error term $R(x)$ to $p(x)$, we get

$$f(x) = y_0 + \frac{u\Delta y_0}{1!} + u(u-1)\frac{\Delta^2 y_0}{2!} + \cdots$$

$$+ u(u-1)\ldots(u - \overline{n-1})\frac{\Delta^n y_0}{n!} + R(x)$$

where $$R(x)=(x-x_0)(x-x_1)\ldots(x-x_n)\frac{f^{n+1}(\xi)}{(n+1)!}$$

$$=u(u-1)\ldots(u-n)\frac{h^{n+1}f^{n+1}(\xi)}{(n+1)!},\quad \xi\epsilon[x_0,x_n]$$

Points to Remember

(i) Newton's forward interpolation formula is applicable only when the arguments are **equally spaced.**

(ii) For better accuracy, x_0 should be chosen so that $\boldsymbol{u=\frac{x-x_0}{h}}$ is very small, i.e., $\boldsymbol{0<u<1}$.

(iii) Using this method, we interpolate the value of y at a point $\boldsymbol{x=a}$ when $\boldsymbol{a\in[x_0,x_1]}$ and extrapolate the value of y at a point slightly backward from x_0.

EXAMPLE 3.1 Using Newton's forward interpolation, find a cubic polynomial $f(x)$ that takes the following data, and hence calculate $f(0.5)$.

x	0	1	2	3
$f(x)$	1	0	1	10

Solution: The forward difference table (diagonal) is given below:

x	$f(x)$	$\Delta f(x)$	$\Delta^2 f(x)$	$\Delta^3 f(x)$
0	**1**			
		−1		
1	0		**2**	
		1		**6**
2	1		8	
		9		
3	10			

As the interpolating point is $0.5\in[0,1]$, we set $x_0=0$.
Hence, we have $f(x_0)=1$, $\Delta f(x_0)=-1$, $\Delta^2 f(x_0)=2$, $\Delta^3 f(x_0)=6$.
The length of differencing $h=1$.
Newton's forward interpolating polynomial is given by

$$f(x)\approx p(u)$$

$$=f(x_0)+\frac{u\Delta f(x_0)}{1!}+u(u-1)\frac{\Delta^2 f(x_0)}{2!}$$

$$+u(u-1)(u-2)\frac{\Delta^3 f(x_0)}{3!}+\cdots$$

where $$u=\frac{(x-x_0)}{h}=\frac{(x-0)}{1}=x$$

Therefore, using these values in the above formula, we have

$$p(u)=1+u\times(-1)+\frac{u(u-1)}{2}\times 2+\frac{u(u-1)(u-2)}{6}\times 6$$

$$=u^3-2u^2+1$$

Therefore,

$$f(x) \approx p(x) = x^3 - x^3 - 2x^2 + 1$$

$$\because \quad u = x$$

which is the required polynomial.

Now, putting $x = 0.5$ in this polynomial, we get

$$f(0.5) \approx 0.625$$

EXAMPLE 3.2 Applying suitable interpolation formula, find $f(1.02)$ when $f(x)$ takes the following data:

x	1.00	1.10	1.20	1.30
$f(x)$	0.8415	0.8912	0.9320	0.9636

Solution: As the arguments given in the table are equidistant and the point of interpolation is at the top of the table, we can apply Newton's forward interpolation formula.

The forward difference table (diagonal) is given below:

x	$f(x)$	$\Delta f(x)$	$\Delta^2 f(x)$	$\Delta^3 f(x)$
1.00	**0.8415**			
		0.0497		
1.10	0.8912		**−0.0089**	
		0.0408		**−0.0003**
1.20	0.9320		−0.0092	
		0.0316		
1.30	0.9636			

As the interpolating point is 1.02 ϵ [1.00, 1.10], we set $x_0 = 1.00$ and hence, we get

$$f(x_0) = 0.8415$$
$$\Delta f(x_0) = 0.0497$$
$$\Delta^2 f(x_0) = -0.0089$$
$$\Delta^3 f(x_0) = -0.0003$$

Length of differencing $h = 0.10$.

Newton's forward interpolating polynomial is given by

$$f(x) \approx p(u)$$
$$= f(x_0) + \frac{u\Delta f(x_0)}{1!} + u(u-1)\frac{\Delta^2 f(x_0)}{2!}$$
$$+ u(u-1)(u-2)\frac{\Delta^3 f(x_0)}{3!} + \cdots$$

where $$u = \frac{(x - x_0)}{h} = \frac{(1.02 - 1.00)}{0.10} = 0.2$$

Therefore, using these values in the above formula, we have

$$f(1.02) \approx 0.8415 + 0.2 \times (0.0497) + \frac{0.2 \times (0.2-1)}{2}$$

$$\times (-0.0089) + \frac{0.2 \times (0.2-1) \times (0.2-2)}{6} \times (-0.0003)$$

$$= 0.8521, \text{ up to 4 decimal places.}$$

EXAMPLE 3.3 Applying suitable interpolation formula, find $\sqrt{2}$ correct to 4 significant figures from the following table:

x	1.9	2.1	2.3	2.5	2.7
$\sqrt{x}$	1.3784	1.4491	1.5166	1.5811	1.6432

Solution: As the arguments given in the table are equidistant and the point of interpolation is at the top of the table, we can apply Newton's forward interpolation formula.

The forward difference table (diagonal) is given below:

x	$f(x)$	$\Delta f(x)$	$\Delta^2 f(x)$	$\Delta^3 f(x)$	$\Delta^4 f(x)$
1.9	**1.3784**				
		0.0707			
2.1	1.4491		**−0.0032**		
		0.0675		**0.0002**	
2.3	1.5166		−0.0030		**0.0004**
		0.0645		0.0006	
2.5	1.5811		−0.0024		
		0.0621			
2.7	1.6432				

Here, we see that the point of interpolation is 2ϵ [1.9, 2.1] so $x_0 = 1.9$ and

$$f(x_0) - 1.3784, \; \Delta f(x_0) = 0.0707$$

$$\Delta^3 f(x_0) = 0.0002$$

$$\Delta^2 f(x_0) = -0.0032$$

$$\Delta^4 f(x_0) = 0.0004$$

Here, the length of differencing $h = 0.2$.

Newton's forward interpolating polynomial is given by

$$f(x) \approx p(u$$

$$f(x_0) + \frac{u \Delta f(x_0)}{1!} + u(u-1)\frac{\Delta^2 f(x_0)}{2!} + u(u-1)(u-2)\frac{\Delta^3 f(x_0)}{3!}$$

$$+ u(u-1)(u-2)(u-3)\frac{\Delta^4 f(x_0)}{4!}$$

where
$$u = \frac{(x - x_0)}{h} = \frac{(2-1.9)}{0.2}$$
$$= 0.5$$

Therefore, we have

$$f(2) \approx 1.3784 + 0.5 \times (0.0707) + \frac{0.5 \times (0.5-1)}{2} \times (-0.0032)$$
$$+ \frac{0.5 \times (0.5-1) \times (0.5-2)}{6} \times (0.0002)$$
$$+ \frac{0.5 \times (0.5-1) \times (0.5-2) \times (0.5-3)}{24} \times (0.0004)$$
$$= 1.414, \text{ correct to 4 significant figures.}$$

3.3.2 Newton's Backward Interpolation

If the arguments $x_0, x_1, \ldots, x_n$ are equally spaced and the point, say x, is very close to x_n, i.e., at the bottom of the table and within the interval (x_0, x_n), then we apply Newton's backward interpolation.

Interpolating Polynomial Let a function $y = f(x)$ be known only for a set of $(n + 1)$ distinct equispaced points $x_0, x_1, \ldots, x_n$ such that $x_i = x_0 + ih$, $(i = 1, 2, \ldots, n)$, where h is the length of each subinterval $[x_i, x_{i+1}]$ and the corresponding values of the function are $y_0 = f(x_0), y_1 = f(x_1), \ldots, y_n = f(x_n)$. As there are $(n + 1)$ values of $y = f(x)$, we can assume a polynomial $p(x)$ of degree not greater than n, such that

$$p(x_i) = f(x_i) = y_i \quad (i = 1, 2, \ldots, n)$$

With this assumption, the interpolating polynomial $p(x)$ can be written as

$$f(x) \approx p(x)$$
$$= A_0 + A_1(x - x_n) + A_2(x - x_n)(x - x_{n-1})$$
$$+ \cdots + A_n(x - x_n)\ldots(x - x_1) \tag{3.4}$$

where $A_0, A_1, \ldots, A_n$ are constants to be determined.
We have $h = x_1 - x_0 = x_2 - x_1 = \ldots = x_n - x_n - 1$.
Putting $x = x_n$ in (3.4), we get

$$p(x_n) = f(x_n) = A_0$$

$\Rightarrow$
$$A_0 = y_n,$$

where
$$y_n = f(x_n)$$

Putting $x = x_n - 1$ in (3.4), we get

$$p(x_{n-1}) = f(x_{n-1}) = A_0 + A_1(x_{n-1} - x_n)$$

$\Rightarrow$
$$y_{n-1} = y_n - hA_1$$

$\Rightarrow$
$$A_1 = \frac{\nabla y_n}{h}$$

Similarly, putting $x = x_{n-2}, x_{n-3}, \ldots, x_0$ successively in (3.4), we determine the values of $A_2, A_3, \ldots, A_n$, respectively, as

$$A_2 = \frac{\nabla^2 y_n}{2!h^2}, A_3 = \frac{\nabla^3 y_n}{3!h^3}, \ldots, A_n = \frac{\nabla^n y_n}{n!h^n}$$

Now, using these values in (3.4), we have

$$p(x) = y_n + (x - x_n)\cdot\frac{\nabla y_n}{h} + (x - x_n)(x - x_{n-1})\cdot\frac{\nabla^2 y_n}{2!h^2}$$

$$+\cdots+(x - x_n)\ldots(x - x_1)\cdot\frac{\nabla^n y_n}{n!h^n} \qquad (3.5)$$

Let $$u = \frac{(x - x_n)}{h}$$

Using this in (3.5), we have

$$f(x) \approx p(uh + x_n)$$

$$= y_n + \frac{u\nabla y_n}{1!} + u(u+1)\frac{\nabla^2 y_n}{2!}$$

$$+\cdots+u(u+1)\ldots\left(u+\overline{n-1}\right)\frac{\nabla^n y_n}{n!} \qquad (3.6)$$

Equation (3.6) is known as **Newton's backward interpolating polynomial.**
The remainder term for this formula is given by

$$R(x) = (x - x_0)(x - x_1)\ldots(x - x_n)\frac{f^{n+1}(\xi)}{(n+1)!}$$

$$= u(u+1)\ldots(u+n)\frac{h^{n+1}f^{n+1}(\xi)}{(n+1)!}, \quad \xi \in [x_0, x_n]$$

Points to Remember

(i) Newton's backward interpolation formula is applicable only when the arguments are **equally spaced.**

(ii) x_n should be chosen so that $u = \dfrac{x - x_n}{h}$ is very small, i.e., $-1 < u < 0$.

(iii) This method is used to interpolate the value of y at a point $x = a$ when $a \in [x_{n-1}, x_n]$ and to extrapolate the value of y at a point slightly forward from x_n.

EXAMPLE 3.4 Using Newton's backward interpolation, find a cubic polynomial $f(x)$ which takes the following data, and hence calculate $f(2.5)$.

x	0	1	2	3
$f(x)$	1	0	1	10

Solution: The backward difference table (diagonal) is given below:

x	$f(x)$	$\nabla f(x)$	$\nabla^2 f(x)$	$\nabla^3 f(x)$
0	1			
		−1		
1	0		2	
		1		**6**
2	1		**8**	
		9		
3	**10**			

As the point of interpolation is 2.5 ϵ [2, 3], we set $x_n = x_3 = 3$.
Hence, our required values are $f(x_3) = 10$, $\nabla f(x_3) = 9$, $\nabla^2 f(x_3) = 8$, $\nabla^3 f(x_3) = 6$.
Also, the length of differencing $h = 1$.
Newton's backward interpolating polynomial is given by

$$f(x) \approx p(u)$$

$$= f(x_3) + \frac{u\nabla f(x_3)}{1!} + u(u+1)\frac{\nabla^2 f(x_3)}{2!}$$

$$+u(u+1)(u+2)\frac{\nabla^3 f(x_3)}{3!} + \cdots$$

where $$u = \frac{(x - x_n)}{h} = \frac{(x-3)}{1} = x - 3$$

Therefore, using these values in the above formula, we have

$$f(x) \approx p(u) = 10 + u \times (9) + \frac{u(u+1)}{2} \times 8 + \frac{u(u+1)(u+2)}{6} \times 6$$

$$= u^3 + 7u^2 + 15u + 10$$

Therefore

$$f(x) \approx p(x) = (x-3)^3 + 7(x-3)^2 + 15(x-3) + 10 = x^3 - 2x^2 + 1$$

which is the required polynomial.
Now, putting $x = 2.5$ in this polynomial, we get

$$f(2.5) \approx 4.125$$

EXAMPLE 3.5 For a few values of x, the corresponding values of $f(x)$ are as follows:

x	4	5	6	7
$f(x)$	3.11	2.96	2.85	2.70

Using the appropriate interpolation formula, find the value of $f(6.5)$.

Solution: As the arguments given in the table are equidistant and the point of interpolation is at the bottom of the table, we can apply Newton's backward interpolation formula.

The backward difference table (diagonal) is given below:

x	$f(x)$	$\nabla f(x)$	$\nabla^2 f(x)$	$\nabla^3 f(x)$
4	3.11			
		−0.15		
5	2.96		0.04	
		−0.11		**−0.08**
6	2.85		**−0.04**	
		−0.15		
7	**2.70**			

Here, the point of interpolation, $x = 6.5$. Hence $x_n = x_3 = 7$, $f(x_3) = 2.70$, $\nabla f(x_3) = -0.15$, $\nabla^2 f(x_3) = -0.04$, $\nabla^3 f(x_3) = -0.08$

The length of differencing $h = 1$.

Newton's backward interpolating polynomial is given by

$$f(x) \approx p(u) = f(x_3) + \frac{u\nabla f(x_3)}{1!} + u(u+1)\frac{\nabla^2 f(x_3)}{2!} + u(u+1)(u+2)\frac{\nabla^3 f(x_3)}{3!} + \cdots$$

where $$u = \frac{(x - x_3)}{h} = \frac{(6.5-7)}{1} = -0.5$$

Therefore, using these values in the above formula, we have

$$f(6.5) \approx 2.70 + (-0.5)\times(-0.15) + \frac{(-0.5)(-0.5+1)}{2}\times(-0.04)$$

$$+ \frac{(-0.5)(-0.5+1)(-0.5+2)}{6}\times(-0.08)$$

$$= 2.785$$

EXAMPLE 3.6 Applying suitable interpolation formula, find $y(19)$ correct to 4 significant figures from the following table:

x:	0	5	10	15	20
y:	1.0	1.6	3.8	8.2	15.4

Solution: As the arguments given in the table are equidistant and the point of interpolation is at the bottom of the table, we can apply Newton's backward interpolation formula.

The backward difference table (diagonal) is given below:

x	y	∇y	$\nabla^2 y$	$\nabla^3 y$	$\nabla^4 y$
0	1.0				
		0.6			
5	1.6		1.6		
		2.2		0.6	
10	3.8		2.2		**0.0**
		4.4		**0.6**	
15	8.2		**2.8**		
		7.2			
20	**15.4**				

The interpolating point $x = 19$. So, $x_n = x_4 = 20$, $x = 19$, length of differencing $h = 5$.

$$y_4 = 15.4, \nabla y_4 = 7.2, \nabla^2 y_4 = 2.8, \nabla^3 y_4 = 0.6, \nabla^4 y_4 = 0.0.$$

Newton's backward interpolating polynomial is given by

$$y \approx p(u) = y_4 + \frac{u \nabla y_4}{1!} + u(u+1)\frac{\nabla^2 y_4}{2!} + u(u+1)(u+2)\frac{\nabla^3 y_4}{3!}$$
$$+ u(u+1)(u+2)(u+3)\frac{\nabla^4 y_4}{4!} + \cdots$$

where $$u = \frac{(x - x_n)}{h} = \frac{(19-20)}{5} = -0.2$$

Therefore, we have

$$y(19) \approx 15.4 + (-0.2)\times(7.2) + \frac{(-0.2)\times(-0.2+1)}{2}\times(2.8)$$
$$+ \frac{(-0.2)\times(-0.2+1)\times(-0.2+2)}{6}\times(0.6)$$
$$+ \frac{(-0.2)\times(-0.2+1)\times(-0.2+2)\times(-0.2+3)}{24}\times(0.0)$$
$$= 13.71, \text{ correct to 4 significant figures.}$$

3.3.3 Central Difference Interpolation

We see that Newton's forward or backward interpolation formulae for equispaced arguments is used to find the best approximation of $y = f(x)$ when the interpolating point is, respectively, at the top or bottom of the table. However, if the interpolating point lies near the middle of the table then both the methods fail to give the best result. In this case, we use central difference formulae, which are discussed below.

Gauss's Forward Interpolation Let $y = f(x)$ be defined at $(2n + 1)$ distinct equidistant points $x_{-n}, x_{-(n-1)}, x_{-(n-2)}, \ldots, x_{-1}, x_0, x_1, \ldots, x_{n-2}, x_{n-1}, x_n$, i.e., $y_i = f(x_i)$, where $x_i = x_0 + ih$ ($i = 0, \pm 1, \pm 2, \ldots, \pm n$). Let $p(x)$ be Gauss's forward interpolating polynomial of degree not more than $2n$, which coincides with the values y_i at the points x_i. Then, the polynomial is given by

$$p(x) = y_0 + u\Delta y_0 + \frac{u(u-1)}{2!}\Delta^2 y_{-1} + \frac{(u+1)u(u-1)}{3!}\Delta^3 y_{-1}$$
$$+ \frac{(u+1)u(u-1)(u-2)}{4!}\Delta^4 y_{-2} + \frac{(u+2)(u+1)u(u-1)(u-2)}{5!}\Delta^5 y_{-2} + \cdots$$

where $$u = \frac{x - x_0}{h}$$

EXAMPLE 3.7 Use Gauss's forward interpolation to calculate $f(2.5)$ from the following table:

x	0	1	2	3	4
$f(x)$	1	11	22	43	75

Solution: The central difference table is given by:

x	$y = f(x)$	Δy	$\Delta^2 y$	$\Delta^3 y$	$\Delta^4 y$
$x_{-2} = 0$	$y_{-2} = 1$				
		$\Delta y_{-2} = 10$			
$x_{-1} = 1$	$y_{-1} = 11$		$\Delta^2 y_{-2} = 1$		
		$\Delta y_{-1} = 11$		$\Delta^3 y_{-2} = 9$	
$x_0 = 2$	$y_0 = 22$		$\Delta^2 y_{-1} = 10$		$\Delta^4 y_{-2} = -8$
		$\Delta y_0 = 21$		$\Delta^3 y_{-1} = 1$	
$x_1 = 3$	$y_1 = 43$		$\Delta^2 y_0 = 11$		
		$\Delta y_1 = 32$			
$x_2 = 4$	$y_2 = 75$				

Here, $x_0 = 2, h = 1, x = 2.5$

and so

$$u = \frac{x - x_0}{h} = \frac{2.5 - 2}{1} = 0.5$$

Gauss's forward interpolating polynomial is

$$f(x) \approx p(x) = y_0 + u\Delta y_0 + \frac{u(u-1)}{2!}\Delta^2 y_{-1} + \frac{(u+1)u(u-1)}{3!}\Delta^3 y_{-1} + \frac{(u+1)u(u-1)(u-2)}{4!}\Delta^4 y_{-2} + \frac{(u+2)(u+1)u(u-1)(u-2)}{5!}\Delta^5 y_{-2} + \cdots$$

where

$$u = \frac{x - x_0}{h}$$

Therefore

$$f(x) \approx 22 + 0.5 \times 21 + \frac{0.5(0.5-1)}{2!} \times 10 + \frac{(0.5+1)0.5(0.5-1)}{3!} \times 1 + \frac{(0.5+1)0.5(0.5-1)(0.5-2)}{4!} \times (-8)$$

$$= 22 + 10.5 - 1.25 - 0.0625 - 0.1875 = 31$$

Gauss's Backward Interpolation Let $y = f(x)$ be defined at $(2n + 1)$ distinct equidistant points $x_{-n}, x_{-(n-1)}, x_{-(n-2)}, \ldots, x_{-1}, x_0, x_1, \ldots, x_{n-2}, x_{n-1}, x_n$, i.e., $y_i = f(x_i)$ where $x_i = x_0 + ih$ ($i = 0, \pm 1, \pm 2, \ldots, \pm n$). Then, Gauss's backward interpolating polynomial $p(x)$ is given by

$$p(x) = y_0 + u\Delta y_{-1} + \frac{(u+1)u}{2!}\Delta^2 y_{-1} + \frac{(u+1)u(u-1)}{3!}\Delta^3 y_{-2} + \frac{(u+2)(u+1)u(u-1)}{4!}\Delta^4 y_{-2} + \frac{(u+2)(u+1)u(u-1)(u-2)}{5!}\Delta^5 y_{-2} + \cdots$$

where $$u = \frac{x - x_0}{h}$$

EXAMPLE 3.8 Use Gauss's backward interpolation to calculate $f(1.5)$ from the following table:

x	0	1	2	3	4
$f(x)$	1	11	22	43	75

Solution: The central difference table is given by:

x	$y = f(x)$	Δy	$\Delta^2 y$	$\Delta^3 y$	$\Delta^4 y$
$x_{-2} = 0$	$y_{-2} = 1$				
		$\Delta y_{-2} = 10$			
$x_{-1} = 1$	$y_{-1} = 11$		$\Delta^2 y_{-2} = 1$		
		$\boldsymbol{\Delta y_{-1} = 11}$		$\boldsymbol{\Delta^3 y_{-2} = 9}$	
$x_0 = 2$	$\boldsymbol{y_0 = 22}$		$\boldsymbol{\Delta^2 y_{-1} = 10}$		$\boldsymbol{\Delta^4 y_{-2} = -8}$
		$\Delta y_0 = 21$		$\Delta^3 y_{-1} = 1$	
$x_1 = 3$	$y_1 = 43$		$\Delta^2 y_0 = 11$		
		$\Delta y_1 = 32$			
$x_2 = 4$	$y_2 = 75$				

Here, $x_0 = 2, h = 1, x = 1.5$

and so
$$u = \frac{x - x_0}{h} = \frac{1.5 - 3}{1}$$
$$= -0.5$$

Gauss's backward interpolating polynomial is

$$f(x) \approx p(x) = y_0 + u\Delta y_{-1} + \frac{(u+1)u}{2!}\Delta^2 y_{-1} + \frac{(u+1)u(u-1)}{3!}\Delta^3 y_{-2}$$
$$+ \frac{(u+2)(u+1)u(u-1)}{4!}\Delta^4 y_{-2} + \frac{(u+2)(u+1)u(u-1)(u-2)}{5!}\Delta^5 y_{-2} + \cdots$$

where $$u = \frac{x - x_0}{h}$$

Therefore

$$f(x) \approx 22 + (-0.5)\times 11 + \frac{(0.5+1)(-0.5)}{2!}\times 10 + \frac{(-0.5+1)(-0.5)(-0.5-1)}{3!}$$
$$\times 9 + \frac{(-0.5+2)(-0.5+1)(-0.5)(-0.5-1)}{4!}\times(-8)$$

$$= 22 - 5.5 - 1.25 + 0.5625 - 0.1875 = 15.625$$

Stirling's Formula Let $y = f(x)$ be defined at $(2n + 1)$ distinct equidistant points $x_{-n}, x_{-(n-1)}, x_{-(n-2)}, \ldots, x_{-1}, x_0, x_1, \ldots, x_{n-2}, x_{n-1}, x_n$, i.e., $y_i = f(x_i)$ where $x_i = x_0 + ih$ $(i = 0, \pm 1, \pm 2, \ldots, \pm n)$. Then, Stirling's interpolating polynomial $p(x)$ is given by

$$p(x) = y_0 + u\frac{(\Delta y_0 + \Delta y_{-1})}{2} + \frac{u^2}{2!}\Delta^2 y_{-1} + \frac{u(u^2-1)}{3!}\frac{(\Delta^3 y_{-2} + \Delta^3 y_{-1})}{2}$$
$$+ \frac{u^2(u^2-1)}{4!}\Delta^4 y_{-2} + \frac{u^2(u^2-1)(u^2-2^2)}{5!}\frac{(\Delta^5 y_{-3} + \Delta^5 y_{-2})}{2} + \cdots$$

where
$$u = \frac{x - x_0}{h}$$

EXAMPLE 3.9 Use Stirling's formula to evaluate e^x when $x = 0.644$, given the following data:

x	0.61	0.62	0.63	0.64	0.65	0.66	0.67
e^x	1.840431	1.858928	1.877610	1.896481	1.915541	1.934792	1.954237

Solution: The forward difference table is:

x	$y = e^x$	Δy	$\Delta^2 y$	$\Delta^3 y$	$\Delta^4 y$
$x_{-3} = 0.61$	$y_{-3} = 1.840431$				
		$\Delta y_{-3} = 0.018497$			
$x_{-2} = 0.62$	$y_{-2} = 1.858928$		$\Delta^2 y_{-3} = 0.000185$		
		$\Delta y_{-2} = 0.018682$		$\Delta^3 y_{-3} = 0.000004$	
$x_{-1} = 0.63$	$y_{-1} = 1.877610$		$\Delta^2 y_{-2} = 0.000189$		$\Delta^4 y_{-3} = 0.000004$
		$\Delta y_{-1} = \mathbf{0.018871}$		$\Delta^3 y_{-2} = \mathbf{0}$	
$x_0 = 0.64$	$y_0 = \mathbf{1.896481}$		$\Delta^2 y_{-1} = \mathbf{0.000189}$		$\Delta^4 y_{-2} = \mathbf{0.000002}$
		$\Delta y_0 = \mathbf{0.019060}$		$\Delta^3 y_{-1} = \mathbf{0.000002}$	
$x_1 = 0.65$	$y_1 = 1.915541$		$\Delta^2 y_0 = 0.000191$		$\Delta^4 y_{-1} = 0.000001$
		$\Delta y_1 = 0.019251$		$\Delta^3 y_0 = 0.000003$	
$x_2 = 0.66$	$y_2 = 1.934792$		$\Delta^2 y_1 = 0.000194$		
		$\Delta y_2 = 0.01945$			
$x_3 = 0.67$	$y_3 = 1.954237$				

Here
$$u = \frac{x - x_0}{h} = \frac{0.644 - 0.64}{0.01}$$
$$= 0.4$$

Therefore, using Stirling's formula, we get

$$e^{0.644} \approx y_0 + \frac{u}{1!}\cdot\frac{(\Delta y_0 + \Delta y_{-1})}{2} + \frac{u^2}{2!}\cdot\Delta^2 y_{-1} + \frac{u(u^2-1)}{3!}\cdot\frac{(\Delta^3 y_{-2} + \Delta^3 y_{-1})}{2} + \frac{u^2(u^2-1)}{4!}\cdot\Delta^4 y_{-2}$$
$$= 1.896481 + \frac{0.4}{1!}\times\frac{(0.019060 + 0.018871)}{2} + \frac{0.4^2}{2!}\times 0.000189$$
$$+ \frac{0.4(0.4^2-1)}{3!}\times\frac{(0 + 0.000002)}{2} + \frac{0.4^2(0.4^2-1)}{4!}\times 0.000002$$
$$= 1.896481 + 0.0075862 + 0.00001512 - 0.00000001$$
$$= 1.904082$$

Bessel's Formula Let $y = f(x)$ be defined at $(2n + 1)$ distinct equidistant points $x_{-n}, x_{-(n-1)}, x_{-(n-2)}, \ldots, x_{-1}, x_0, x_1, \ldots, x_{n-2}, x_{n-1}, x_n$, i.e., $y_i = f(x_i)$ where $x_i = x_0 + ih$ $(i = 0, \pm 1, \pm 2, \ldots, \pm n)$. Then, Bessel's interpolating polynomial $p(x)$ is given by

$$p(x) = \frac{y_0 + y_1}{2} + v\Delta y_0 + \frac{\left(v^2 - \frac{1}{4}\right)}{2!}\frac{(\Delta^2 y_0 + \Delta^2 y_{-1})}{2} + \frac{v(v^2 - \frac{1}{4})}{3!}\Delta^3 y_{-1}$$
$$+ \frac{\left(v^2 - \frac{1}{4}\right)\left(v^2 - \frac{9}{4}\right)}{4!}\frac{(\Delta^4 y_{-1} + \Delta^4 y_{-2})}{2} + \frac{v\left(v^2 - \frac{1}{4}\right)\left(v^2 - \frac{9}{4}\right)}{5!}\Delta^5 y_{-2} + \cdots$$

where
$$v = u - \frac{1}{2}$$

and
$$u = \frac{x - x_0}{h}$$

EXAMPLE 3.10 Use Bessel's formula to evaluate $f(x)$ when $x = 3.7$, given the following data:

x	0	1	2	3	4	5	6
$f(x)$	1	7	15	35	80	190	420

Solution: The forward difference table is:

x	$y - f(x)$	Δy	$\Delta^2 y$	$\Delta^3 y$	$\Delta^4 y$	$\Delta^5 y$	$\Delta^6 y$
$x_{-3} = 0$	$y_{-3} = 1$						
		$\Delta y_{-3} = 6$					
$x_{-2} = 1$	$y_{-2} = 7$		$\Delta^2 y_{-3} = 2$				
		$\Delta y_{-2} = 8$		$\Delta^3 y_{-3} = 10$			
$x_{-1} = 2$	$y_{-1} = 15$		$\Delta^2 y_{-2} = 12$		$\Delta^4 y_{-3} = 3$		
		$\Delta y_{-1} = \mathbf{20}$		$\Delta^3 y_{-2} = 13$		$\Delta^5 y_{-3} = 24$	
$x_0 = 3$	$\mathbf{y_0 = 35}$		$\mathbf{\Delta^2 y_{-1} = 25}$		$\mathbf{\Delta^4 y_{-2} = 27}$		$\Delta^6 y_{-3} = -36$
		$\mathbf{\Delta y_0 = 45}$		$\mathbf{\Delta^3 y_{-1} = 40}$		$\mathbf{\Delta^5 y_{-2} = -12}$	
$x_1 = 4$	$\mathbf{y_1 = 80}$		$\mathbf{\Delta^2 y_0 = 65}$		$\mathbf{\Delta^4 y_{-1} = 15}$		
		$\Delta y_{-1} = 110$		$\Delta^3 y_0 = 55$			
$x_2 = 5$	$y_2 = 190$		$\Delta^2 y_1 = 120$				
		$\Delta y_2 = 230$					
$x_3 = 6$	$y_3 = 420$						

Here, $x_0 = 3$, $h = 1$, $x = 3.7$

and so
$$u = \frac{x - x_0}{h} = \frac{3.7 - 3}{1}$$
$$= 0.7$$

Therefore, using Bessel's formula, we get

$$f(x) \approx p(x) = \frac{y_0 + y_1}{2} + v\Delta y_0 + \frac{\left(v^2 - \frac{1}{4}\right)}{2!}\frac{(\Delta^2 y0 + \Delta^2 y-1)}{2} + \frac{v(v^2 - \frac{1}{4})}{3!}\Delta^3 y_{-1}$$
$$+ \frac{\left(v^2 - \frac{1}{4}\right)\left(v^2 - \frac{9}{4}\right)}{4!}\frac{(\Delta^4 y_{-1} + \Delta^4 y_{-2})}{2} + \frac{v\left(v^2 - \frac{1}{4}\right)\left(v^2 - \frac{9}{4}\right)}{5!}\Delta^5 y_{-2} + \cdots$$

where
$$v = u - \frac{1}{2} = 0.7 - 0.5$$
$$= 0.2$$

Therefore

$$f(3.7) \approx \frac{35+80}{2} + 0.2 \times 45 + \frac{\left((0.2)^2 - \frac{1}{4}\right)}{2!}\frac{(65+25)}{2} + \frac{0.2\left((0.2)^2 - \frac{1}{4}\right)}{3!} \times 40$$
$$+ \frac{\left((0.2)^2 - \frac{1}{4}\right)\left((0.2)^2 - \frac{9}{4}\right)}{4!}\frac{(15+27)}{2} + \frac{0.2\left((0.2)^2 - \frac{1}{4}\right)\left((0.2)^2 - \frac{9}{4}\right)}{5!} \times (-12)$$
$$= 57.5 + 9 - 4.725 - 0.28 + 0.4061 - 0.0093$$
$$= 61.8918$$

Laplace–Everett Formula Let $y = f(x)$ be defined at $(2n + 1)$ distinct equidistant points $x_{-n}, x_{(n-1)}, x_{-(n-2)}, \ldots, x_{-1}, x_0, x_1, \ldots, x_{n-2}, x_{n-1}, x_n$, i.e., $y_i = f(x_i)$ where $x_i = x_0 + ih$ $(i = 0, \pm 1, \pm 2, \ldots, \pm n)$. Then, Laplace–Everett's interpolating polynomial $p(x)$ is given by

$$p(x) = uy_1 + \frac{(u+1)u(u-1)}{3!}\Delta^2 y_0 + \frac{(u+2)(u+1)u(u-1)(u-2)}{5!}\Delta^4 y_{-1} + \cdots + wy_0$$
$$+ \frac{(w+1)w(w-1)}{3!}\Delta^2 y_{-1} + \frac{(w+2)(w+1)w(w-1)(w-2)}{5!}\Delta^4 y_{-2} + \cdots$$

where
$$w = 1 - u$$

and
$$u = \frac{x - x_0}{h}$$

EXAMPLE 3.11 Use Laplace–Everett's formula to evaluate $f(x)$ when $x = 2.5$, given the following data:

x	0	1	2	3	4	5
$f(x)$	1.2314	1.4251	1.6010	1.8101	2.0102	2.2500

Solution: The forward difference table is:

x	$y = f(x)$	Δy	$\Delta^2 y$	$\Delta^3 y$	$\Delta^4 y$
$x_{-2} = 0$	$y_{-2} = 1.2314$				
		$\Delta y_{-2} = 0.1937$			
$x_{-1} = 1$	$y_{-1} = 1.4251$		$\Delta^2 y_{-2} = 0.0178$		
		$\Delta y_{-1} = 0.1759$		$\Delta^3 y_{-2} = 0.0510$	
$x_0 = 2$	$\mathbf{y_0 = 1.6010}$		$\mathbf{\Delta^2 y_{-1} = 0.0332}$		$\mathbf{\Delta^4 y_{-2} = 0.0932}$
		$\Delta y_0 = 0.2091$		$\Delta^3 y_{-1} = -0.0422$	
$x_1 = 3$	$\mathbf{y_1 = 1.8101}$		$\mathbf{\Delta^2 y_0 = -0.0090}$		$\mathbf{\Delta^4 y_{-1} = 0.0909}$
		$\Delta y_1 = 0.2001$		$\Delta^3 y_0 = 0.0487$	
$x_2 = 4$	$y_2 = 2.0102$		$\Delta^2 y_1 = 0.0397$		
		$\Delta y_2 = 0.2398$			
$x_3 = 5$	$y_3 = 2.2500$				

Here, $x_0 = 2, h = 1, x = 2.5$

and so
$$u = \frac{x - x_0}{h} = \frac{2.5-2}{1}$$
$$= 0.5$$

Therefore, using Laplace–Everett's formula, we get

$$f(x) \approx p(x) = uy_1 + \frac{(u+1)u(u-1)}{3!}\Delta^2 y_0 + \frac{(u+2)(u+1)u(u-1)(u-2)}{5!}\Delta^4 y_{-1} + \cdots$$
$$+ wy_0 + \frac{(w+1)w(w-1)}{3!}\Delta^2 y_{-1} + \frac{(w+2)(w+1)w(w-1)(w-2)}{5!}\Delta^4 y_{-2} + \cdots$$

where
$$w = 1 - u = 1 - 0.5 = 0.5$$

Therefore

$$f(2.5) \approx 0.5 \times 1.8101 + \frac{(0.5+1)0.5(0.5-1)}{3!} \times (-0.0090)$$
$$+ \frac{(0.5+2)(0.5+1)0.5(0.5-1)(0.5-2)}{5!} \times 0.0909$$
$$+ 0.5 \times 1.6010 + \frac{(0.5+1)0.5(0.5-1)}{3!} \times 0.0332$$
$$+ \frac{(0.5+2)(0.5+1)0.5(0.5-1)(0.5-2)}{5!} \times (-0.0932)$$
$$= (0.90505 + 0.0005625 + 0.001065234)$$
$$+ (0.8005 - 0.002075 - 0.001092187)$$
$$= 1.7072$$

3.4 INTERPOLATION WITH UNEQUAL INTERVALS

When the arguments $x_0, x_1, \ldots, x_n$ are not equally spaced, we cannot apply Newton's forward or backward formula. In addition, if the interpolating point lies anywhere within the interval $[x_0, x_n]$, then application of the previous two methods may produce significant error as the condition $0 < |u| < 1$ may not be satisfied. In such cases, we can apply Lagrange's interpolation and Newton's divided difference interpolation.

3.4.1 Lagrange's Interpolation

When the arguments $x_0, x_1, \ldots, x_n$ are not equally spaced and the point, say x, lies anywhere within the interval (x_0, x_n) then we generally apply Lagrange's interpolation. Howerver, it should be noted that Lagrange's interpolation formula is applicable even when the arguments are equidistant.

Interpolating Polynomial Let a function $y = f(x)$ be known only for a set of $(n + 1)$ distinct equispaced or unequispaced points $x_0, x_1, \ldots, x_n$ and the corresponding values of the function are $y_0 = f(x_0), y = f(x_1), \ldots, y_n = f(x_n)$. As there are $(n + 1)$ values of $y = f(x)$, we can assume a polynomial $p(x)$ of degree not greater than n such that

$$p(x_i) = f(x_i) = y_i \ (i = 1, 2, \ldots, n)$$

With this assumption, the interpolating polynomial $p(x)$ can be written as

$$f(x) \approx p(x) = A_0(x-x_1)(x-x_2)\dots(x-x_n) + A_1(x-x_0)(x-x_2)\dots(x-x_n) + \cdots + A_n(x-x_0)(x-x_1)\dots(x-x_{n-1}) \quad (3.7)$$

where, $A_0, A_1, \dots, A_n$ are constants to be determined.
Putting $x = x_0, x_1, \dots x_n$ successively in (3.7), we get

$$A_0 = \frac{f(x_0)}{(x_0-x_1)(x_0-x_2)\dots(x_0-x_n)} = \frac{y_0}{(x_0-x_1)(x_0-x_2)\dots(x_0-x_n)}, \text{ as } y_0 = f(x_0)$$

$$A_1 = \frac{f(x_1)}{(x_1-x_0)(x_1-x_2)\dots(x_1-x_n)} = \frac{y_1}{(x_1-x_0)(x_1-x_2)\dots(x_1-x_n)}, \text{ as } y_1 = f(x_1)$$

$$A_n = \frac{f(x_n)}{(x_n-x_0)(x_n-x_1)\dots(x_n-x_{n-1})} = \frac{y_n}{(x_n-x_0)(x_n-x_1)\dots(x_n-x_{n-1})}, \text{ as } y_n = f(x_n)$$

Putting the values of $A_0, A_1, \dots, A_n$ in (3.7), we get

$$f(x) \approx p(x) = \frac{(x-x_1)(x-x_2)\dots(x-x_n)}{(x_0-x_1)(x_0-x_2)\dots(x_0-x_n)} y_0 + \frac{(x-x_0)(x-x_2)\dots(x-x_n)}{(x_1-x_0)(x_1-x_2)\dots(x_1-x_n)} y_1 + \cdots + \frac{(x-x_0)(x-x_1)\dots(x-x_{n-1})}{(x_n-x_0)(x_n-x_1)\dots(x_n-x_{n-1})} y_n \quad (3.8)$$

which is known as **Lagrange's interpolating polynomial.**
The remainder term for this formula is given by

$$R(x) = (x-x_0)(x-x_1)\dots(x-x_n)\frac{f^{n+1}(\xi)}{(n+1)!}, \ \xi \epsilon [x_0, x_n]$$

Points to Remember

(i) Lagrange's interpolation formula is applicable to both **equispaced** and **unequispaced** arguments.
(ii) The interpolating point may lie anywhere in the interval $[x_0, x_n]$.
(iii) The restriction of this formula is due to the increment in degree of the polynomial. For any increase in the degree of the interpolating polynomial by adding a new interpolating point, the whole calculation has to be recomputed.

EXAMPLE 3.12 Using Lagrange's interpolation, find a cubic polynomial $y(x)$ which takes the following data, and hence calculate $y(10)$.

x	5	6	9	11
$y(x)$	12	13	14	16

Solution: Lagrange's interpolating polynomial is given by

$$y(x) \approx p(x) = \frac{(x-x_1)(x-x_2)(x-x_3)}{(x_0-x_1)(x_0-x_2)(x_0-x_3)} y_0 + \frac{(x-x_0)(x-x_2)(x-x_3)}{(x_1-x_0)(x_1-x_2)(x_1-x_3)} y_1 + \frac{(x-x_0)(x-x_1)(x-x_3)}{(x_2-x_0)(x_2-x_1)(x_2-x_3)} y_2 + \frac{(x-x_0)(x-x_1)(x-x_2)}{(x_3-x_0)(x_3-x_1)(x_3-x_2)} y_3 + \cdots$$

From the given table, we have

$$x_0 = 5, x_1 = 6, x_2 = 9, x_3 = 11 \text{ and } y_0 = 12, y_1 = 13, y_2 = 14, y_3 = 16$$

Therefore, using these values in equation (3.15), we get

$$y(x) \approx p(x) = \frac{(x-6)(x-9)(x-11)}{(5-6)(5-9)(5-11)} \times 12 + \frac{(x-5)(x-9)(x-11)}{(6-5)(6-9)(6-11)} \times 13 + \frac{(x-5)(x-6)(x-11)}{(9-5)(9-6)(9-11)} \times 14$$
$$+ \frac{(x-5)(x-6)(x-9)}{(11-5)(11-6)(11-9)} \times 16$$
$$= \frac{x^3 - 26x^2 + 219x - 594}{(-24)} \times 12 + \frac{x^3 - 25x^2 + 199x - 495}{15} \times 13$$
$$+ \frac{x^3 - 22x^2 + 151x - 330}{(-24)} \times 14 + \frac{x^3 - 20x^2 + 129x - 270}{60} \times 16$$
$$= 3x^3 - 70x^2 + 557x - 690$$

which is the required polynomial.

Now, putting $x = 10$ in this polynomial, we get

$$y(10) \approx 14.667, \text{ correct to 5 significant figures.}$$

EXAMPLE 3.13 Using the appropriate interpolation formula, find the polynomial $y(x)$ and hence calculate $y(5.5)$ for the given data:

x	0	2	3	5	7
$y(x)$	1	47	97	251	477

Solution: As the arguments given in the table are not equidistant, we should apply Lagrange's interpolation formula.

Lagrange's interpolating polynomial is given by

$$y(x) \approx p(x) = \frac{(x-x_1)(x-x_2)(x-x_3)}{(x_0-x_1)(x_0-x_2)(x_0-x_3)} y_0 + \frac{(x-x_0)(x-x_2)(x-x_3)}{(x_1-x_0)(x_1-x_2)(x_1-x_3)} y_1$$
$$+ \frac{(x-x_0)(x-x_1)(x-x_3)}{(x_2-x_0)(x_2-x_1)(x_2-x_3)} y_2 + \frac{(x-x_0)(x-x_1)(x-x_2)}{(x_3-x_0)(x_3-x_1)(x_3-x_2)} y_3 + \cdots$$

From the given table, we have

$$x_0 = 0, x_1 = 2, x_2 = 3, x_3 = 5, x_4 = 7 \text{ and } y_0 = 1, y_1 = 47, y_2 = 97, y_3 = 251, y_4 = 477$$

Therefore, using these values in the above formula, we get

$$p(x) = \frac{(x-2)(x-3)(x-5)(x-7)}{(0-2)(0-3)(0-5)(0-7)} \times 1 + \frac{(x-0)(x-3)(x-5)(x-7)}{(2-0)(2-3)(2-5)(2-7)} \times 47$$
$$+ \frac{(x-0)(x-2)(x-5)(x-7)}{(3-0)(3-2)(3-5)(3-7)} \times 97 + \frac{(x-0)(x-2)(x-3)(x-7)}{(5-0)(5-2)(5-3)(5-7)} \times 251$$
$$+ \frac{(x-0)(x-2)(x-3)(x-5)}{(7-0)(7-2)(7-3)(7-5)} \times 477$$

$$= 9x^2 + 5x - 1$$

which is the required polynomial.
Now, putting $x = 5.5$ in this polynomial, we get

$$y(5.5) \approx 298.75, \text{ correct to 5 significant figures.}$$

EXAMPLE 3.14 Construct Lagrange's interpolating polynomial using the following data and hence calculate $y(46)$.

x	40	45	50	55
$y = f(x)$	15.22	13.99	12.62	11.13

Solution: Lagrange's interpolating polynomial is given by

$$y(x) \approx p(x) = \frac{(x-x_1)(x-x_2)(x-x_3)}{(x_0-x_1)(x_0-x_2)(x_0-x_3)} y_0 + \frac{(x-x_0)(x-x_2)(x-x_3)}{(x_1-x_0)(x_1-x_2)(x_1-x_3)} y_1$$

$$+ \frac{(x-x_0)(x-x_1)(x-x_3)}{(x_2-x_0)(x_2-x_1)(x_2-x_3)} y_2$$

$$+ \frac{(x-x_0)(x-x_1)(x-x_2)}{(x_3-x_0)(x_3-x_1)(x_3-x_2)} y_3 + \cdots$$

From the given table, we have

$$x_0 = 40, x_1 = 45, x_2 = 50, x_3 = 55 \text{ and } y_0 = 15.22, y_1 = 13.99, y_2 = 12.62, y_3 = 11.13$$

Therefore, using these values in the above formula, we get

$$p(x) = \frac{(x-45)(x-50)(x-55)}{(40-45)(40-50)(40-55)} \times 15.22 + \frac{(x-40)(x-50)(x-55)}{(45-40)(45-50)(45-55)} \times 13.99$$

$$+ \frac{(x-40)(x-45)(x-55)}{(50-40)(50-45)(50-55)} \times 12.62$$

$$+ \frac{(x-40)(x-45)(x-50)}{(55-40)(55-45)(55-50)} \times 11.13$$

$$= 2.7 \times 10^{-5} x^3 - 6.4 \times 10^{-3} x^2 + 15.3 \times 10^{-2} x + 17.62$$

which is the required polynomial.
Now, putting $x = 46$ in this polynomial, we get

$$y(46) \approx 13.744, \text{ correct to 5 significant figures.}$$

3.4.2 Newton's Divided Difference Formula

This method is complementary to Lagrange's interpolation. As we know, whenever a new point is added to an existing set of data, the calculation for Lagrange's interpolating polynomial has to be made afresh.

In such cases, we can apply Newton's divided difference interpolation to overcome the disadvantage of Lagrange's interpolation.

Interpolating Polynomial From the definition of divided differences, we have

$$[x, x_0] = [x_0, x] = \frac{y - y_0}{x - x_0}$$

$$\Rightarrow \qquad y = y_0 + (x - x_0)[x, x_0] \tag{3.9}$$

Again

$$[x, x_0, x_1] = \frac{[x, x_0] - [x_0, x_1]}{x - x_1}$$

$$\Rightarrow \qquad [x, x_0] = [x_0, x_1] + (x - x_1)[x, x_0, x_1]$$

Using this in (3.9), we get

$$y = y_0 + (x - x_0)[x_0, x_1] + (x - x_0)(x - x_1)[x, x_0, x_1]$$

Proceeding in this way, we get

$$\begin{aligned} y = y_0 &+ (x - x_0)[x_0, x_1] + (x - x_0)(x - x_1)[x_0, x_1, x_2] \\ &+ (x - x_0)(x - x_1)(x - x_2)[x_0, x_1, x_2, x_3] + \cdots \\ &+ (x - x_0)(x - x_1)\ldots(x - x_{n-1})[x_0, x_1, \ldots, x_n] \\ &+ (x - x_0)(x - x_1)\ldots(x - x_n)[x, x_0, x_1, \ldots, x_n] \end{aligned} \tag{3.10}$$

The last term of (3.10) is the remainder term. Hence, we have

$$\begin{aligned} y \approx p(x) = y_0 &+ (x - x_0)[x_0, x_1] + (x - x_0)(x - x_1)[x_0, x_1, x_2] \\ &+ (x - x_0)(x - x_1)(x - x_2)[x_0, x_1, x_2, x_3] + \cdots \\ &+ (x - x_0)(x - x_1)\ldots(x - x_{n-1})[x_0, x_1, \ldots, x_n] \end{aligned} \tag{3.11}$$

where $p(x_i) = f(x_i)$ for all $i = 0, 1, 2, \ldots, n$ is the required polynomial with the remainder term

$$R(x) = (x - x_0)(x - x_1)\ldots(x - x_n)[x, x_0, x_1, \ldots, x_n]$$

Note: Some authors denote these divided differences as $f[x_0, x_1]$, $f[x_0, x_1, x_2]$, … instead of denoting $[x_0, x_1]$, $[x_0, x_1, x_2]$, …, respectively.

EXAMPLE 3.15 Using Newton's divided difference formula, find a cubic polynomial $y(x)$ which takes the following data, and hence calculate $y(2)$.

x	0	1	4	5
$y(x)$	8	11	68	123

Solution: The divided difference table is given below:

x	$y(x)$	1st order	2nd order	3rd order
0	**8**			
		3		
1	11		**4**	
		19		**1**
4	68		9	
		55		
5	123			

Newton's divided difference formula is given by

$$y(x) \approx p(x) = y_0 + (x - x_0)[x_0, x_1] + (x - x_0)(x - x_1)[x_0, x_1, x_2]$$
$$+ (x - x_0)(x - x_1)(x - x_2)[x_0, x_1, x_2, x_3] + \cdots$$

From the above table, we have

$x_0 = 0, x_1 = 1, x_2 = 4, x_3 = 5$ and $y_0 = 8$, $[x_0, x_1] = 3$, $[x_0, x_1, x_2] = 4$, $[x_0, x_1, x_2, x_3] = 1$

Therefore, using these values in the above formula, we get

$$p(x) = 8 + (x - 0) \times 3 + (x - 0)(x - 1) \times 4 + (x - 0)(x - 1)(x - 4) \times 1$$
$$= x^3 - x^2 + 3x + 8$$

which is the required polynomial.

Now, putting $x = 2$ in this polynomial, we get

$$y(2) \approx 18$$

EXAMPLE 3.16 Use Newton's divided difference formula to find $f(5)$ from the given data:

x	**0**	**2**	**3**	**4**	**7**	**8**
$f(x)$	4	26	58	112	466	668

Solution: The divided difference table is given below:

x	$f(x)$	1st order	2nd order	3rd order	4th order
0	**4**				
		11			
2	26		**7**		
		32		**1**	
3	58		11		**0**
		54		1	
4	112		16		0
		118		1	
7	466		21		
		202			
8	668				

Now, using Newton's divided difference formula

$$f(x) \approx p(x) = f(x_0) + (x - x_0)[x_0, x_1] + (x - x_0)(x - x_1)[x_0, x_1, x_2] + \cdots$$

We get, from the above table

$$f(5) \approx 4 + (5-0) \times 11 + (5-0)(5-2) \times 7 + (5-0)(5-2)(5-3) \times 1$$
$$= 194$$

EXAMPLE 3.17 Using Newton's divided difference formula, find $f(3.4)$ from the following data:

x	2.5	2.8	3.0	3.1	3.6
$f(x)$	12.1825	16.4446	20.0855	22.1980	36.5982

Solution: The divided difference table is shown below:

x	$f(x)$	1st order	2nd order	3rd order	4th order
2.5	**12.1825**				
		14.207			
2.8	16.4446		**7.995**		
		18.2045		**2.9**	
3.0	20.0855		9.735		**0.83782**
		21.125		3.8216	
3.1	22.1980		12.7923		
		28.8004			
3.6	36.5982				

Now, using Newton's divided difference formula

$$f(x) \approx p(x) = f(x_0) + (x - x_0)[x_0, x_1] + (x - x_0)(x - x_1)[x_0, x_1, x_2] + \cdots$$

We get, from the above table

$$f(5) \approx 12.1825 + (3.4 - 2.5) \times 14.207 + (3.4 - 2.5)(3.4 - 2.8) \times 7.995$$
$$+ (3.4 - 2.5)(3.4 - 2.8)(3.4 - 3.0) \times 2.9$$
$$+ (3.4 - 2.5)(3.4 - 2.8)(3.4 - 3.0)(3.4 - 3.1) \times 0.83782$$
$$= 29.967, \text{ correct to 5 significant digits.}$$

SOLVED EXAMPLES

EXAMPLE 3.18 Use Newton's forward interpolation to compute $f(2.1)$ from the following table:

x	0	2	4	6	8	10
$f(x)$	1	5	17	37	45	51

Solution: The forward difference table (diagonal) is given below:

x	$f(x)$	$\Delta f(x)$	$\Delta^2 f(x)$	$\Delta^3 f(x)$	$\Delta^4 f(x)$	$\Delta^5 f(x)$
0	1					
		4				
2	**5**		8			
		12		0		
4	17		**8**		−20	
		20		**−20**		50
6	37		−12		**30**	
		8		10		
8	45		−2			
		6				
10	51					

As the interpolating point is 2.1 ϵ [2, 4], we set $x_0 = 2$.

So $$f(x_0)=5,\ \Delta f(x_0)=12,\ \Delta^2 f(x_0)=8,\ \Delta^3 f(x_0)=-20,\ \Delta^4 f(x_0)=30$$

The step length $h = 2$.

Newton's forward interpolation formula is given by

$$f(x)\approx p(u)=f(x_0)+\frac{u\Delta f(x_0)}{1!}+u(u-1)\frac{\Delta^2 f(x_0)}{2!}+u(u-1)(u-2)\frac{\Delta^3 f(x_0)}{3!}$$
$$+u(u-1)(u-2)(u-3)\frac{\Delta^4 f(x_0)}{4!}+\cdots$$

where $$u=\frac{(x-x_0)}{h}=\frac{(2.1-2)}{2}$$
$$=0.05$$

Therefore, we get

$$f(2.1)\approx 5+0.05\times(12)+\frac{0.05\times(0.05-1)}{2}\times(8)+\frac{0.05\times(0.05-1)\times(0.05-2)}{6}\times(-20)$$
$$+\frac{0.05\times(0.05-1)\times(0.05-2)\times(0.05-3)}{24}\times(30)$$
$$=4.7597,\ \text{correct to 4 significant figures.}$$

EXAMPLE 3.19 From the following table of the function $(x)=\frac{1}{x}$, calculate 1/2.72, correct to 4 decimal places using suitable interpolation formula.

x	2.7	2.8	2.9
$f(x)$	0.3704	0.3571	0.3448

Solution: As the interpolating point is 2.72 ϵ [2.7, 2.8], i.e., at the top of the table, and the arguments are equidistant, we apply Newton's forward interpolation formula. The forward difference table (diagonal) is given below:

x	$f(x)$	$\Delta f(x)$	$\Delta^2 f(x)$
2.7	**0.3704**		
		−0.0133	
2.8	0.3571		**0.0010**
		−0.0123	
2.9	0.3448		

Here, we set $x_0 = 2.7$.

So $$f(x_0) = 0.3704,\ \Delta f(x_0) = -0.0133,\ \Delta^2 f(x_0) = 0.0010$$

The step length $h = 0.1$.
Newton's forward interpolation formula is given by

$$f(x) \approx p(u) = f(x_0) + \frac{u\Delta f(x_0)}{1!} + u(u-1)\frac{\Delta^2 f(x_0)}{2!} + u(u-1)(u-2)\frac{\Delta^3 f(x_0)}{3!}$$
$$+u(u-1)(u-2)(u-3)\frac{\Delta^4 f(x_0)}{4!} + \cdots$$

where $$u = \frac{(x-x_0)}{h} = \frac{(2.72-2.7)}{0.1}$$
$$= 0.2$$

Therefore, we get

$$f(2.72) \approx 0.3704 + 0.2\times(-0.0133) + \frac{0.2\times(0.2-1)}{2!}\times(0.0010)$$

$$= 0.3677, \text{ correct to 4 decimal places.}$$

EXAMPLE 3.20 Use Newton's forward interpolation to compute $f(1.3)$ from the following table, correct to 2 decimal places.

x	0	1	2	3	4
$f(x)$	1	1.5	2.2	3.1	4.3

Solution: The forward difference table (diagonal) is given below:

x	$f(x)$	$\Delta f(x)$	$\Delta^2 f(x)$	$\Delta^3 f(x)$
0	1			
		0.5		
1	**1.5**		0.2	
		0.7		0
2	2.2		**0.2**	
		0.9		**0.1**
3	3.1		0.3	
		1.2		
4	4.3			

As the interpolating point is 1.3 ∊ [1, 2], we set $x_0 = 1$.

So $$f(x_0) = 1.5,\ \Delta f(x_0) = 0.7,\ \Delta^2 f(x_0) = 0.2,\ \Delta^3 f(x_0) = 0.1$$

The step length $h = 1$.

Newton's forward interpolation formula is given by

$$f(x) \approx p(u) = f(x_0) + \frac{u\Delta f(x_0)}{1!} + u(u-1)\frac{\Delta^2 f(x_0)}{2!} + u(u-1)(u-2)\frac{\Delta^3 f(x_0)}{3!}$$
$$+u(u-1)(u-2)(u-3)\frac{\Delta^4 f(x_0)}{4!} + \cdots$$

where $$u = \frac{(x - x_0)}{h} = \frac{(1.3-1)}{1} = 0.3$$
$$= 0.3$$

Therefore, we get

$$f(1.3) \approx 1.5 + 0.3\times(0.7) + \frac{0.3\times(0.3-1)}{2}\times(0.2) + \frac{0.3\times(0.3-1)\times(0.3-2)}{6}\times(0.1)$$
$$= 1.689 \approx 1.69, \text{ correct to 2 decimal places.}$$

EXAMPLE 3.21 Use suitable interpolation formulae to compute $f(0.23)$ and $f(0.29)$, correct to 4 decimal places, from the following table:

x	0.20	0.22	0.24	0.26	0.28	0.30
$f(x)$	1.6596	1.6698	1.6804	1.6912	1.7024	1.7139

Solution: The difference table (diagonal) is

x	$f(x)$	$\Delta f(x)$ or $\nabla f(x)$	$\Delta^2 f(x)$ or $\nabla^2 f(x)$
0.20	1.6596		
		0.0102	
0.22	**1.6698**		0.0004
		0.0106	
0.24	1.6804		**0.0002**
		0.0108	
0.26	1.6912		0.0004
		0.0112	
0.28	1.7024		**0.0003**
		0.0115	
0.30	**1.7139**		

Here, we ignore the differences of higher order as they are very small.

Part I: As the interpolating point is 0.23 ϵ [0.22, 0.24], we set $x_0 = 0.22$ and apply Newton's forward interpolation formula.

We have $f(x_0) = 1.6698$, $\Delta f(x_0) = 0.0106$, $\Delta^2 f(x_0) = 0.0002$

The step length $h = 0.02$.

Newton's forward interpolation formula is given by

$$f(x) \approx p(u) = f(x_0) + \frac{u\Delta f(x_0)}{1!} + u(u-1)\frac{\Delta^2 f(x_0)}{2!} + \cdots$$

where
$$u = \frac{(x - x_0)}{h} = \frac{(0.23 - 0.22)}{0.02} = 0.5$$
$$= 0.5$$

Therefore, we get

$$f(0.23) \approx 1.6698 + 0.5 \times (0.0106) + \frac{0.5 \times (0.5 - 1)}{2} \times (0.0002)$$
$$= 1.675075 \approx 1.6751, \text{ correct up to 4 decimal places.}$$

Part II: As the interpolating point is 0.29 ϵ [0.28, 0.30], we apply Newton's backward interpolation formula.

We have $f(x_n) = 1.7139$, $\nabla f(x_n) = 0.0115$, $\nabla^2 f(x_n) = 0.0003$

Newton's backward interpolation formula is given by

$$f(x) \approx p(u) = f(x_n) + \frac{u\nabla f(x_0)}{1!} + u(u+1)\frac{\nabla^2 f(x_n)}{2!} + \cdots$$

where
$$u = \frac{(x - x_n)}{h} = \frac{(0.29 - 0.30)}{0.02}$$
$$= -0.5$$

Therefore, we get

$$f(0.29) \approx 1.7139 - 0.5 \times (0.0115) + \frac{-0.5 \times (-0.5 + 1)}{2} \times (0.0003)$$
$$= 1.70811 \approx 1.7081, \text{ correct to 4 decimal places.}$$

EXAMPLE 3.22 From the following table, estimate the number of students who obtained marks between 40 and 45.

x	30–40	40–50	50–60	60–70	70–80
f(*x*)	31	42	51	35	31

Solution: Let $f(x)$ denote the number of students with marks above 30 and below 80. Then, the forward difference table is given by:

Marks obtained	x	$f(x)$	$\Delta f(x)$	$\Delta^2 f(x)$	$\Delta^3 f(x)$	$\Delta^4 f(x)$
below 40	**40**	**31**				
			42			
below 50	50	73		**9**		
			51		**−25**	
below 60		124		−16		**37**
			35		12	
below 70	70	159		−4		
			31			
below 80	80	190				

As we see, the point of interpolation is 45 ϵ [40, 50], so $x_0 = 40$

and
$$u = \frac{45-40}{10} = 0.5$$

Therefore, using Newton's interpolation formula, we get

$$f(45) \approx 31 + 0.5\times(42) + \frac{0.5\times(0.5-1)}{2}\times(9) + \frac{0.5\times(0.5-1)\times(0.5-2)}{6}\times(-25)$$
$$+\frac{0.5\times(0.5-1)\times(0.5-2)\times(0.5-3)}{24}\times(37)$$
$$= 47.868 \approx 48$$

Therefore, the number of students who obtained marks between 40 and 45 is given by

$$f(45) - f(40) \approx 48 - 31 = 17$$

EXAMPLE 3.23 Find the equation of the cubic curve that passes through the points (0, −5), (1, −10), (2, −9), (3, 4) and (4, 35).

Solution: As the arguments are equispaced, we can apply Newton's forward interpolation to find the curve.

The forward difference table is given by:

x	$f(x)$	$\Delta f(x)$	$\Delta^2 f(x)$	$\Delta^3 f(x)$	$\Delta^4 f(x)$
0	**−5**				
		−5			
1	−10		**6**		
		1		**6**	
2	−9		12		**0**
		13		6	
3	4		18		
		31			
4	35				

Here, we set $x_0 = 0$.

Therefore $\quad f(x_0) = -5, \Delta f(x_0) = -5, \Delta^2 f(x_0) = 6, \Delta^3 f(x_0) = 6, \Delta^4 f(x_0) = 0$

The step length $h = 1$.
Newton's forward interpolation formula is given by

$$f(x) \approx p(u) = f(x_0) + \frac{u\Delta f(x_0)}{1!} + u(u-1)\frac{\Delta^2 f(x_0)}{2!}$$

$$+u(u-1)(u-2)\frac{\Delta^3 f(x_0)}{3!} + u(u-1)(u-2)(u-3)\frac{\Delta^4 f(x_0)}{4!} + \cdots$$

where $\quad u = \frac{(x-x_0)}{h} = \frac{(x-0)}{1} = x$

Therefore, we get

$$p(u) = -5 + u\times(-5) + \frac{u\times(u-1)}{2}\times(6) + u(u-1)(u-2)\times\frac{6}{3!} + u(u-1)(u-2)(u-3)\times\frac{0}{4!}$$

$$= -5 - 5u + 3u^2 - 3u + u^3 - 3u^2 + 2u$$

$$= u^3 - 6u - 5$$

Hence, the required cubic curve is given by $f(x) = x^3 - 6x - 5$

EXAMPLE 3.24 Given the following data, find $f(x)$ assuming it to be a polynomial of degree 3 in x.

x	0	1	2	3
$f(x)$	1	2	11	34

Solution: As the arguments are equispaced, we can apply Newton's forward interpolation to find the curve. The forward difference table (diagonal) is given below:

x	$f(x)$	$\Delta f(x)$	$\Delta^2 f(x)$	$\Delta^3 f(x)$
0	**1**			
		1		
1	2		**8**	
		9		**6**
2	11		14	
		23		
3	34			

Here, we set $x_0 = 0$.

So $\quad f(x_0) = 1, \Delta f(x_0) = 1, \Delta^2 f(x_0) = 8, \Delta^3 f(x_0) = 6$

The step length $h = 1$.
Newton's forward interpolation formula is given by

$$f(x) \approx p(u) = f(x_0) + \frac{u\Delta f(x_0)}{1!} + u(u-1)\frac{\Delta^2 f(x_0)}{2!} + u(u-1)(u-2)\frac{\Delta^3 f(x_0)}{3!} + \cdots$$

where $u = \dfrac{(x - x_0)}{h} = \dfrac{(x-0)}{1}$

$= x$

Therefore, we get

$$p(u) = 1 + u \times (1) + \frac{u \times (u-1)}{2} \times (8) + u(u-1)(u-2) \times \frac{6}{3!}$$

$$= 1 + u + 4u^2 - 4u + u^3 - 3u^2 + 2u$$

$$= u^3 + u^2 - u + 1$$

Hence, the required cubic curve is given by $f(x) = x^3 + x^2 - x + 1$

EXAMPLE 3.25 Given the following table, find $f(1.5)$ and $f(7.5)$.

x	1	2	3	4	5	6	7	8
$f(x)$	1	8	27	64	125	216	343	512

Solution: The difference table (diagonal) is:

x	$f(x)$	$\Delta f(x)$ or $\nabla f(x)$	$\Delta^2 f(x)$ or $\nabla^2 f(x)$	$\Delta^3 f(x)$ or $\nabla^3 f(x)$
1	**1**			
		7		
2	8		**12**	
		19		**6**
3	27		18	
		37		6
4	64		24	
		61		6
5	125		30	
		91		6
6	216		36	
		127		**6**
7	343		**42**	
		169		
8	**512**			

Part I: As the interpolating point is $1.5 \in [1, 2]$, we set $x_0 = 1$ and apply Newton's forward interpolation formula.

We have $f(x_0) - 1$, $\Delta f(x_0) = 7$, $\Delta^2 f(x_0) = 12$, $\Delta^3 f(x_0) = 6$

The step length $h = 1$.

Newton's forward interpolation formula is given by

$$f(x) \approx p(u) = f(x_0) + \frac{u\Delta f(x_0)}{1!} + u(u-1)\frac{\Delta^2 f(x_0)}{2!} + u(u-1)(u-2)\frac{\Delta^3 f(x_0)}{3!} + \cdots$$

where $u = \dfrac{(x - x_0)}{h} = \dfrac{(1.5-1)}{1}$

$= 0.5$

Therefore, we get

$$f(1.5) \approx 11 + 0.5 \times (7) + \frac{0.5 \times (0.5-1)}{2} \times (12) + \frac{0.5 \times (0.5-1)(0.5-2)}{6} \times (6)$$

$$= 3.375$$

Part II: As the interpolating point is 7.5 ϵ [7, 8], we apply Newton's backward interpolation formula.

We have $f(x_n) = 512$, $\nabla f(x_n) = 169$, $\nabla^2 f(x_n) = 42$, $\nabla^3 f(x_n) = 6$

Newton's backward interpolation formula is given by

$$f(x) \approx p(u) = f(x_n) + \frac{u \nabla f(x_0)}{1!} + u(u+1)\frac{\nabla^2 f(x_n)}{2!} + \cdots$$

where $$u = \frac{(x - x_n)}{h} = \frac{(7.5-8)}{1}$$

$$= -0.5$$

Therefore, we get

$$f(7.5) \approx 512 - 0.5 \times (169) + \frac{-0.5 \times (-0.5+1)}{2} \times (42) + \frac{-0.5 \times (-0.5+1)(-0.5+2)}{6} \times (6)$$

$$= 421.875$$

EXAMPLE 3.26 The values of sin x are given below for different values of x. Form a difference table and from that table, find sin 32°.

x	30°	35°	40°	45°	50°	55°
sin x	0.5000	0.5736	0.6428	0.7071	0.7660	0.8192

Solution: The forward difference table (diagonal) is given below:

x	sin x	Δsin x	Δ²sin x	Δ³sin x	Δ⁴sin x	Δ⁵sin x
30°	**0.5000**					
		0.0736				
35°	0.5736		**−0.0044**			
		0.0692		**−0.0005**		
40°	0.6428		−0.0049		**0**	
		0.0643		−0.0005		**0.0002**
45°	0.7071		−0.0054		0.0002	
		0.0589		−0.0003		
50°	0.7660		−0.0057			
		0.0532				
55°	0.8192					

As the interpolating point is 32° ϵ [30°, 35°], we set $x_0 = 30°$.

So $$\sin(x_0)=0.5000,\ \Delta\sin(x_0)=0.0736,\ \Delta^2\sin(x_0)=-0.0044$$

$$\Delta^3\sin(x_0)=-0.0005,\ \Delta^4\sin(x_0)=0$$

The step length $h = 5°$.

Newton's forward interpolation formula is given by

$$f(x)\approx p(u)=f(x_0)+\frac{u\Delta f(x_0)}{1!}+u(u-1)\frac{\Delta^2 f(x_0)}{2!}+u(u-1)(u-2)\frac{\Delta^3 f(x_0)}{3!}$$

$$+u(u-1)(u-2)(u-3)\frac{\Delta^4 f(x_0)}{4!}+\cdots$$

where $$u=\frac{(x-x_0)}{h}=\frac{(32^0-30^0)}{5^0}$$

$$=0.4$$

and $$f(x)=\sin x$$

Therefore, we have

$$\sin 32°\approx 0.5000+0.4\times(0.0736)+\frac{0.4\times(0.4-1)}{2}\times(-0.0044)+\frac{0.4\times(0.4-1)\times(0.4-2)}{6}\times(-0.0005)$$

$$+\frac{0.4\times(0.4-1)\times(0.4-2)\times(0.4-3)(0.4-4)}{120}\times(0.0002)$$

≈ 0.5299, correct to 4 significant figures.

EXAMPLE 3.27 Apply Lagrange's interpolation formula to find $f(x)$ at $x = 5$ if $f(1) = 2$, $f(2) = 4$, $f(3) = 8$, $f(4) = 16$ and $f(7) = 128$.

Solution: Lagrange's interpolating polynomial is given by

$$y(x)\approx p(x)=\frac{(x-x_1)(x-x_2)(x-x_3)(x-x_4)}{(x_0-x_1)(x_0-x_2)(x_0-x_3)(x_0-x_4)}y_0+\frac{(x-x_0)(x-x_2)(x-x_3)(x-x_4)}{(x_1-x_0)(x_1-x_2)(x_1-x_3)(x_1-x_4)}y_1+\cdots$$

From the given table, we have

$$x_0=1,\ x_1=2,\ x_2=3,\ x_3=4,\ x_4=7 \text{ and } y_0=2,\ y_1=4,\ y_2=8,\ y_3=16,\ y_4=128$$

The interpolating point $x = 5$.

Therefore, using these values, we get

$$y(5)\approx\frac{(5-2)(5-3)(5-4)(5-7)}{(1-2)(1-3)(1-4)(1-7)}\times 2+\frac{(5-1)(5-3)(5-4)(5-7)}{(2-1)(2-3)(2-4)(2-7)}\times 4$$

$$+\frac{(5-1)(5-2)(5-4)(5-7)}{(3-1)(3-2)(3-4)(3-7)}\times 8+\frac{(5-1)(5-2)(5-3)(5-7)}{(4-1)(4-2)(4-3)(4-7)}\times 16$$

$$+\frac{(5-1)(5-2)(5-3)(5-4)}{(7-1)(7-2)(7-3)(7-4)}\times 128$$

$= 32.93333$, correct to 5 decimal places.

EXAMPLE 3.28 Using Lagrange's interpolation formula, find the value of $f(x)$ for $x = 0$, from the following table:

x	−2	−1	2	4
$f(x)$	−9	−1	11	69

Solution: Lagrange's interpolating polynomial is given by

$$y(x) \approx p(x) = \frac{(x-x_1)(x-x_2)(x-x_3)}{(x_0-x_1)(x_0-x_2)(x_0-x_3)} y_0 + \frac{(x-x_0)(x-x_2)(x-x_3)}{(x_1-x_0)(x_1-x_2)(x_1-x_3)} y_1 + \cdots$$

From the given table, we have

$$x_0 = -2, x_1 = -1, x_2 = 2, x_3 = 4 \text{ and}$$
$$y_0 = -9, y_1 = -1, y_2 = 11, y_3 = 69$$

The interpolating point $x = 0$.

Therefore, using these values, we get

$$y(0) \approx \frac{(0+1)(0-2)(0-4)}{(-2+1)(-2-2)(-2-4)} \times (-9) + \frac{(0+2)(0-2)(0-4)}{(-1+2)(-1-2)(-1-4)} \times (-1) + \frac{(0+1)(0+2)(0-4)}{(2+1)(2+2)(2-4)} \times 11 + \frac{(0+1)(0+2)(0-2)}{(4+1)(4+2)(4-2)} \times 69$$

$$\approx 1$$

EXAMPLE 3.29 Use a suitable interpolation formula to find $f(102)$, correct to 2 decimal places, from the following table:

x	93.0	96.2	100.0	104.2	108.7
$f(x)$	11.38	12.80	14.70	17.07	19.91

Solution: As the given arguments are not equispaced, we use Lagrange's interpolation. Now, Lagrange's interpolating polynomial is given by

$$y(x) \approx p(x) = \frac{(x-x_1)(x-x_2)(x-x_3)}{(x_0-x_1)(x_0-x_2)(x_0-x_3)} y_0 + \frac{(x-x_0)(x-x_2)(x-x_3)}{(x_1-x_0)(x_1-x_2)(x_1-x_3)} y_1 + \cdots$$

From the given table, we have

$$x_0 = 93.0, x_1 = 96.2, x_2 = 100.0, x_3 = 104.2, x_4 = 108.7$$
$$y_0 = 11.38, y_1 = 12.80, y_2 = 14.70, y_3 = 17.07, y_4 = 19.91$$

The interpolating point $x = 102$.

Therefore, using these values, we get

$$y(102) \approx \frac{(102-96.2)(102-100.0)(102-104.2)(102-108.7)}{(93.0-96.2)(93.0-100.0)(93.0-104.2)(93.0-108.7)} \times 11.38$$

$$+\frac{(102-93.0)(102-100.0)(102-104.2)(102-108.7)}{(96.2-93.0)(96.2-100.0)(96.2-104.2)(96.2-108.7)} \times 12.80$$

$$+\frac{(102-93.0)(102-96.2)(102-104.2)(102-108.7)}{(100.0-93.0)(100.0-96.2)(100.0-104.2)(100.0-108.7)} \times 14.70$$

$$+\frac{(102-93.0)(102-96.2)(102-100.0)(102-108.7)}{(104.2-93.0)(104.2-96.2)(104.2-100.0)(104.2-108.7)} \times 17.07$$

$$+\frac{(102-93.0)(102-96.2)(102-100.0)(102-104.2)}{(108.7-93.0)(108.7-96.2)(108.7-100.0)(108.7-104.2)} \times 19.91$$

$$= 15.87$$

EXAMPLE 3.30 Find by Lagrange's formula the interpolation polynomial that corresponds to the following data:

x	-1	0	2	5
$f(x)$	9	5	3	15

Solution: Lagrange's interpolating polynomial is given by

$$y(x) \approx p(x) = \frac{(x-x_1)(x-x_2)(x-x_3)}{(x_0-x_1)(x_0-x_2)(x_0-x_3)} y_0 + \frac{(x-x_0)(x-x_2)(x-x_3)}{(x_1-x_0)(x_1-x_2)(x_1-x_3)} y_1 + \cdots$$

From the given table, we have

$$x_0 = -1, x_1 = 0, x_2 = 2, x_3 = 5 \text{ and } y_0 = 9, y_1 = 5, y_2 = 3, y_3 = 15$$

Therefore, using these values, we get

$$y(0) \approx \frac{(x-0)(x-2)(x-5)}{(-1-0)(-1-2)(-1-5)} \times 9 + \frac{(x+1)(x-2)(x-5)}{(0+1)(0-2)(0-5)} \times 5$$

$$+\frac{(x+1)(x-0)(x-5)}{(2+1)(2-0)(2-5)} \times 3 + \frac{(x+1)(x-0)(x-2)}{(5+1)(5-0)(5-2)} \times 15$$

$$= x^2 - 3x + 5$$

EXAMPLE 3.31 Using Newton's divided difference formula, find a cubic polynomial $y(x)$ which takes the following data:

x	2	4	7	8
$y(x)$	3	43	778	1515

Solution: The divided difference table is given below:

x	$y(x)$	1st order	2nd order	3rd order
2	**3**			
		20		
4	43		**45**	
		245		**13**
7	778		123	
		737		
8	1515			

Newton's divided difference formula is given by

$$y(x) \approx p(x) = y_0 + (x - x_0)[x_0, x_1] + (x - x_0)(x - x_1)[x_0, x_1, x_2]$$
$$+ (x - x_0)(x - x_1)(x - x_2)[x_0, x_1, x_2, x_3] + \cdots$$

From the above table, we have

$$x_0 = 2, x_1 = 4, x_2 = 7, x_3 = 8 \text{ and } y_0 = 3, [x_0, x_1] = 20, [x_0, x_1, x_2] = 45, [x_0, x_1, x_2, x_3] = 13$$

Therefore, using these values in the above formula, we get

$$p(x) = 3 + (x - 2) \times 20 + (x - 2)(x - 4) \times 45 + (x - 2)(x - 4)(x - 7) \times 13$$
$$= 13x^3 - 124x^2 + 400x - 405, \text{ which is the required polynomial.}$$

EXAMPLE 3.32 Use Newton's divided difference interpolation formula to find $f(12)$ from the given data:

x	11	13	14	18	19	21
$f(x)$	1342	2210	2758	5850	6878	9282

Solution: The divided difference table is given below:

x	$f(x)$	1st order	2nd order	3rd order	4th order
11	**1342**				
		434			
13	2210		**38**		
		548		**1**	
14	2758		45		**0**
		773		1	
18	5850		51		0
		1028		1	
19	6878		58		
		1202			
21	9282				

Now, using Newton's divided difference formula

$$f(x) \approx p(x) = f(x_0) + (x - x_0)[x_0, x_1] + (x - x_0)(x - x_1)[x_0, x_1, x_2] + \cdots$$

We get, from the above table

$$f(12) \approx 1342 + (12-11)\times 434 + (12-11)(12-13)\times 38 + (12-11)(12-13)(12-14)\times 1$$

$$= 1740$$

MULTIPLE CHOICE QUESTIONS

1. In interpolation, the value of x lies
   (a) between the smallest and the largest values of x
   (b) outside the range of max. and min. values of x
   (c) may be anything
   (d) half of the smallest and largest values of x
2. Divided difference formula is used for
   (a) equispaced points (b) unequally spaced points
   (c) both (a) and (b) (d) none of these
3. The restriction on the interpolation points for Newton's forward and backward formula is
   (a) should not be so large (b) should be in arithmetic progression
   (c) should be in geometric progression (d) should be positive
4. Lagrange's interpolation formula deals with
   (a) equispaced arguments only (b) unequispaced arguments only
   (c) both (a) and (b) (d) none of these
5. In Newton's forward interpolation, the interval should be
   (a) equally spaced (b) not equally spaced
   (c) may be equally spaced (d) both (a) and (b)
6. Newton's forward interpolation formula is used
   (a) near the beginning of the tabulated values
   (b) near the middle of the tabulated values
   (c) near the end of the tabulated values
   (d) for any of these
7. Newton's backward interpolation formula is used
   (a) near the beginning of the tabulated values
   (b) near the middle of the tabulated values
   (c) near the end of the tabulated values
   (d) for any of these
8. Lagrange's interpolation formula is used
   (a) near the beginning of the tabulated values
   (b) near the middle of the tabulated values
   (c) near the end of the tabulated values
   (d) all of these
9. In Lagrange's polynomial, the sum of the Lagrangian coefficients is
   (a) 1 (b) −1 (c) 0 (d) none
10. For $n = 1$, Lagrange's interpolation formula becomes
    (a) an equation of a straight line (b) an equation of a parabola
    (c) an equation of a hyperbola (d) none of these

11. Newton's forward or backward interpolation is used for equispaced arguments.
(a) true (b) false

12. In Newton's forward interpolation formula, the value of $u = \dfrac{x - x_0}{h}$ lies between
(a) 1 and 2 (b) −1 and 0 (c) 0 and 1 (d) −1 and 1

13. In Newton's backward interpolation formula, the value of $u = \dfrac{x - x_n}{h}$ lies between
(a) 1 and 2 (b) −1 and 0 (c) 0 and 1 (d) −1 and 1

14. The process of computing the value of a function inside the given range is called extrapolation.
(a) true (b) false

15. If $y = f(x)$ be defined at $(n + 1)$ distinct interpolating points, the degree of the Lagrangian function is
(a) at least n (b) at most n (c) $n + 1$ (d) n

16. The restriction on the interpolation points for Lagrange's formula is
(a) should have unequal spacing
(b) should have equal spacing
(c) both (a) and (b)
(d) none of these

17 In Newton's forward or backward interpolation, the value of h changes from interval to interval.
(a) true (b) false

18 If the point of interpolation is slightly outside the given range, we apply interpolation.
(a) true (b) false

19 For a given set of values of x and $f(x)$, the interpolating polynomial is
(a) unique (b) not unique (c) a constant (d) none of these

20 The construction of an interpolation polynomial for a function $f(x)$ is not recommended if
(a) $f(x)$ is not a polynomial (b) $f(x)$ is not derivable somewhere
(c) $f(x)$ has abrupt changes (d) graph of $f(x)$ is unknown

Answers

1. (a) 2. (c) 3. (b) 4. (c) 5. (a) 6. (a) 7. (c) 8. (d) 9. (a) 10. (a) 11. (a)
12. (c) 13. (b) 14. (b) 15. (b) 16. (c) 17. (b) 18. (b) 19. (a) 20. (c)

SHORT ANSWER QUESTIONS

1. What is interpolation?
2. What is the difference between interpolation and extrapolation? Give suitable examples.
3. Write down the restrictions in the application of Newton's forward interpolation formula.
4. Mention the inefficiency of Lagrange's interpolation formula.
5. Derive Newton's forward interpolation formula for $n + 1$ arguments.
6. Obtain Newton's backward interpolation formula for $n + 1$ arguments.
7. Derive Lagrange's interpolation formula for $n + 1$ arguments.

8. Derive Newton's divided difference interpolation formula for $n + 1$ arguments.
9. Find the forward interpolating polynomial for the function $f(x)$ where $f(0) = 1, f(1) = 1, f(2) = 1$ and $f(3) = 2$.
10. The function $y = \sin x$ is tabulated as given.

x	0	$\pi/4$	$\pi/2$
$y = \sin x$	0.0	0.70711	1.0

Find the value of $\sin(\pi/3)$ using Lagrange's interpolation formula correct to 5 decimal places.

11. Using Lagrange's interpolation formula, find the form of y(x) from the following data:

x	0	1	3	4
$f(x)$	−12	0	12	24

12. From the following table, express y as a function of x and then find y at $x = 0.5$.

x	0	1	2	3	4
y	3	6	11	18	27

13. Use Lagrange's interpolation formula to find the value of $f(x)$ for $x = 0$, given the following table:

x	−1	−2	2	4
$f(x)$	−1	−9	11	69

14. Find Lagrange's interpolation polynomial passing through the set of points and use it to find out y at $x = 1.5$.

x	0	1	2
y	4	3	6

15. Find Lagrange's interpolation polynomial which corresponds to the following data:

x	−1	0	2	5
y	9	5	3	15

16. The following table gives the viscosity of oil as a function of temperature. Use Lagrange's formula to find the viscosity of oil at a temperature of 140°.

Temperature	110	130	160	190
Viscosity	10.8	8.1	5.5	4.8

17. Use Lagrange's interpolation formula to find $f(12)$.

x	5	6	9	11	14
$f(x)$	12	13	14	16	20

18. Find the polynomial $f(x)$ and hence calculate $f(5.5)$ for the given data:

x	0	2	3	5	7
$f(x)$	1	47	97	251	477

19. Given the following data, find $\log_{10} 656$.

x	654	658	659
$\log_{10} x$	2.8156	2.8182	2.8189

20. Estimate $f(0.9)$ using Newton's divided difference formula, given the following table:

x	0	1	2	4
$f(x)$	5	14	41	98

21. Find the cubic polynomial that takes the following values:

x	0	1	2	3
$f(x)$	1	2	1	10

22. Construct Lagrange's interpolation polynomial using the following data:

x	40	45	50	55
$f(x)$	15.22	13.99	12.62	11.13

Answers

9. $f(x) \approx -\frac{1}{6}\left(x^3 + 3x^2 - 16x + 6\right)$
10. 0.85076, correct to 5 decimal places
11. $y(x) \approx x^3 - 6x^2 + 17x - 12$
12. $y \approx x^2 + 2x + 3$; 4.250, correct to 3 decimal places
13. 1.000, correct to 3 decimal places
14. $p(x) = 2x^2 - 3x + 4$; 4.000, correct to 3 decimal places
15. $p(x) = -x^3 + 7x^2 - 6x - 5$
16. 7.033, correct to 3 decimal places
17. 17.483, correct to 3 decimal places
18. $f(x) \approx 9x^2 + 5x + 1$; 300.750, correct to 3 decimal places
19. 2.8170, correct to 4 decimal places
20. 12.08, correct to 2 decimal places
21. $y(x) \approx 2x^3 - 7x^2 + 6x + 1$
22. $y(x) \approx 2.7 \times 10^{-5} x^3 - 6.4 \times 10^{-3} x^2 + 15.3 \times 10^{-2} x + 17.62$

LONG ANSWER QUESTIONS

1. Using Newton's divided difference formula, calculate $y(3.4)$.

x	2.5	2.8	3.0	3.1	3.6
y	12.1825	16.4446	20.0855	22.1980	36.5982

2. Apply Lagrange's interpolation formula to find $f(x)$ at $x = 5$ if $f(1) = 2, f(2) = 4, f(3) = 8, f(4) = 16$ and $f(7) = 128$.

3. Given the following table of function $f(x) = 1/x$, find $\frac{1}{2.27}$ using a suitable interpolation formula. Find an estimate of the error.

x	2.7	2.8	2.9
$f(x)$	0.3704	0.3571	0.3448

4. The area A of a circle of diameter d is given for the following values:

d:	80	85	90	95	100
A:	5026	5674	6362	7088	7854

Find the approximate value for the area of the circle of diameter 103.

5. Use Newton's divided difference formula to approximate $f(0.05)$, from the following table:

x	0.0	0.2	0.4	0.6	0.8
$f(x)$	1.00000	1.22140	1.49182	1.82212	2.22554

6. The following table gives values of $\tan x$ for $0.10 \leq x \leq 0.30$. Find $y(0.12)$.

x	0.10	0.15	0.20	0.25	0.30
$y = \tan x$	0.1003	0.1511	0.2027	0.2553	0.3093

7. Compute the values of $f(3.5)$ and $f(7.5)$ using Newton's interpolation from the following table:

x	3	4	5	6	7	8
$f(x)$	27	64	125	216	343	512

8. Find the value of $\log_{10} 2.91$, using the following table:

x	2.0	2.2	2.4	2.6	2.8	3.0
$\log_{10} x$	0.30103	0.34242	0.38021	0.41497	0.44716	0.47721

9. Values of x (in degrees) and $\sin x$ are given in the following table.

x (in degrees)	15	20	25	30	35	40
$f(x)$	0.2588190	0.3420201	0.4226183	0.5	0.5735764	0.6427876

Determine the value of sin 38° by Newton's backward difference interpolation formula.

10. Values of $\sin x$ are given below, for different values of x. Form a difference table and from this table, find sin 52°.

x	30°	35°	40°	45°	50°	55°
$y = \sin x$	0.5000	0.5736	0.6428	0.7071	0.7660	0.8192

11. Use Newton's divided difference formula to find $f(5)$ from the following data:

x	0	2	3	4	7	8
$f(x)$	4	26	58	112	466	668

12. The following data gives the melting point of an alloy of zinc and lead. Θ is the temperature and x is the percentage of lead. Using a suitable interpolation formula, find Θ when $x = 48$.

x	40	50	60	70	80	90
Θ	184	204	226	250	276	304

13. If $y(10) = 35.3$, $y(15) = 32.4$, $y(20) = 29.2$, $y(25) = 26.1$, $y(30) = 23.2$ and $y(35) = 20.5$, find $y(12)$ using Newton's forward interpolation formula.

14. Find the third degree curve $y = f(x)$ passing through the points (2,3), (4,43), (7,778) and (8,1515) using Newton's divided difference formula.

15. Calculate $f(1.135)$ using a suitable formula.

x	1.140	1.145	1.150	1.155	1.160	1.165
$f(x)$	0.13103	0.13541	0.13976	0.14410	0.14842	0.15272

16. Using Newton's divided difference formula, estimate $f(8)$, given the following table:

x	4	5	7	10	11	13
$f(x)$	48	100	294	900	1210	2028

17. From the following table, estimate the number of students who obtained marks between 40 and 45.

x	30–40	40–50	50–60	60–70	70–80
$f(x)$	31	42	51	35	31

18. From the following table, find the value of $f(12)$ by Newton's divided difference interpolation formula.

x	11	13	14	18	19	21
$f(x)$	1342	2210	2758	5850	6878	9282

19. The following table gives the distance in nautical miles of the visible horizon for the height (given in feet) above the earth's surface.

x	100	150	200	250	300	350	400
$f(x)$	10.63	13.03	15.04	16.81	18.42	19.9	21.27

Use Newton's forward interpolation formula to estimate y when $x = 120$ *ft*.

20. The ordinates $f(x)$ of a normal curve in terms of standard deviation x are given in the following table:

x	1.00	1.02	1.04	1.06	1.08
$f(x)$	0.2420	0.2371	0.2323	0.2275	0.2227

Find the ordinate for $x = 1.07$.

21. From the following table, estimate the value of $e^{0.15}$.

x	0.1	0.2	0.3	0.4	0.5
e^x	1.10517	1.22140	1.34986	1.49182	1.64872

Answers

1. 29.9671, correct to 4 decimal places
2. 32.93333, correct to 5 decimal places
3. 0.4390, correct to 4 decimal places
4. 8335
5. 1.05072, correct to 5 decimal places
6. 0.1205, correct to 4 decimal places
7. 43; 422
8. 0.46392
9. 0.6156612, correct to 7 decimal places
10. 0.7880, correct to 4 decimal places
11. 194
12. 200
13. 34.2, correct to 1 decimal place
14. $y(x) \approx 13x^3 - 124x^2 + 400x - 405$
15. 0.12653, correct to 5 decimal places
16. 628
17. 17
18. 1740
19. 11.65, correct to 2 decimal places
20. 0.2251, correct to 4 decimal places
21. 1.16183, correct to 5 decimal places

4 Numerical Integration

4.1 INTRODUCTION

The definite integral

$$\int_a^b f(x)\,dx$$

can be evaluated by some analytical methods when the integrand $f(x)$ is either explicitly known or is in a simple known form. However, there are a lot of practical problems involving integrals that cannot be solved by our limited analytical methods. For example, if we want to evaluate the integrals

$$\int_1^3 e^{2x}\log_e 5x\,dx \quad \text{or} \quad \int_2^4 \frac{\cos x}{x+\sin x}\,dx$$

then we may experience difficulties in the application of analytical methods. Moreover, if $f(x)$ is in a very complicated form, mathematical analysis may fail in the efficient computation of the integral.

Numerical integration is a technique used to determine the approximate value of the integral $\int_a^b f(x)\,dx$, where $[a, b]$ is finite even when the integrand $f(x)$ is not known explicitly, and where $f(x)$ is replaced by a suitable interpolating polynomial $p(x)$. This approximation is usually known as **numerical integration** or **quadrature**. The error of integration E is given by

$$\int_a^b f(x)\,dx - \int_a^b p(x)\,dx$$

and is called the **error of approximation.**

Degree of Precision A quadrature formula is said to have a degree of precision n (a positive integer) if the following two conditions are satisfied:

- The error is zero for an arbitrary polynomial of degree $\leq n$
- The error is non-zero for an arbitrary polynomial of degree $n + 1$

4.2 GENERAL QUADRATURE FORMULA FOR EQUIDISTANT ORDINATES

Let us consider the integral $= \int_a^b f(x)\,dx$, where $f(x)$ is defined at every point within the interval $[a, b]$. We divide the interval into n number of equal subintervals, each of width $h = (b - a)/n$, so that $a = x_0$, $x_1 = x_0 + h$, …, $x_n = x_0 + nh = b$. Let $y_0 = f(x_0)$, $y_1 = f(x_1)$, …, $y_n = f(x_n)$. Then

$$I = \int_a^b f(x)\,dx = \int_{x_0}^{x_0+nh} f(x)\,dx = h \cdot \int_0^n f(u)\,du$$

where $u = (x - x_0)/h$

Now, replacing the integrand $f(u)$ by Newton's forward interpolating polynomial, we get

$$I \approx h \cdot \int_0^n \left[y_0 + \frac{u\Delta y_0}{1!} + u(u-1)\frac{\Delta^2 y_0}{2!} + \cdots + u(u-1)\ldots(u-\overline{n-1})\frac{\Delta^n y_0}{n!} \right] du$$

$$= h \cdot \int_0^n \left[y_0 + u\Delta y_0 + (u^2 - u)\frac{\Delta^2 y_0}{2!} + (u^3 - 3u^2 + 2u)\frac{\Delta^3 y_0}{3!} + \cdots \right] du$$

$$= h\left[uy_0 + \frac{u^2\Delta y_0}{2} + \left(\frac{u^3}{3} - \frac{u^2}{2}\right)\frac{\Delta^2 y_0}{2!} + \left(\frac{u^4}{4} - u^3 + u^2\right)\frac{\Delta^3 y_0}{3!} + \cdots \right]_0^n$$

$$= nh\left[y_0 + \frac{n\Delta y_0}{2} + \left(\frac{n^2}{3} - \frac{n}{2}\right)\frac{\Delta^2 y_0}{2!} + \left(\frac{n^3}{4} - n^2 + n\right)\frac{\Delta^3 y_0}{3!} + \cdots \right] \quad (4.1)$$

This equation is known as the Gauss–Legendre general quadrature formula. We now derive three quadrature formulae by putting $n = 1, 2, 3$ in equation (4.1).

4.3 TRAPEZOIDAL RULE

Putting $\boldsymbol{n} = \mathbf{1}$ in equation (4.1) and taking the curve $y = p(x)$ through only **two points** (x_0, y_0) and (x_1, y_1) as a polynomial of degree **1** so that differences of order higher than 1 vanish, for the first subinterval $[x_0, x_1]$, we get

$$I_{0,1} = \int_{x_0}^{x_1} f(x)\,dx = \int_{x_0}^{x_0+h} f(x)\,dx \approx 1 \cdot h\left[y_0 + \frac{1 \cdot \Delta y_0}{2} \right]$$

$$= \frac{h}{2}\left[2y_0 + 1\cdot(y_1 - y_0)\right] = \frac{h}{2}\left[y_0 + y_1\right] \quad (4.2)$$

This is known as the **trapezoidal rule** for two points.
Similarly, for the 2nd interval $[x_1, x_2]$

$$I_{1,2} \approx \frac{h}{2}\left[y_1 + y_2\right]$$

Similarly

$$I_{2,3} \approx \frac{h}{2}\left[y_2 + y_3\right] \text{ for the 3rd interval } [x_2, x_3]$$

$$I_{3,4} \approx \frac{h}{2}\left[y_3 + y_4\right] \text{ for the 4th interval } [x_3, x_4]$$

....

....

$$I_{n-1,n} \approx \frac{h}{2}\left[y_{n-1} + y_n\right] \text{ for the } n^{th} \text{ interval } [x_{n-1}, x_n]$$

Therefore, we get

$$I = \int_{x_0}^{x_0+nh} f(x)dx = \int_{x_0}^{x_0+h} f(x)dx + \int_{x_0+h}^{x_0+2h} f(x)dx + \cdots + \int_{x_0+(n-1)h}^{x_0+nh} f(x)dx$$

$$= I_{0,1} + I_{1,2} + \cdots + I_{n-1,n} \approx \frac{h}{2}[y_0 + y_1] + \frac{h}{2}[y_1 + y_2] + \cdots + \frac{h}{2}[y_{n-1} + y_n]$$

$$= \frac{h}{2}[y_0 + y_n + 2(y_1 + y_2 + \cdots + y_{n-1})] \tag{4.3}$$

The quadrature formula (4.3) is known as the **composite trapezoidal rule.**

Note:

- Equation (4.2) shows that the interpolating polynomial $p(x)$ is developed taking into account only two arguments, x_0, x_1, and their corresponding values, y_0, y_1. So, according to the fundamental theorem of differences, $p(x)$ should be of degree 1, i.e., $p(x) = ax + b$; a, b are constants.
- This formula is applicable for any finite number (odd or even) of subintervals.
- Accuracy in evaluating the integral using the trapezoidal rule can be improved by increasing the number of subintervals.

4.3.1 Geometrical Significance of Trapezoidal Rule

From equation (4.2), we see that the right-hand side represents the area of a trapezium formed by the **straight line** joining the points (x_0, y_0) and (x_1, y_1); the x-axis and the ordinates $y_0 = f(x_0)$ and $y_1 = f(x_1)$ are parallel sides; and the distance between them is the subinterval length h. Again, the right-hand side of (4.3) is the sum of the areas of n trapeziums. In this case, the formula replaces the curve $y = f(x)$ by $\boldsymbol{n}$ **straight lines** joining the points (x_0, y_0) and (x_1, y_1); (x_1, y_1) and (x_2, y_2); ...; (x_{n-1}, y_{n-1}) and (x_n, y_n). This is shown in Fig. 4.1.

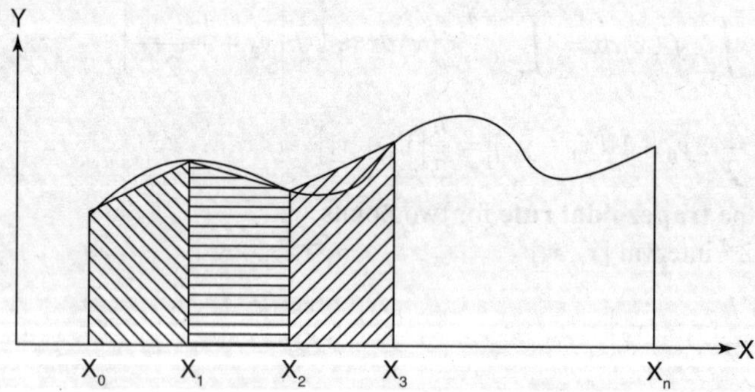

Fig. 4.1 *Trapezoidal rule*

4.3.2 Error in Trapezoidal Rule

Local Error Let $y = f(x)$ be differentiable successively in the interval (x_{i-1}, x_i), $i = 1, 2, \ldots, n$. Then, expanding y in Taylor's series around $x = x_{i-1}$, we get

$$I_{i-1,i} = \int_{x_{i-1}}^{x_i} y dx = \int_{x_{i-1}}^{x_i} \left[y_{i-1} + (x - x_{i-1}) y'_{i-1} + \frac{(x - x_{i-1})^2}{2!} y''_{i-1} + \cdots \right] dx$$

$$= hy_{i-1} + \frac{h^2}{2} y'_{i-1} + \frac{h^3}{6} y''_{i-1} + \cdots$$

where $\quad x_i - x_{i-1} = h$ (4.4)

Now, using the trapezoidal rule in the interval (x_{i-1}, x_i), we get

$$I_{i-1,i} \approx \frac{h}{2}[y_{i-1} + y_i] = \frac{h}{2}[y_{i-1} + y(x_{i-1} + h)]$$

$$= \frac{h}{2}\left[y_{i-1} + y_{i-1} + \frac{h}{1!} y'_{i-1} + \frac{h^2}{2!} y''_{i-1} + \frac{h^3}{3!} y'''_{i-1} + \cdots\right]$$

$$= hy_{i-1} + \frac{h^2}{2} y'_{i-1} + \frac{h^3}{4} y''_{i-1} + \cdots \tag{4.5}$$

Therefore, using (4.4) and (4.5), the error term in the interval (x_{i-1}, x_i) can be given by

$$E_i = I_{i-1,i} - \frac{h}{2}[y_{i-1} + y_i] = -\frac{h^3}{12} y''_{i-1} + \cdots$$

Neglecting the derivatives of order higher than 2, we have

$$E_i \approx -\frac{h^3}{12} y''_{i-1} \tag{4.6}$$

which is the error term for the interval (x_{i-1}, x_i) and is known as the **local error**.

Therefore, local error in the trapezoidal rule is of **order h^3**, i.e., $O(h^3)$.

Global Error Using (4.6), we obtain the total error in n number of subintervals as

$$\sum_{i=1}^{n} E_i \approx -\frac{h^3}{12}[y''_0 + y''_1 + \cdots + y''_n] \approx -\frac{h^3}{12} ny''(\xi)$$

$$= -\frac{h^3}{12}\frac{(b-a)}{h} y''(\xi) = -\frac{(b-a)}{12} h^2 y''(\xi)$$

where $\quad y''(\xi) = max\{y''_0, y''_1, \ldots, y''_n\}$

and each $y''_j, j = 0,1,2,\ldots,n$ is approximated by $y''(\xi)$

This error, committed in the composite trapezoidal rule, is known as the **global error**.

Therefore, global error in the trapezoidal rule is of **order h^2**, i.e., $O(h^2)$.

Note: Since both local and global errors involve 2nd order derivatives, if the polynomial is of degree 1, no error is committed in both forms of the trapezoidal rule as the 2nd order derivative of the polynomial vanishes. For this reason, the degree of precision of the trapezoidal rule is **one**.

4.4 SIMPSON'S 1/3 RULE

Putting $n = 2$ in equation (4.1) and taking the curve $y = p(x)$ through only **three points (x_0, y_0), (x_1, y_1)** and **(x_2, y_2)** as a polynomial of degree **2** so that differences of order higher than 2 vanish, for the first two consecutive subintervals $[x_0, x_1]$ and $[x_1, x_2]$, we get

$$I_{0,2} = \int_{x_0}^{x_2} f(x)dx = \int_{x_0}^{x_0+2h} f(x)dx \approx 2 \cdot h\left[y_0 + \frac{2.\Delta y_0}{2} + \left(\frac{4}{3} - 1\right)\Delta^2 y_0\right]$$

$$= \frac{2 \cdot h}{6}\left[6y_0 + 6\cdot(y_1 - y_0) + (y_2 - 2y_1 + y_0)\right]$$

$$= \frac{h}{3}\left[y_0 + 4y_1 + y_2\right] \qquad (4.7)$$

This is known as **Simpson's 1/3 rule** for three points.

Similarly, for the next two consecutive subintervals $[x_2, x_3]$ and $[x_3, x_4]$,

$$I_{2,4} \approx \frac{h}{3}\left[y_2 + 4y_3 + y_4\right]$$

Similarly

$$I_{4,6} \approx \frac{h}{3}\left[y_4 + 4y_5 + y_6\right] \text{ for intervals } [x_4, x_5] \text{ and } [x_5, x_6]$$

$$I_{6,8} \approx \frac{h}{3}\left[y_6 + 4y_7 + y_8\right] \text{ for intervals } [x_6, x_7] \text{ and } [x_7, x_8]$$

....

....

$$I_{n-2,n} \approx \frac{h}{3}\left[y_{n-2} + 4y_{n-1} + y_n\right]$$

for intervals $[x_{n-2}, x_{n-1}]$ and $[x_{n-1}, x_n]$

Therefore, we get

$$I = \int_{x_0}^{x_0+nh} f(x)\,dx = \int_{x_0}^{x_0+2h} f(x)\,dx + \int_{x_0+2h}^{x_0+4h} f(x)\,dx + \cdots + \int_{x_0+(n-2)h}^{x_0+nh} f(x)\,dx$$

$$= I_{0,2} + I_{2,4} + \cdots + I_{n-2,n} \approx \frac{h}{3}\left[y_0 + 4y_1 + y_2\right]$$

$$+ \frac{h}{3}\left[y_2 + 4y_3 + y_4\right] + \cdots + \frac{h}{3}\left[y_{n-2} + 4y_{n-1} + y_n\right]$$

$$= \frac{h}{3}\left[y_0 + y_n + 2(y_2 + y_4 + \cdots + y_{n-2}) + 4(y_1 + y_3 + \cdots + y_{n-1})\right] \qquad (4.8)$$

The quadrature formula (4.8) is known as the **composite Simpson's 1/3 rule.**

Note

- Equation (4.7) shows that the interpolating polynomial $p(x)$ is constructed using only three arguments, x_0, x_1, x_2 and their corresponding values, y_0, y_1, y_2. So, according to the fundamental theorem of differences, $p(x)$ should be of degree 2, i.e., $p(x) = ax^2 + bx + c$; a, b, c are constants.
- As for each integration the formula considers two consecutive subintervals at a time, this formula is applicable for even number of subintervals only.
- Accuracy in evaluating the integral using Simpson's 1/3 rule can be improved by increasing the number of subintervals.

4.4.1 Geometrical Significance of Simpson's 1/3 Rule

From equation (4.7), we see that the curve $y = f(x)$ is replaced by the second degree polynomial, i.e., a **parabola** passing through (x_0, y_0), (x_1, y_1) and (x_2, y_2). Therefore, the area bounded by the curve $y = f(x)$, the x-axis and the ordinates $y_0 = f(x_0)$ and $y_2 = f(x_2)$ is approximated by the area bounded by the parabola passing through (x_0, y_0), (x_1, y_1) and (x_2, y_2); the x-axis and the ordinates $y_0 = f(x_0)$ and $y_2 = f(x_2)$. Thus, in composite Simpson's 1/3 rule, the formula replaces the curve $y = f(x)$ by ***n*/2 parabolas** passing through (x_0, y_0), (x_1, y_1) and (x_2, y_2); (x_2, y_2), (x_3, y_3) and (x_4, y_4); …; (x_{n-2}, y_{n-2}), (x_{n-1}, y_{n-1}) and (x_n, y_n). This is shown in Fig. 4.2.

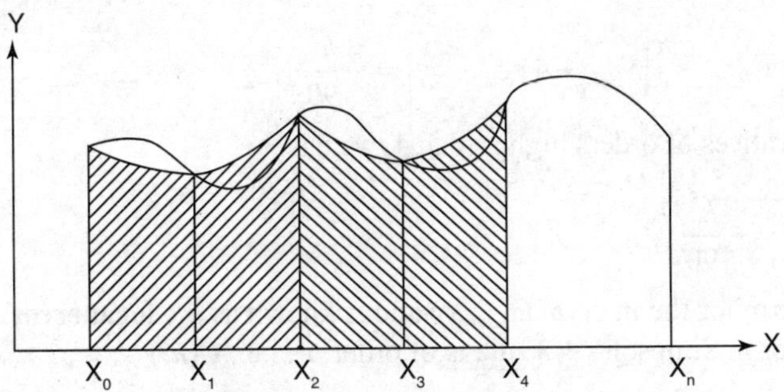

Fig. 4.2 *Simpson's 1/3 rule*

4.4.2 Error in Simpson's 1/3 Rule

Local Error Let $y = f(x)$ be differentiable successively in the intervals (x_{i-2}, x_{i-1}) and (x_{i-1}, x_i), $i = 2, 4, \ldots, n$. Then, expanding y in Taylor's series around $x = x_{i-2}$, we get

$$I_{i-2,i} = \int_{x_{i-2}}^{x_i} y dx$$

$$= \int_{x_{i-2}}^{x_i} \left[y_{i-2} + (x - x_{i-2}) y'_{i-2} + \frac{(x - x_{i-2})^2}{2!} y''_{i-2} + \frac{(x - x_{i-2})^3}{3!} y'''_{i-2} + \frac{(x - x_{i-2})^4}{4!} y^{iv}_{i-2} + \cdots \right] dx$$

$$= 2hy_{i-2} + \frac{4h^2}{2} y'_{i-2} + \frac{8h^3}{6} y''_{i-2} + \frac{16h^4}{24} y'''_{i-2} + \frac{32h^4}{120} y^{iv}_{i-2} + \cdots,$$

where $x_i - x_{i-2} = 2h$

$$= 2hy_{i-2} + 2h^2 y'_{i-2} + \frac{4h^3}{3} y''_{i-2} + \frac{2h^4}{3} y'''_{i-2} + \frac{4h^5}{15} y^{iv}_{i-2} + \cdots \tag{4.9}$$

Now, using Simpson's 1/3 rule for the intervals (x_{i-2}, x_{i-1}) and (x_{i-1}, x_i), we get

$$I_{i-2,i} \approx \frac{h}{3}[y_{i-2} + 4y_{i-1} + y_i] = \frac{h}{3}[y_{i-2} + 4y(x_{i-2} + h) + y(x_{i-2} + 2h)]$$

$$= \frac{h}{3}\left[y_{i-2} + 4\left\{ y_{i-2} + \frac{h}{1!} y'_{i-2} + \frac{h^2}{2!} y''_{i-2} + \frac{h^3}{3!} y'''_{i-2} + \frac{h^4}{4!} y^{iv}{}_{i-2} + \cdots \right\} + \left\{ y_{i-2} + \frac{2h}{1!} y'_{i-2} + \frac{4h^2}{2!} y''_{i-2} + \frac{8h^3}{3!} y'''_{i-2} + \frac{16h^4}{4!} y^{iv}{}_{i-2} + \cdots \right\} \right]$$

$$= 2hy_{i-2} + 2h^2 y'_{i-2} + \frac{4h^3}{3} y''_{i-2} + \frac{2h^4}{3} y'''_{i-2} + \frac{5h^5}{18} y^{iv}_{i-2} + \cdots \tag{4.10}$$

Therefore, using (4.9) and (4.10), the error term in the interval (x_{i-2}, x_i) can be given by

$$E_i = I_{i-1,i} - \frac{h}{3}\left[y_{i-2} + 4y_{i-1} + y_i\right] = -\frac{h^5}{90} y^{iv}_{i-2} + \cdots$$

Neglecting the derivatives of orders higher than 4, we have

$$E_i \approx = -\frac{h^5}{90} y^{iv}_{i-2} \tag{4.11}$$

which is the error term for the interval (x_{i-2}, x_i) and is known as the **local error**.
Therefore, local error in Simpson's 1/3 rule is of order $\boldsymbol{h^5}$, i.e., $\boldsymbol{O(h^5)}$.

Global Error Using (4.11), we obtain the total error in $n/2$ pairs of subintervals as

$$\sum_{i=2}^{n} E_i = E_2 + E_4 + \cdots + E_n \approx -\frac{h^5}{90}\left[y_0^{iv} + y_2^{iv} + \cdots + y_{n-2}^{iv}\right]$$

$$\approx -\frac{h^5}{90}\left(\frac{n}{2}\right) y^{iv}(\xi)$$

$$= -\frac{h^5}{180}\frac{(b-a)}{h} y^{iv}(\xi) = -\frac{(b-a)}{180} h^4 y^{iv}(\xi)$$

where $y^{iv}(\xi) = max\left\{y_0^{iv}, y_2^{iv}, \ldots, y_{n-2}^{iv}\right\}$ and each $y_j^{iv}, j = 0, 2, \ldots n-2$ is approximated by $y^{iv}(\xi)$.

This error, committed in composite Simpson's 1/3 rule, is known as the **global error**.
Therefore, global error in composite Simpson's 1/3 rule is of **order** $\boldsymbol{h^4}$, i.e., $\boldsymbol{O(h^4)}$.

Note: Since both local and global errors involve 4th order derivatives, if the polynomial is of degree 3, no error is committed in both forms of Simpson's 1/3 rule as the 4th order derivative of the polynomial vanishes. For this reason, the degree of precision of Simpson's 1/3 rule is **three**.

4.5 SIMPSON'S 3/8 RULE

Putting $n = 3$ in (4.1) and taking the curve $y = p(x)$ through only **four points** (x_0, y_0), (x_1, y_1), (x_2, y_2) and (x_3, y_3) as a polynomial of degree **3** so that differences of orders higher than 3 vanish, for the first three consecutive subintervals $[x_0, x_1]$, $[x_1, x_2]$ and $[x_2, x_3]$, we get

$$I_{0,3} = \int_{x_0}^{x_3} f(x)\,dx = \int_{x_0}^{x_0+3h} f(x)\,dx \approx 3\cdot h\left[y_0 + \frac{3.\Delta y_0}{2} + \frac{4}{3}\Delta^2 y_0 + \frac{1}{8}\Delta^3 y_0\right]$$

$$= \frac{3\cdot h}{8}\left[8y_0 + 12(y_1 - y_0) + 6(y_2 - 2y_1 + y_0) + (y_3 - 3y_2 + 3y_1 - y_0)\right]$$

$$= \frac{3h}{8}\left[y_0 + 3y_1 + 3y_2 + y_3\right] \tag{4.12}$$

This is known as **Simpson's 3/8 rule** for four points.
Similarly, for next three consecutive subintervals $[x_3, x_4]$, $[x_4, x_5]$ and $[x_5, x_6]$,

$$I_{3,6} \approx \frac{3h}{8}[y_3 + 3y_4 + 3y_5 + y_6]$$

Similarly

$I_{6,9} \approx \frac{3h}{8}[y_6 + 3y_7 + 3y_8 + y_9]$ for intervals $[x_6, x_7]$, $[x_7, x_8]$ and $[x_8, x_9]$,

$I_{9,12} \approx \frac{3h}{8}[y_9 + 3y_{10} + 3y_{11} + y_{12}]$ for intervals $[x_9, x_{10}]$, $[x_{10}, x_{11}]$ and $[x_{11}, x_{12}]$,

....

....

$I_{n-3,n} \approx \frac{3h}{8}[y_{n-3} + 3y_{n-2} + 3y_{n-1} + y_n]$ for intervals $[x_{n-3}, x_{n-2}]$, $[x_{n-2}, x_{n-1}]$ and $[x_{n-1}, x_n]$

Therefore, we get

$$I = \int_{x_0}^{x_0+nh} f(x)dx = \int_{x_0}^{x_0+3h} f(x)dx + \int_{x_0+3h}^{x_0+6h} f(x)dx + \cdots + \int_{x_0+(n-3)h}^{x_0+nh} f(x)dx$$

$$= I_{0,3} + I_{3,6} + \cdots + I_{n-3,n}$$

$$\approx \frac{3h}{8}[y_0 + 3y_1 + 3y_2 + y_3] + \frac{3h}{8}[y_3 + 3y_4 + 3y_5 + y_6] + \cdots + \frac{3h}{8}[y_{n-3} + 3y_{n-2} + 3y_{n-1} + y_n]$$

$$= \frac{3h}{8}[y_0 + y_n + 3(y_1 + y_2 + y_4 + y_5 + \cdots + y_{n-2} + y_{n-1}) + 2(y_3 + y_6 + \cdots + y_{n-3})] \tag{4.13}$$

The quadrature formula (4.13) is known as the **composite Simpson's 3/8 rule**.

4.6 BOOLE'S RULE

Putting $n = 4$ in (4.1) and taking the curve $y = p(x)$ through only **five points** (x_0, y_0), (x_1, y_1), (x_2, y_2), (x_3, y_3) and (x_4, y_4) as a polynomial of degree **4** so that differences of orders higher than 4 vanish, for the first four consecutive subintervals $[x_0, x_1]$, $[x_1, x_2]$, $[x_2, x_3]$ and $[x_3, x_4]$, we get

$$I_{0,4} = \int_{x_0}^{x_4} f(x)dx = \int_{x_0}^{x_0+4h} f(x)dx \approx 4.h\left[y_0 + 2\Delta y_0 + \frac{5}{3}\Delta^2 y_0 + \frac{2}{3}\Delta^3 y_0 + \frac{7}{90}\Delta^4 y_0\right]$$

$$-\frac{2h}{45}[7y_0 + 32y_1 + 12y_2 + 32y_3 + 7y_4] \tag{4.14}$$

This is known as **Boole's rule** for five points.

Thus, the composite rule is given by

$$I \approx \frac{2h}{45}[7(y_0 + y_n) + 32(y_1 + y_3 + y_5 + y_7 + \cdots) + 12(y_2 + y_6 + \cdots) + 14(y_4 + y_8 + \cdots)] \tag{4.15}$$

4.7 WEDDLE'S RULE

Putting $n = 6$ in (4.1) and taking the curve $y = p(x)$ through only **seven points** (x_0, y_0), (x_1, y_1), (x_2, y_2), (x_3, y_3), (x_4, y_4), (x_5, y_5) and (x_6, y_6) as a polynomial of degree **6** so that differences of orders higher than 6 vanish, we get

$$I_{0,6} = \int_{x_0}^{x_6} f(x)dx = \int_{x_0}^{x_0+6h} f(x)dx$$

$$\approx 6\cdot h\left[y_0 + 3\Delta y_0 + \frac{9}{2}\Delta^2 y_0 + 4\Delta^3 y_0 + \frac{41}{20}\Delta^4 y_0 + \frac{11}{20}\Delta^5 y_0 + \frac{41}{840}\Delta^6 y_0\right]$$

$$= \frac{6h}{20}\left[20y_0 + 60\Delta y_0 + 90\Delta^2 y_0 + 80\Delta^3 y_0 + 41\Delta^4 y_0 + 11\Delta^5 y_0 + \frac{41}{42}\Delta^6 y_0\right]$$

$$= \frac{6h}{20}\left[20y_0 + 60\Delta y_0 + 90\Delta^2 y_0 + 80\Delta^3 y_0 + 41\Delta^4 y_0 + 11\Delta^5 y_0 + \Delta^6 y_0\right]$$

considering $\frac{41}{42} \approx 1$

$$= \frac{3h}{10}\left[y_0 + 5y_1 + y_2 + 6y_3 + y_4 + 5y_5 + y_6\right] \tag{4.16}$$

This is known as **Weddle's rule** for seven points.

Thus, the composite form of Weddle's rule is given by

$$\int_a^b f(x)dx = \int_{x_0}^{x_0+nh} f(x)dx$$

$$\approx \frac{3h}{10}[(y_0 + y_n) + 5(y_1 + y_7 + y_{13} + \cdots) + (y_2 + y_8 + y_{14} + \cdots)$$
$$+ 6(y_3 + y_9 + y_{15} + \cdots) + (y_4 + y_{10} + y_{16} + \cdots)$$
$$+ 5(y_5 + y_{11} + y_{17} + \cdots) + 2(y_6 + y_{12} + y_{18} + \cdots)] \tag{4.17}$$

$$= \frac{3h}{10}[(y_0 + y_n) + 5(y_1 + y_5 + y_7 + y_{11} + \cdots) + (y_2 + y_4 + y_8 + y_{10} + \cdots)$$
$$+ 6(y_3 + y_9 + y_{15} + \cdots) + 2(y_6 + y_{12} + y_{18} + \cdots)] \tag{4.18}$$

4.8 NEWTON–COTE'S QUADRATURE FORMULA

Let us consider the definite integral $= \int_a^b f(x)dx$, where $f(x)$ is defined at every point within the interval $[a, b]$. We divide the interval into n number of equal subintervals, each of length $h = (b - a)/n$, so that $a = x_0, x_1 = x_0 + h, \ldots, x_n = x_0 + nh = b$. Let $y_0 = f(x_0), y_1 = f(x_1), \ldots, y_n = f(x_n)$. We replace $f(x)$ with a suitable polynomial $p(x) = \sum_{i=0}^{n} L_i(x)y_i$.

where $$L_i(x) = \frac{(x-x_0)(x-x_1)\ldots(x-x_{i-1})(x-x_{i+1})\ldots(x-x_n)}{(x_i-x_0)(x_i-x_2)\ldots(x_i-x_{i-1})(x_i-x_{i+1})\ldots(x_i-x_n)}$$

Now $x_i = x_0 + ih$ and $x = x_0 + uh$ and so $dx = hdu$

and $$L_i(x) = \frac{u(u-1)\ldots(u-\overline{i-1})(u-\overline{i+1})\ldots(u-n)}{i(i-1)\ldots1(-1)\ldots(i-n)}$$

Therefore $$\int_a^b f(x)dx = \int_{x_0}^{x_0+nh} f(x)dx \approx \int_{x_0}^{x_n} p(x)dx = h\int_0^n\left[\sum_{i=0}^{n} L_i(x)y_i\right]du$$

$$= nh\left[\frac{1}{n}\sum_{i=0}^{n} y_i\int_0^n L_i(x)du\right] = nh\sum_{i=0}^{n} C_i^n y_i = (b-a)\sum_{i=0}^{n} C_i^n y_i \quad (4.19)$$

where $$C_i^n = \frac{1}{n}\int_0^n L_i(x)du \text{ ,} 0 \le i \le n \quad (4.20)$$

are known as Cote's coefficients and (4.19) is called the **Newton–Cote's formula for integration.**

Properties of Cote's Coefficients

(i) $\sum_{i=0}^{n} C_i^n = 1$

Proof: Let $f(x) = 1$. Then

$$\int_a^b f(x)dx = (b-a)\sum_{i=0}^{n} C_i^n y_i$$

gives

$$\int_a^b 1\, dx = (b-a)\sum_{i=0}^{n} C_i^n$$

$$\Rightarrow \quad b-a = (b-a)\sum_{i=0}^{n} C_i^n$$

$$\Rightarrow \quad \sum_{i=0}^{n} C_i^n = 1$$

(ii) $C_i^n = C_{n-i}^n$

Proof: From equation (4.20), we get

$$C_i^n = \frac{1}{n}\int_0^n L_i(x)du = \frac{1}{n}\frac{(-1)^{n-i}}{i!(n-i)!}\int_0^n \frac{u(u-1)\ldots(u-\overline{i-1})(u-\overline{i+1})\ldots(u-n)}{u-i}du$$

$$= \frac{1}{n}\frac{(-1)^{2n-i}}{i!(n-i)!}\int_0^n \frac{u'(u'-1)\ldots(u'-\overline{i-1})(u'-\overline{i+1})\ldots(u'-n)}{u'-(n-i)}du'$$

where $u' = n - u$

$$= \frac{1}{n}\frac{(-1)^{i}}{i!(n-i)!}\int_0^n \frac{u'(u'-1)\ldots(u'-\overline{i-1})(u'-\overline{i+1})\ldots(u'-n)}{u'-(n-i)}du' = C_{n-i}^n$$

4.8.1 Trapezoidal Rule

Putting $n = 1$ in (4.19), we get

$$\int_{x_0}^{x_1} f(x)dx \approx \int_{x_0}^{x_1} p(x)dx = 1.h\sum_{i=0}^{1} C_i^1 y_i = h\left[C_0^1 y_0 + C_1^1 y_1\right]$$

Now, from (4.20), we get

$$C_0^1 = \frac{1}{1}\int_0^1 \frac{(u-1)}{-1}du = \left[u - \frac{u^2}{2}\right]_0^1 = \frac{1}{2}$$

and $$C_1^1 = \frac{1}{1}\int_0^1 u du = \left[\frac{u^2}{2}\right]_0^1 = \frac{1}{2}$$

Therefore $$\int_{x_0}^{x_1} f(x)dx \approx h\left[\frac{1}{2}y_0 + \frac{1}{2}y_1\right] = \frac{h}{2}[y_0 + y_1]$$

4.8.2 Simpson's 1/3 Rule

Putting $n = 2$ in (4.19), we get

$$\int_{x_0}^{x_2} f(x)dx \approx \int_{x_0}^{x_2} p(x)dx = 2\cdot h\sum_{i=0}^{2} C_i^2 y_i = 2h\left[C_0^2 y_0 + C_1^2 y_1 + C_2^2 y_2\right]$$

Now, from (4.20), we get

$$C_0^2 = \frac{1}{2}\int_0^2 \frac{(u-1)(u-2)}{(-1)(-2)}du = \frac{1}{6}$$

$$C_1^2 = \frac{1}{2}\int_0^2 \frac{u(u-2)}{1(-1)}du = \frac{4}{6}$$

and $$C_2^2 = \frac{1}{2}\int_0^2 \frac{u(u-1)}{2.1}du = \frac{1}{6}$$

Therefore $$\int_{x_0}^{x_2} f(x)dx \approx 2h\left[\frac{1}{6}y_0 + \frac{4}{6}y_1 + \frac{1}{6}y_2\right] = \frac{h}{3}[y_0 + 4y_1 + y_2]$$

SOLVED EXAMPLES

EXAMPLE 4.1 Evaluate $\int_0^1 \frac{dx}{1+x}$ by Simpson's composite rule taking 11 ordinates, and hence find the value of $\log_e 2$, correct to 5 significant figures.

Solution: Let $y = f(x) = \dfrac{1}{1+x}$.

We are given $a = 0$, $b = 1$, $n = 10$ (as the number of ordinates is 11).

So $$h = \frac{b-a}{n} = \frac{1-0}{10} = 0.1$$

The values of x and $y = f(x)$ from $a = 0$ to $b = 1$ are given in the following table:

Values of x_i	Values of $y_i = f(x_i)$		
	$i = 0, 10$	$i = 2, 4, 6, 8$	$i = 1, 3, 5, 7, 9$
$x_0 = 0$	$y_0 = 1$	–	–
$x_1 = 0.1$	–	–	$y_1 = 0.9091$
$x_2 = 0.2$	–	$y_2 = 0.8333$	–
$x_3 = 0.3$	–	–	$y_3 = 0.7692$
$x_4 = 0.4$	–	$y_4 = 0.7143$	–
$x_5 = 0.5$	–	–	$y_5 = 0.6667$
$x_6 = 0.6$	–	$y_6 = 0.6250$	–
$x_7 = 0.7$	–	–	$y_7 = 0.5882$
$x_8 = 0.8$	–	$y_8 = 0.5556$	–
$x_9 = 0.9$	–	–	$y_9 = 0.5263$
$x_{10} = 1$	$y_{10} = 0.5$	–	–
Total	1.5	2.7282	3.4595

Simpson's composite rule is given by

$$\int_a^b f(x)\,dx \approx \frac{h}{3}\left[y_0 + y_n + 2(y_2 + y_4 + \cdots + y_{n-2}) + 4(y_1 + y_3 + \cdots + y_{n-1})\right]$$

Using this, we get

$$\int_0^1 \frac{dx}{1+x} \approx \frac{0.1}{3}[1.5 + 2\times 2.7282 + 4\times 3.4595]$$
$$= 0.6931467$$

Again $\quad \int_0^1 \frac{dx}{1+x} = \left[\log_e(1+x)\right]_0^1 = \log_e 2$

Therefore $\quad \log_e 2 \approx 0.6931467 \approx 0.69315$, correct to 5 significant figures.

EXAMPLE 4.2 Find the value of $\log 2^{1/3}$ from $\int_0^1 \frac{x^2\,dx}{1+x^3}$ using Simpson's 1/3 rule with $h = 0.25$.

Solution: Let $y = f(x) = \frac{x^2}{1+x^3}$.

We are given $a = 0$, $b = 1$, $h = 0.25$.

The values of x and $y = f(x)$ from $a = 0$ to $b = 1$ are given in the following table:

Values of x_i	Values of $y_i = f(x_i)$		
	$i = 0, 4$	$i = 2$	$i = 1, 3$
$x_0 = 0$	$y_0 = 0$	–	–
$x_1 = 0.25$	–	–	$y_1 = 0.06154$
$x_2 = 0.50$	–	$y_2 = 0.22222$	–
$x_3 = 0.75$	–	–	$y_3 = 0.39560$
$x_4 = 1$	$y_4 = 0.50000$		–
Total	0.50000	0.22222	0.45714

Simpson's composite rule is given by

$$\int_a^b f(x)\,dx \approx \frac{h}{3}\left[y_0 + y_n + 2(y_2 + y_4 + \cdots + y_{n-2}) + 4(y_1 + y_3 + \cdots + y_{n-1})\right]$$

Using this, we get

$$\int_0^1 \frac{x^2 dx}{1+x^3} \approx \frac{0.25}{3}\left[0 + 0.5 + 2\times 0.22222 + 4(0.06154 + 0.39560)\right]$$

$$= 0.23108$$

Again $\quad \displaystyle\int_0^1 \frac{x^2 dx}{1+x^3} = \frac{1}{3}\left[\log(1+x^3)\right]_0^1 = \frac{1}{3}\log 2 = \log 2^{1/3}$

Therefore $\quad \log 2^{1/3} \approx 0.23108$

EXAMPLE 4.3 Evaluate $\displaystyle\int_0^1 \frac{dx}{1+x^2}$ by Simpson's 1/3 rule taking $n = 6$, and hence compute an approximate value of π, correct to 5 significant figures.

Solution: Let $y = f(x) = \dfrac{1}{1+x^2}$.

We are given $a = 0$, $b = 1$, $n = 6$.

So $\quad h = \dfrac{b-a}{n} = \dfrac{1-0}{6} = 1/6$

The values of x and $y = f(x)$ from $a = 0$ to $b = 1$ are given in the following table:

Values of x_i	Values of $y_i = f(x_i)$
$x_0 = 0$	$y_0 = 1$
$x_1 = 1/6$	$y_1 = 36/37$
$x_2 = 1/3$	$y_2 = 9/10$

Cont'd

Table cont'd

Values of x_i	Values of $y_i = f(x_i)$
$x_3 = 1/2$	$y_3 = 4/5$
$x_4 = 2/3$	$y_4 = 9/13$
$x_5 = 5/6$	$y_5 = 36/61$
$x_6 = 1$	$y_6 = 1/2$

Simpson's composite rule is given by

$$\int_a^b f(x)\,dx \approx \frac{h}{3}\left[y_0 + y_n + 2(y_2 + y_4 + \cdots + y_{n-2}) + 4(y_1 + y_3 + \cdots + y_{n-1})\right]$$

Using this, we get

$$\int_0^1 \frac{dx}{1+x^2} \approx \frac{1/6}{3}\left[1 + 1/2 + 2(9/10 + 9/13) + 4(36/37 + 4/5 + 36/61)\right]$$

$$= 0.7854$$

Again $$\int_0^1 \frac{dx}{1+x^2} = \left[\tan^{-1} x\right]_0^1 = \tan^{-1}(1) = \pi/4$$

Therefore $\pi/4 \approx 0.7854 \Rightarrow \pi \approx 3.1416$, correct to 5 significant figures.

EXAMPLE 4.4 Compute the value of π from the formula $\frac{\pi}{4} = \int_0^1 \frac{dx}{1+x^2}$ using the composite trapezoidal rule with 10 subintervals and compare it with the exact value.

Solution: Let $y = f(x) = \frac{1}{1+x^2}$.

We are given $a = 0$, $b = 1$, $n = 10$.

So $$h = \frac{b-a}{n} = \frac{1-0}{10} = 0.1.$$

The values of x and $y = f(x)$ from $a = 0$ to $b = 1$ are given in the following table:

Values of x_i	Values of $y_i = f(x_i)$
$x_0 = 0$	$y_0 = 1$
$x_1 = 0.1$	$y_1 = 0.9901$
$x_2 = 0.2$	$y_2 = 0.9615$
$x_3 = 0.3$	$y_3 = 0.9174$
$x_4 = 0.4$	$y_4 = 0.8621$
$x_5 = 0.5$	$y_5 = 0.8$
$x_6 = 0.6$	$y_6 = 0.7353$

Cont'd

Table cont'd

Values of x_i	Values of $y_i = f(x_i)$
$x_7 = 0.7$	$y_7 = 0.6711$
$x_8 = 0.8$	$y_8 = 0.6098$
$x_9 = 0.9$	$y_9 = 0.5525$
$x_{10} = 1$	$y_{10} = 0.5$

The trapezoidal composite rule is given by

$$\int_a^b f(x)\,dx \approx \frac{h}{2}\left[y_0 + y_n + 2(y_1 + y_2 + y_3 + \cdots + y_{n-1})\right]$$

Using this, we get

$$\int_0^1 \frac{dx}{1+x^2} \approx \frac{0.1}{2}[1+0.5+2(0.9901+0.9615+0.9174+0.8621+0.8$$

$$+0.7353+0.6711+0.6098+0.5525)]$$

$$= 0.7854$$

Therefore, from the given condition, we have

$$\pi/4 \approx 0.7854 \Rightarrow \pi \approx 3.1416$$

Now, if we consider 3.14159 as the exact value of π, then the percentage error is given by

$$\left|\frac{3.14159-3.1416}{3.14159}\right| \times 100\% = 0.00032\%$$

EXAMPLE 4.5 Evaluate $\int xe^x\,dx$ where the interval is $(0, -1)$ using the trapezoidal rule, taking $n = 6$.

Solution: Let $y = f(x) = xe^x$.

We are given $a = -1$, $b = 0$, $n = 6$.

So

$$h = \frac{b-a}{n} = \frac{0-(-1)}{6} = 1/6$$

The values of x and $y = f(x)$ from $a = -1$ to $b = 0$ are given in the following table:

Values of x_i	Values of $y_i = f(x_i)$
$x_0 = -1$	$y_0 = -0.3678$
$x_1 = -5/6$	$y_1 = -0.3622$
$x_2 = -2/3$	$y_2 = -0.3405$
$x_3 = -1/2$	$y_3 = -0.3033$
$x_4 = -1/3$	$y_4 = -0.2388$
$x_5 = -1/6$	$y_5 = -0.1411$
$x_6 = 0$	$y_6 = 0$

The trapezoidal composite rule is given by

$$\int_a^b f(x)\,dx \approx \frac{h}{2}\left[y_0 + y_n + 2(y_1 + y_2 + y_3 + \cdots + y_{n-1})\right]$$

Using this, we get

$$\int_0^1 \frac{dx}{1+x^2} \approx \frac{1/6}{2}\left[-0.3678 + 0 + 2(-0.3622 - 0.3405 - 0.3033 - 0.2388 - 0.1411)\right]$$

$$\approx -0.2616$$

EXAMPLE 4.6 Calculate using Simpson's 1/3 rule the value of the integral $\int_0^1 \frac{xdx}{1+x}$, correct to 3 significant figures taking 6 intervals.

Solution: Let $y = f(x) = \frac{x}{1+x}$.

We are given $a = 0$, $b = 1$, $n = 6$.

So $$h = \frac{b-a}{n} = \frac{1-0}{6} = 1/6$$

The values of x and $y = f(x)$ from $a = 0$ to $b = 1$ are given in the following table:

Values of x_i	Values of $y_i = f(x_i)$
$x_0 = 0$	$y_0 = 0$
$x_1 = 1/6$	$y_1 = 1/7$
$x_2 = 1/3$	$y_2 = 1/4$
$x_3 = 1/2$	$y_3 = 1/3$
$x_4 = 2/3$	$y_4 = 2/5$
$x_5 = 5/6$	$y_5 = 5/11$
$x_6 = 1$	$y_6 = 1/2$

Simpson's composite rule is given by

$$\int_a^b f(x)\,dx \approx \frac{h}{3}\left[y_0 + y_n + 2(y_2 + y_4 + \cdots + y_{n-2}) + 4(y_1 + y_3 + \cdots + y_{n-1})\right]$$

Using this, we get

$$\int_0^1 \frac{xdx}{1+x} \approx \frac{1/6}{3}\left[0 + 1/2 + 2(1/4 + 2/5) + 4(1/7 + 1/3 + 5/11)\right]$$

≈ 0.307, correct to 3 significant figures.

EXAMPLE 4.7 Calculate the approximate value of $\int_0^{\pi/2} \sin x dx$ by composite trapezoidal rule using 11 ordinates. Also compare it with the actual value of the integral.

Solution: Let $y = f(x) = \sin x$.

We are given $a = 0$, $b = \pi/2$, $n = 10$ (as the number of ordinates is 11).

So
$$h = \frac{b-a}{n} = \frac{\pi/2-0}{10} = \pi/20$$

The values of x and $y = f(x)$ from $a = 0$ to $b = \pi/2$ are given in the following table:

Values of x_i	**Values of $y_i = f(x_i)$**
0	0.0
$\pi/20$	0.15643
$\pi/10$	0.30902
$3\pi/20$	0.45400
$\pi/5$	0.58778
$\pi/4$	0.70711
$3\pi/10$	0.80902
$7\pi/20$	0.89101
$2\pi/5$	0.95106
$9\pi/20$	0.98769
$\pi/2$	1.0

The trapezoidal composite rule is given by
$$\int_a^b f(x)\,dx \approx \frac{h}{2}\left[y_0 + y_n + 2(y_1 + y_2 + y_3 + \cdots + y_{n-1})\right]$$

Using this, we get
$$\int_0^{\pi/2} \sin x\,dx \approx \frac{\pi/20}{2}[0+1+2(0.15643+0.30902+0.454+0.58778$$
$$+0.70711+0.80902+0.89101+0.95106+0.98769)]$$
$$\approx 0.998$$

Now
$$\int_0^{\pi/2} \sin x\,dx = -[\cos x]_0^{\pi/2} = 1$$

So
$$\text{the percentage error} = \left|\frac{1-0.998}{1}\right| \times 100\% = 0.2\%$$

EXAMPLE 4.8 Compute $I = \int \frac{x}{\sin x}dx$ where the interval is (0, 1/2) using Simpson's 1/3 rule with $h = 1/4$.

Solution: Let $y = f(x) = \frac{x}{\sin x}$.

We are given $a = 0$, $b = 1/2$, $h = 1/4$.

Here, we see that the given integration is an **improper** one as the integrand $f(x) = \dfrac{x}{\sin x}$ is not defined at $x = 0$. But the **limiting value** of it at that point is 1. Hence, considering this, the values of x and $y = f(x)$ from $a = 0$ to $b = 1$ are given in the following table:

Values of x_i	Values of $y_i = f(x_i)$
0	1
1/4	1.0105
1/2	1.0429

Simpson's composite rule is given by

$$\int_a^b f(x)\,dx \approx \frac{h}{3}\left[y_0 + y_n + 2(y_2 + y_4 + \cdots + y_{n-2}) + 4(y_1 + y_3 + \cdots + y_{n-1})\right]$$

Using this, we get

$$\int_o^{1/2} \frac{x}{\sin x}\,dx \approx \frac{1/4}{3}[1 + 1.0429 + 4 \times 1.0105]$$

$$\approx 0.51$$

EXAMPLE 4.9 Calculate the value of $\displaystyle\int_{1.2}^{1.6}\left(x + \frac{1}{x}\right)dx$, correct to 3 significant figures, by the trapezoidal rule, taking 4 intervals.

Solution: Let $y = f(x) = (x + 1/x)$.

We are given $a = 1.2$, $b = 1.6$, $n = 4$.

So $$h = \frac{b-a}{n} = \frac{1.6 - 1.2}{4} = 0.1$$

The values of x and $y = f(x)$ from $a = 0$ to $b = 1$ are given in the following table.

Values of x_i	Values of $y_i = f(x_i)$
1.2	2.03333
1.3	2.06923
1.4	2.11429
1.5	2.16667
1.6	2.22500

The trapezoidal composite rule is given by

$$\int_a^b f(x)\,dx \approx \frac{h}{2}\left[y_0 + y_n + 2(y_1 + y_2 + y_3 + \cdots + y_{n-1})\right]$$

Using this, we get

$$\int_{1.2}^{1.6} (x+1/x)\,dx \approx \frac{0.1}{2}\left[2.03333+2.22500+2(2.06923+2.11429+2.16667)\right]$$

$$\approx 0.848, \text{ correct to 3 significant figures.}$$

EXAMPLE 4.10 Evaluate $\int_0^1 (4x-3x^2)dx$, taking 10 intervals by Simpson's 1/3 rule.

Solution: Let $y = f(x) = 4x - 3x^2$.

We are given $a = 0$, $b = 1$, $n = 10$.

So $$h = \frac{b-a}{n} = \frac{1-0}{10} = 0.1$$

The values of x and $y = f(x)$ from $a = 0$ to $b = 1$ are given in the following table:

Values of x_i	Values of $y_i = f(x_i)$
0	0
0.1	0.37
0.2	0.68
0.3	0.93
0.4	1.12
0.5	1.25
0.6	1.32
0.7	1.33
0.8	1.28
0.9	1.17
1	1

Simpson's composite rule is given by

$$\int_a^b f(x)\,dx \approx \frac{h}{3}\left[y_0 + y_n + 2(y_2 + y_4 + \cdots + y_{n-2}) + 4(y_1 + y_3 + \cdots + y_{n-1})\right]$$

Using this, we get

$$\int_0^1 (4x-3x^2)dx \approx \frac{0.1}{3}\left[0+1+2(0.68+1.12+1.32+1.28)\right.$$

$$\left.+4(0.37+0.93+1.25+1.33+1.17)\right]$$

$$= 1$$

EXAMPLE 4.11 Apply the trapezoidal rule to integrate $\int_0^1 \sqrt{1-x^2}\,dx$, taking 10 subintervals.

Solution: Let $y = f(x) = \sqrt{1-x^2}$.

We are given $a = 0$, $b = 1$, $n = 10$.

So
$$h = \frac{b-a}{n} = \frac{1-0}{10} = 0.1$$

The values of x and $y = f(x)$ from $a = 0$ to $b = 1$ are given in the following table:

Values of x_i	**Values of $y_i = f(x_i)$**
0	1
0.1	0.995
0.2	0.980
0.3	0.954
0.4	0.917
0.5	0.866
0.6	0.8
0.7	0.7141
0.8	0.6
0.9	0.4359
1	0

The trapezoidal composite rule is given by

$$\int_a^b f(x)\,dx \approx \frac{h}{2}\left[y_0 + y_n + 2\left(y_1 + y_2 + y_3 + \cdots + y_{n-1}\right)\right]$$

Using this, we get

$$\int_0^1 \sqrt{1-x^2}\,dx \approx \frac{0.1}{2}[1+0+2(0.995+0.980+0.954+0.917+0.866$$
$$+0.8+0.7141+0.6+0.4359)]$$
$$\approx 0.776$$

EXAMPLE 4.12 Evaluate $\int_0^1 xe^x\,dx$ using the trapezoidal rule, taking $n = 6$.

Solution: Let $y = f(x) = xe^x$.

We are given $a = 0$, $b = 1$, $n = 6$.

So
$$h=\frac{b-a}{n}=\frac{1-0}{6}=1/6$$

The values of x and $y=f(x)$ from $a=0$ to $b=1$ are given in the following table:

Values of x_i	Values of $y_i=f(x_i)$
0	0
1/6	0.1969
1/3	0.4652
1/2	0.8244
2/3	1.2985
5/6	1.9175
1	2.7183

The trapezoidal composite rule is given by

$$\int_a^b f(x)\,dx \approx \frac{h}{2}\left[y_0+y_n+2(y_1+y_2+y_3+\cdots+y_{n-1})\right]$$

Using this, we get

$$\int_0^1 xe^x\,dx \approx \frac{1/6}{2}\left[0+2.7183+2(0.1969+0.4652+0.8244+1.2985+1.9175)\right]$$
$$\approx 1.0103$$

EXAMPLE 4.13 A curve passes through points as given in the following table. Find the area bounded by the curve, x-axis, $x=1$ and $x=9$.

x	1	2	3	4	5	6	7	8	9
y	1342	2210	2758	5850	6878	9282	1.2	2.1	2.3

Solution: If $y=f(x)$ be the curve for the above problem, the area is given by $\int_1^9 f(x)\,dx$.

From the given table, it is clear that the x-values are equidistant with constant step length 1. Hence, we can apply the composite trapezoidal rule by assigning $a=1$, $b=9$, $h=1$.

Now, the trapezoidal composite rule is given by

$$\int_a^b f(x)\,dx \approx \frac{h}{2}\left[y_0+y_n+2(y_1+y_2+y_3+\cdots+y_{n-1})\right]$$

Using this, we get

$$\int_1^9 f(x)\,dx \approx \frac{1}{2}\left[1342+2.3+2(2210+2758+5850+6878+9282+1.2+2.1)\right]$$
$$=27653.45$$

EXAMPLE 4.14 A curve passes through points as given in the following table. Find the area bounded by the curve, x-axis, $x = 1$ and $x = 9$.

x	1	2	3	4	5	6	7	8	9
y	0.2	0.7	1	1.3	1.5	1.7	1.9	2.1	2.3

Solution: If $y = f(x)$ be the curve for the above problem, the area is given by $\int_1^9 f(x)dx$.

From the given table, it is clear that the x-values are equidistant with constant step length 1. Hence, we can apply the composite trapezoidal rule by assigning $a = 1$, $b = 9$, $h = 1$.

Now, the trapezoidal composite rule is given by

$$\int_a^b f(x)dx \approx \frac{h}{2}\left[y_0 + y_n + 2(y_1 + y_2 + y_3 + \cdots + y_{n-1})\right]$$

Using this, we get

$$\int_1^9 f(x)dx \approx \frac{1}{2}\left[0.2 + 2.3 + 2(0.7 + 1 + 1.3 + 1.5 + 1.7 + 1.9 + 2.1)\right]$$

$$= 11.45$$

EXAMPLE 4.15 Find the value of $\int_0^{\pi/2} \sqrt{1 - 0.162\sin^2 x}\,dx$ using Simpson's 1/3 rule, taking 6 equal subintervals.

Solution: Let $y = f(x) = \sqrt{1 - 0.162\sin^2 x}$.

We are given $a = 0$, $b = \pi/2$, $n = 6$.

So $$h = \frac{b-a}{n} = \frac{\pi/2 - 0}{6} = \pi/12$$

The values of x and $y = f(x)$ from $a = 0$ to $b = \pi/2$ are given in the following table:

Values of x_i	Values of $y_i = f(x_i)$
0	1.0
$\pi/12$	0.9945
$\pi/6$	0.9806
$\pi/4$	0.9586
$\pi/3$	0.9373
$5\pi/12$	0.9213
$\pi/2$	0.9154

Simpson's composite rule is given by

$$\int_a^b f(x)dx \approx \frac{h}{3}\left[y_0 + y_n + 2(y_2 + y_4 + \cdots + y_{n-2}) + 4(y_1 + y_3 + \cdots + y_{n-1})\right]$$

Using this, we get

$$\int_0^{\pi/2} \sqrt{1-0.162\sin^2 x}\,dx \approx \frac{\pi/12}{3}\left[1 + 0.9154 + 2(0.9806 + 0.9373)\right.$$

$$\left. + 4(0.9945 + 0.9586 + 0.9213)\right]$$

$$\approx 1.505, \text{ correct to 3 decimal places.}$$

EXAMPLE 4.16 Find from the following table, the area bounded by the curve and the x-axis from $x = 7.47$ to $x = 7.52$.

x	7.47	7.48	7.49	7.50	7.51	7.52
$f(x)$	1.93	1.95	1.98	2.01	2.03	2.06

Solution: The area bounded by the curve and the x-axis from $x = 7.47$ to $x = 7.52$ is given by $\int_{7.47}^{7.52} f(x)dx$. Now, from the given table, we see that the x-values are equidistant with constant step length 0.01. Hence, we can apply the composite trapezoidal rule by assigning $a = 7.47$, $b = 7.52$, $h = 0.01$.

The trapezoidal composite rule is given by

$$\int_a^b f(x)dx \approx \frac{h}{2}\left[y_0 + y_n + 2(y_1 + y_2 + y_3 + \cdots + y_{n-1})\right]$$

Using this, we get

$$\int_{7.47}^{7.52} f(x)dx \approx \frac{1}{2}\left[1.93 + 2.06 + 2(1.95 + 1.98 + 2.01 + 2.03)\right]$$

$$= 0.09965 \approx 0.0100$$

EXAMPLE 4.17 Evaluate $\int_0^1 \cos x dx$, taking 5 equal intervals. Explain the reason behind your choice of the integration formula used.

Solution: As the number of subintervals (5) is odd, we cannot use Simpson's 1/3 rule. So we must solve the given problem by the trapezoidal rule.

We are given $a = 0$, $b = 1$, $n = 5$.

So
$$h = \frac{b-a}{n} = \frac{1-0}{5} = 0.2$$

The values of x and $y = f(x)$ from $a = 0$ to $b = 1$ are given in the following table:

Values of x_i	Values of $y_i = f(x_i)$
0	1.0000
0.2	0.9801
0.4	0.9211
0.6	0.8253
0.8	0.6967
1	0.5403

The trapezoidal composite rule is given by

$$\int_a^b f(x)dx \approx \frac{h}{2}\left[y_0 + y_n + 2(y_1 + y_2 + y_3 + \cdots + y_{n-1})\right]$$

Using this, we get

$$\int_0^1 \cos x dx \approx \frac{0.2}{2}\left[1.0000 + 0.5403 + 2(0.9801 + 0.9211 + 0.8253 + 0.6967)\right]$$

$$\approx 0.8387$$

EXAMPLE 4.18 Evaluate $\int_1^5 \log_{10} x dx$, taking 8 subintervals, correct to 4 decimal places by (i) Trapezoidal rule (ii) Simpson's 1/3 rule.

Solution: Let $y = f(x) = \log_{10}x$.

We are given $a = 0$, $b = 5$, $n = 8$.

So $$h = \frac{b-a}{n} = \frac{5-1}{8} = 0.5$$

The values of x and $y = f(x)$ from $a = 1$ to $b = 5$ are given in the following table:

Values of x_i	Values of $y_i = f(x_i)$
1.0	0.0000
1.5	0.1761
2.0	0.3010
2.5	0.3979
3.0	0.4771
3.5	0.5441
4.0	0.6021
4.5	0.6532
5.0	0.6990

(i) The trapezoidal composite rule is given by

$$\int_a^b f(x)dx \approx \frac{h}{2}\left[y_0 + y_n + 2(y_1 + y_2 + y_3 + \cdots + y_{n-1})\right]$$

Using this, we get

$$\int_1^5 \log_{10} x dx \approx \frac{0.5}{2}[0.0000 + 0.6990 + 2(0.1761 + 0.3010 + 0.3979$$
$$+ 0.4771 + 0.5441 + 0.6021 + 0.6532)]$$
$$\approx 1.751, \text{ correct to 3 decimal places.}$$

(ii) Simpson's 1/3 rule is given by

$$\int_a^b f(x)dx \approx \frac{h}{3}\left[y_0 + y_n + 2(y_2 + y_4 + \cdots + y_{n-2}) + 4(y_1 + y_3 + \cdots + y_{n-1})\right]$$

Using this, we get

$$\int_1^5 \log_{10} x dx \approx \frac{0.5}{3}[0.0000 + 0.6990 + 2(0.3010 + 0.4771 + 0.6021)$$
$$+ 4(0.1761 + 0.3979 + 0.5441 + 0.6532)]$$
$$\approx 1.757, \text{ correct to 3 decimal places.}$$

EXAMPLE 4.19 Evaluate $\int_0^6 \frac{dx}{(1+x)^2}$ by Simpson's 1/3 rule taking $n = 6$, and hence compute an approximate value of π, correct to 5 significant figures.

Solution: Let $y = f(x) = \frac{1}{(1+x)^2}$.

We are given $a = 0$, $b = 6$, $n = 6$.

So
$$h = \frac{b-a}{n} = \frac{6-0}{6} = 1$$

The values of x and $y = f(x)$ from $a = 0$ to $b = 6$ are given in the following table:

Values of x_i	Values of $y_i = f(x_i)$
$x_0 = 0$	$y_0 = 1.00000$
$x_1 = 1$	$y_1 = 0.25000$
$x_2 = 2$	$y_2 = 0.11111$
$x_3 = 3$	$y_3 = 0.06250$
$x_4 = 4$	$y_4 = 0.04000$
$x_5 = 5$	$y_5 = 0.02778$
$x_6 = 6$	$y_6 = 0.02041$

Simpson's composite rule is given by

$$\int_a^b f(x)dx \approx \frac{h}{3}\left[y_0 + y_n + 2(y_2 + y_4 + \ldots + y_{n-2}) + 4(y_1 + y_3 + \ldots + y_{n-1})\right]$$

Using this, we get

$$\int_0^6 \frac{dx}{(1+x)^2} \approx \frac{1}{3}[1.00000 + 0.02041 + 2(0.11111 + 0.04000)$$

$$+ 4(0.25000 + 0.06250 + 0.02778)]$$

$$= 0.89458$$

EXAMPLE 4.20 Evaluate $\int_0^6 \frac{dx}{1+x^2}$ by Weddle's rule taking $n = 6$, correct to 5 significant figures.

Solution: Let $y = f(x) = \frac{1}{1+x^2}$.

We are given $a = 0$, $b = 6$, $n = 6$.

So $$h = \frac{b-a}{n} = \frac{6-0}{6} = 1$$

The values of x and $y = f(x)$ from $a = 0$ to $b = 6$ are given in the following table:

Values of x_i	Values of $y_i = f(x_i)$
$x_0 = 0$	$y_0 = 1.00000$
$x_1 = 1$	$y_1 = 0.50000$
$x_2 = 2$	$y_2 = 0.20000$
$x_3 = 3$	$y_3 = 0.10000$
$x_4 = 4$	$y_4 = 0.05882$
$x_5 = 5$	$y_5 = 0.03846$
$x_6 = 6$	$y_6 = 0.02703$

Weddle's rule is given by

$$\int_a^b f(x)dx \approx \frac{3h}{10}[(y_0 + y_n) + 5(y_1 + y_5 + y_7 + y_{11} + \cdots) + (y_2 + y_4 + y_8 + y_{10} + \cdots)$$

$$+ 6(y_3 + y_9 + y_{15} + \cdots) + 2(y_6 + y_{12} + y_{18} + \cdots)]$$

Using this, we get

$$\int_0^6 \frac{dx}{1+x^2} \approx \frac{3\times 1}{10}[(1.00000 + 0.02703) + 5\times(0.50000 + 0.03846)$$

$$+ 0.20000 + 0.05882 + 6\times 0.10000]$$

$$\approx 1.3734, \text{ correct to 5 significant figures.}$$

EXAMPLE 4.21 Evaluate $\int_0^6 \frac{dx}{1+x^3}$ by Weddle's rule taking $n = 6$, correct to 3 decimal places.

Solution: Let $y = f(x) = \frac{1}{1+x^3}$.

We are given $a = 0$, $b = 6$, $n = 6$.

So
$$h = \frac{b-a}{n} = \frac{6-0}{6} = 1$$

The values of x and $y = f(x)$ from $a = 0$ to $b = 6$ are given in the following table:

Values of x_i	Values of $y_i = f(x_i)$
$x_0 = 0$	$y_0 = 1.0000$
$x_1 = 1$	$y_1 = 0.5000$
$x_2 = 2$	$y_2 = 0.1111$
$x_3 = 3$	$y_3 = 0.0357$
$x_4 = 4$	$y_4 = 0.0153$
$x_5 = 5$	$y_5 = 0.0079$
$x_6 = 6$	$y_6 = 0.0046$

Weddle's rule is given by

$$\int_a^b f(x)dx \approx \frac{3h}{10}\left[(y_0 + y_n) + 5(y_1 + y_5 + y_7 + y_{11} + \cdots) + (y_2 + y_4 + y_8 + y_{10} + \cdots)\right.$$
$$\left. + 6(y_3 + y_9 + y_{15} + \cdots) + 2(y_6 + y_{12} + y_{18} + \cdots)\right]$$

Using this, we get

$$\int_0^6 \frac{dx}{1+x^2} \approx \frac{3\times 1}{10}\left[(1.0000 + 0.0046) + 5\times(0.5000 + 0.0079)\right.$$
$$\left. + 0.1111 + 0.0153 + 6\times 0.0357\right]$$
$$\approx 1.165, \text{ correct to 3 decimal places.}$$

EXAMPLE 4.22 Evaluate $\int_0^{1.5} \frac{x^3 dx}{e^x + 1}$ by Weddle's rule taking $n = 6$, correct to 3 decimal places.

Solution: Let $y = f(x) = \frac{x^3}{e^x + 1}$.

We are given $a = 0$, $b = 1.5$, $n = 6$.

So
$$h = \frac{b-a}{n} = \frac{1.5-0}{6} = 0.25$$

The values of x and $y = f(x)$ from $a = 0$ to $b = 1.5$ are given in the following table:

Values of x_i	Values of $y_i = f(x_i)$
$x_0 = 0$	$y_0 = 0.0000$
$x_1 = 0.25$	$y_1 = 0.0068$
$x_2 = 0.50$	$y_2 = 0.0472$
$x_3 = 0.75$	$y_3 = 0.1353$
$x_4 = 1.00$	$y_4 = 0.2689$
$x_5 = 1.25$	$y_5 = 0.4350$
$x_6 = 1.50$	$y_6 = 0.6157$

Weddle's rule is given by

$$\int_a^b f(x)dx \approx \frac{3h}{10}\left[(y_0 + y_n) + 5(y_1 + y_5 + y_7 + y_{11} + \cdots) + (y_2 + y_4 + y_8 + y_{10} + \cdots) + 6(y_3 + y_9 + y_{15} + \cdots) + 2(y_6 + y_{12} + y_{18} + \cdots)\right]$$

Using this, we get

$$\int_0^6 \frac{dx}{1+x^2} \approx \frac{3\times 0.25}{10}\left[(0.0000 + 0.6157) + 5\times(0.0068 + 0.4350) + 0.0472 + 0.2689 + 6\times 0.1353\right]$$

$$\approx 0.296, \text{ correct to 3 decimal places.}$$

EXAMPLE 4.23 Evaluate $\int_0^4 \frac{dx}{1+x^2}$ by Boole's rule taking $n = 8$, correct up to 3 decimal places.

Solution: Let $y = f(x) = \frac{1}{1+x^2}$.

We are given $a = 0$, $b = 4$, $n = 8$.

So

$$h = \frac{b-a}{n} = \frac{4-0}{8} = 0.5$$

The values of x and $y = f(x)$ from $a = 0$ to $b = 4$ are given in the following table:

Values of x_i	Values of $y_i = f(x_i)$
$x_0 = 0.0$	$y_0 = 1.0000$
$x_1 = 0.5$	$y_1 = 0.8000$
$x_2 = 1.0$	$y_2 = 0.5000$
$x_3 = 1.5$	$y_3 = 0.3077$
$x_4 = 2.0$	$y_4 = 0.2000$

Cont'd

Table cont'd

Values of x_i	Values of $y_i = f(x_i)$
$x_5 = 2.5$	$y_5 = 0.1379$
$x_6 = 3.0$	$y_6 = 0.1000$
$x_7 = 3.5$	$y_7 = 0.0755$
$x_8 = 4.0$	$y_8 = 0.0588$

Boole's rule is given by

$$\int_a^b f(x)\,dx \approx \frac{2h}{45}\left[7(y_0 + y_n) + 32(y_1 + y_3 + y_5 + y_7 + \cdots)\right.$$

$$\left. + 12(y_2 + y_6 + \cdots) + 14(y_4 + y_8 + \cdots)\right]$$

Using this, we get

$$\int_0^4 \frac{dx}{1+x^2} \approx \frac{2\times 0.25}{45}\left[7\times(1.0000 + 0.0588) + 32\times(0.8000 + 0.3077 + 0.1379 + 0.0755)\right.$$

$$\left. + 12\times(0.5000 + 0.1000) + 14\times 0.2000\right]$$

≈ 1.326, correct to 3 decimal places.

4.9 ROMBERG'S METHOD

We see from the previous discussion that the value obtained by numerical integration depends on the value of the step length h. So, for a given value of h, we cannot improve the value of the given integration. Herein lies the utility of Romberg's method. Using this method, we generally obtain a better approximation to the exact value of a definite integral. Consider the definite integral

$$I = \int_a^b f(x)\,dx$$

Let I_1, I_2 be the approximate values of I obtained by the trapezoidal rule with two different step-lengths h_1, h_2, respectively, where in general $h_2 = \frac{h_1}{2}$. Let E_1, E_2 be the corresponding errors.

We know that the error committed in the composite trapezoidal rule is of order h^2. So, we have

$$E_1 = kh_1^2 \text{ and } E_2 = kh_2^2$$

Therefore $$I = I_1 + E_1 = I_1 + kh_1^2 \text{ and } I = I_2 + E_2 = I_2 + kh_2^2 = I_2 + k\frac{h_1^2}{4}$$

Therefore $$I_1 + kh_1^2 = I_2 + k\frac{h_1^2}{4} \Rightarrow \frac{kh_1^2}{4} = \frac{(I_2 - I_1)}{3}$$

Therefore $$I = I_2 + k\frac{h_1^2}{4} = I_2 + \frac{(I_2 - I_1)}{3}$$

This formula is known as Romberg's working formula.

EXAMPLE 4.24 Evaluate $\int_0^1 dx/\left(1+x^2\right)$, correct to 4 decimal places by Romberg's method.

Solution: Let $h_1 = 0.5$. The values of the integrand $f(x)=\dfrac{1}{1+x^2}$ are

x	0	0.5	1
$f(x)$	1	0.8	0.5

Therefore, by the trapezoidal rule, we get

$$I_1 = \frac{0.5}{2}[1+0.5+2\times 0.8] = 0.775$$

Again, let $h_2 = \dfrac{h_1}{2} = 0.25$. Then, the values of the integrand are

x	0	0.25	0.5	0.75	1
$f(x)$	1	0.9412	0.8	0.64	0.5

Therefore, by the trapezoidal rule, we get

$$I_2 = \frac{0.25}{2}[1+0.5+2\times(0.9412+0.8+0.64)]$$
$$= 0.7828$$

Therefore, by Romberg's method for I_1 and I_2, we have

$$I = I_2 + \frac{I_2 - I_1}{3} = 0.7828 + \frac{0.7828 - 0.775}{3}$$
$$= 0.7854 \tag{1}$$

Now, let $h_3 = \dfrac{h_2}{2} = 0.125$. Then, the values of the integrand are

x	0	0.125	0.25	0.375	0.5	0.625	0.75	0.875	1
$f(x)$	1	0.9846	0.9412	0.8767	0.8	0.7191	0.64	0.5664	0.5

Therefore, by the trapezoidal rule, we get

$$I_3 = \frac{0.125}{2}[1+0.5+2\times(0.9846+0.9412+0.8767$$
$$+0.8+0.7191+0.64+0.5664)]$$
$$= 0.78475$$

Therefore, by Romberg's method for I_2 and I_3, we have

$$I = I_3 + \frac{I_3 - I_2}{3} = 0.78475 + \frac{0.78475 - 0.7828}{3}$$
$$= 0.7854 \tag{2}$$

From (1) and (2), we see that $I = \int_0^1 dx/(1+x^2) = 0.7854$

EXAMPLE 4.25 Evaluate $\int_0^1 dx/(1+x)$ by Romberg's method.

Solution: Let $h_1 = 0.5$. The values of the integrand $f(x) = \frac{1}{1+x}$ are

x	0	0.5	1
f(x)	1	0.6667	0.5

Therefore, by the trapezoidal rule, we get

$$I_1 = \frac{0.5}{2}[1+0.5+2\times 0.6667]$$
$$= 0.70835$$

Again, let $h_2 = \frac{h_1}{2} = 0.25$. Then, the values of the integrand are

x	0	0.25	0.5	0.75	1
f(x)	1	0.8	0.6667	0.5714	0.5

Therefore, by the trapezoidal rule, we get

$$I_2 = \frac{0.25}{2}[1+0.5+2\times(0.8+0.6667+0.5714)]$$
$$= 0.697025$$

Therefore, by Romberg's method for I_1 and I_2, we have

$$I = I_2 + \frac{I_2 - I_1}{3} = 0.697025 + \frac{0.697025 - 0.70835}{3}$$
$$= 0.69325 \quad (1)$$

Now, let $h_3 = \frac{h_2}{2} = 0.125$. Then, the values of the integrand are

x	0	0.125	0.25	0.375	0.5	0.625	0.75	0.875	1
f(x)	1	0.8889	0.8	0.7273	0.6667	0.6154	0.5714	0.5333	0.5

Therefore, by the trapezoidal rule, we get

$$I_3 = \frac{0.125}{2}[1+0.5+2\times(0.8889+0.8+0.7273+0.6667$$
$$+0.6154+0.5714+0.5333)]$$
$$= 0.694125$$

Therefore, by Romberg's method for I_2 and I_3, we have

$$I = I_3 + \frac{I_3 - I_2}{3} = 0.694125 + \frac{0.694125 - 0.697025}{3}$$

$$= 0.693158 \qquad (2)$$

From (1) and (2), we see that the two values obtained by Romberg's method are not equal. So, let $h_4 = \frac{h_3}{2} = 0.0625$. Then, the values of the integrand are

x	0	0.0625	0.125	0.1875	0.25	0.3125	0.375	0.4375	0.5
$f(x)$	1	0.9412	0.8889	0.8421	0.8	0.7619	0.7273	0.6957	0.6667

x	0.5625	0.625	0.6875	0.75	0.8125	0.875	0.9375	1
$f(x)$	0.64	0.6154	0.5926	0.5714	0.5517	0.5333	0.5161	0.5

Therefore, by the trapezoidal rule, we get

$$I_4 = \frac{0.25}{2}\left[1 + 0.5 + 2\times(0.94120.8889 + 0.8421 + 0.8 + 0.7619 + 0.7273 + 0.6957\right.$$

$$+ 0.6667 + 0.64 + 0.6154 + 0.5926$$

$$\left.+ 0.5714 + 0.5517 + 0.5333 + 0.5161)\right]$$

$$= 0.693394$$

Therefore, by Romberg's method for I_3 and I_4, we have

$$I = I_4 + \frac{I_4 - I_3}{3} = 0.69315 \qquad (3)$$

(2) and (3) show that the results are almost same. Hence, the required value is 0.69315.

MULTIPLE CHOICE QUESTIONS

1. Simpson's 1/3 rule requires the interval to be divided into an odd number of subintervals.
   (a) true (b) false
2. In evaluating $\int_a^b f(x)\,dx$, the error in the trapezoidal rule is of the order
   (a) h^2 (b) h^3 (c) h^4 (d) h
3. The truncation error in the composite trapezoidal rule is of the order
   (a) h^2 (b) h^3 (c) h^4 (d) none of these
4. The truncation error in composite Simpson's 1/3 rule is of the order
   (a) h^2 (b) h^4 (c) h^5 (d) none of these
5. The degree of precision of the trapezoidal rule is
   (a) 1 (b) 2 (c) 3 (d) 4

6. In the trapezoidal rule, if the interval of the integration $\int_2^9 f(x)\,dx$ is divided into equal subintervals taking 8 ordinates, the length of each subinterval is
(a) 2 (b) 0.5 (c) 1 (d) 1.5

7. The order of h in the error expression of Simpson's 1/3 rule is
(a) 2 (b) 4 (c) 3 (d) 5

8. By evaluating $\int_0^1 \frac{dx}{1+x^2}$ by a numerical integration method, we can obtain an approximate value of
(a) $\log_e 2$ (b) π (c) e (d) $\log_{10} 2$

9. In the trapezoidal rule, the portion of the curve is replaced by a
(a) straight line (b) circular path (c) parabolic path (d) none of these

10. The order of h in the error expression of the trapezoidal rule is
(a) 6 (b) 3 (c) 5 (d) 2

11. The degree of precision of Simpson's 1/3 rule is
(a) 1 (b) 2 (c) 3 (d) 5

12. In Simpson's 1/3 rule of finding $\int_a^b f(x)\,dx$, $f(x)$ is approximated by
(a) line segment (b) parabola (c) circular sector (d) part of ellipse

13. For the trapezoidal rule of numerical integration, the number of subintervals should be
(a) even (b) even or odd (c) odd (d) multiple of three

14. The degree of the approximation polynomial in Simpson's 1/3 rule is
(a) 3 (b) 2 (c) 1 (d) (a), (b) or (c)

15. In evaluating $\int_a^b f(x)\,dx$, the error in Simpson's 1/3 rule is of the order
(a) h^3 (b) h^5 (c) h^4 (d) h

16. The degree of the approximating polynomial corresponding to the trapezoidal rule is
(a) 1 (b) 0 (c) 2 (d) 3

17. The degree of the approximating polynomial corresponding to Simpson's 1/3 rule is
(a) 2 (b) 1 (c) 3 (d) 4

18. In evaluating $\int_a^b f(x)\,dx$, the error in Simpson's 1/3 rule is of the order
(a) h^3 (b) h^5 (c) h^4 (d) h

Answers

1. (a) 2. (a) 3. (b) 4. (d) 5. (b) 6. (b) 7. (b) 8. (a) 9. (a) 10. (b)
11. (a) 12. (d) 13. (d) 14. (b) 15. (c) 16. (a) 17. (a) 18. (c)

SHORT ANSWER QUESTIONS

1. What is numerical integration? What is the degree of precision for quadrature formula in numerical integration?
2. Derive the general quadrature formula for numerical integration.
3. Deduce the composite trapezoidal rule from the general quadrature formula.
4. Derive composite Simpson's 1/3 rule from the general quadrature formula.
5. Explain the geometrical significance of the trapezoidal rule.
6. Derive the expression of the error term in the trapezoidal rule.
7. Show that the value of $\int_0^1 x dx$, computed by the trapezoidal rule taking 2 intervals is equal to its exact value.
8. What do you mean by geometrical significance of Simpson's 1/3 rule?
9. Obtain the expression of the error term in composite Simpson's 1/3 rule.
10. Justify whether Simpson's 1/3 rule is exact for $\int_a^b x^3 dx$.
11. Find the approximate value of $\int \frac{dx}{1+x}$ when the interval is (0, 1) and $h = 1/2$ using the trapezoidal rule.
12. Estimate the value of the following integral by Simpson's 1/3 rule taking 4 strips: $\int_1^3 \frac{dx}{x}$
13. Find the value of the integral $\int_0^1 e^x dx$ by the trapezoidal rule with $h = 0.1$.
14. Evaluate $\int xe^x dx$ where the interval is (0, –1) using the trapezoidal rule, taking $n = 6$.
15. Compute $I = \int \frac{xdx}{\sin x}$, where the interval is (0, ½) using Simpson's rule with $h = 1/4$.
16. Calculate the area of the function $f(x) = \sin x$ with limits (0, 90°) by Simpson's 1/3 rule using 11 ordinates.
17. Evaluate $\int_0^1 \frac{dx}{1+x^2}$ by Simpson's 1/3 rule taking $n = 6$.
18. Calculate $\int_0^6 x(1+x)dx$ by the trapezoidal rule with $n = 6$.
19. Evaluate $\int_0^3 \sqrt{x}dx$ by the trapezoidal rule with $n = 3$.
20. Estimate the area bounded by the curve $y = f(x)$, x-axis and the ordinates $x = 0$ and $x = 2$ from the following data using Simpson's 1/3 rule.

x	0.0	0.5	1.0	1.5	2.0
$f(x)$	0.3989	0.3521	0.2420	0.3295	0.0540

Answers

10. Exact. 11. 0.70835 12. 1.100 13. 0.1720 14. −0.2616 15. 0.51 16. 1
17. 0.7854 18. 91 19. 3.2802 20. 0.4772

LONG ANSWER QUESTIONS

1. Establish Simpson's 1/3 rule for numerical integration. Also find the error term in Simpson's 1/3 rule.
2. Find the value of log $2^{1/3}$ from $\int_0^1 \frac{x^2 dx}{1+x^3}$ using Simpson's 1/3 rule, with $h = 0.25$.
3. Evaluate $\int_0^1 \frac{dx}{1+x}$ by Simpson's composite rule taking 11 ordinates, and hence find the value of $\log_e 2$, correct to 5 significant figures.
4. Evaluate $\int_0^1 \frac{dx}{1+x^2}$ using Simpson's 1/3 rule, taking $h = 1/4$. Hence, compute an approximate value of π.
5. Evaluate $\int_0^1 \frac{dx}{1+x^2}$ using Simpson's 1/3 rule, taking $n = 6$.
6. Compute the value of π from the formula $\frac{\pi}{4} = \int_0^1 \frac{dx}{1+x^2}$ using the trapezoidal rule with 10 subintervals.
7. Calculate using Simpson's 1/3 rule, the value of the integral $\int_0^1 \frac{xdx}{1+x}$ to 3 significant figures, taking 6 intervals.
8. A curve passes through points as given in the following table. Find the area bounded by the curve, x-axis, $x = 1$ and $x = 9$.

x	1	2	3	4	5	6	7	8	9
y	1342	2210	2758	5850	6878	9282	1.2	2.1	2.3

9. A curve passes through points as given in the following table. Find the area bounded by the curve, x-axis, $x = 1$ and $x = 9$.

x	1	2	3	4	5	6	7	8	9
y	0.2	0.7	1	1.3	1.5	1.7	1.9	2.1	2.3

10. Calculate the approximate value of $\int_0^{\pi/2} \sin x dx$ by the composite trapezoidal rule using 11 ordinates. Also compare it with the actual value of the integral.

11. Evaluate $\int_0^1 (4x - 3x^2)\,dx$ by the trapezoidal rule, taking 10 equal intervals. Compute the exact integral and find the absolute and relative errors.

12. If

x:	4.0	4.2	4.4	4.6	4.8	5.0	5.2
f(x):	1.3883	1.4351	1.4876	1.5261	1.5686	1.6094	1.6487

Find $\int_{4.0}^{5.2} f(x)dx$ by (i) Trapezoidal rule (ii) Simpson's 1/3 rule

13. Apply the trapezoidal rule to integrate $\int_0^1 \frac{dx}{1+x}$, taking 10 subintervals.

14. Use the composite trapezoidal rule to compute $\int_0^1 e^{-x^2} dx$ correct to 3 significant figures, taking 5 ordinates.

15. Calculate the value of $\int_{1.2}^{1.6} (x + 1/x)\,dx$, correct to 3 significant figures, taking 4 intervals by the trapezoidal rule.

16. Apply the trapezoidal rule to integrate $\int_0^1 \sqrt{1-x^2}\,dx$, taking 10 subintervals.

17. Evaluate $\int_0^1 (4x - 3x^2)dx$ by Simpson's 1/3 rule, taking 10 equal intervals.

18. Compute $\int_1^{1.5} e^x dx$ using Simpson's 1/3 rule, taking 10 equal intervals. The solution should be correct to 5 decimal places.

19. Find the value of $\int_0^{\pi/2} \sqrt{1 - 0.162\sin^2 x}\,dx$ using Simpson's 1/3 rule, taking 6 equal subintervals.

20. Evaluate $\int_1^5 \log_{10} x dx$ using Simpson's 1/3 rule, correct to 3 decimal places, taking 10 equal intervals.

21. Apply Simpson's 1/3 rule to evaluate $\int_{-3}^{3} |x|\,dx$, taking 6 subintervals. Compare this with the exact value and compute the relative error.

22. The following table gives the values of acceleration (*f*) of a particle in *cm/sec*^{2} at interval of time (*t*) in *sec*.

t:	0.0	0.5	1.0	1.5	2.0
f:	0.3989	0.3521	0.2420	0.1295	0.0540

Using this, find the velocity of the particle at $t = 2$ *secs*.

Answers

2. 0.23108, correct to 5 decimal places
3. 0.6931466; 0.69315, correct to 5 decimal places
4. 0.7854; 3.1416, correct to 4 decimal places
5. 0.7854, correct to 4 decimal places
6. 3.1416, correct to 4 decimal places
7. 0.307, correct to 3 decimal places
8. 27653.45, correct to 2 decimal places
9. 11.45, correct to 2 decimal places
10. 0.998, correct to 3 decimal places
11. 0.995, correct to 3 decimal places
12. (i) 1.82906; (ii) 1.82879, correct to 5 decimal places
13. 0.6938, correct to 4 decimal places
14. 0.743, correct to 3 decimal places
15. 0.847935, correct to 6 decimal places
16. 0.77613, correct to 5 decimal places
17. 1.00, correct to 2 decimal places
18. 1.76341, correct to 5 decimal places
19. 1.505, correct to 3 decimal places.
20. 1.757, correct to 3 decimal places
21. 8.667, correct to 3 decimal places; exact value is 9 and relative error is 0.037
22. 0.4772, correct to 4 decimal places

5

Numerical Solutions of Ordinary Differential Equations

5.1 INTRODUCTION

Differential equations arise due to modelling of physical, chemical, biological and social phenomena in Engineering and Science. But thus far, very few of them have been solved completely using analytical methods. In fact, a wide variety of differential equations, especially non-linear, remain unsolved. Therefore, approximate solutions of these equations have been attempted using numerical techniques. In this chapter, we discuss some numerical methods used to solve ordinary differential equations with initial conditions.

Although a number of different numerical techniques are used to solve ordinary differential equations, we can group them under two categories: **single-step method** and **multi-step method**.

Single-step Method When a numerical technique requires only the value of y_i to compute the next value y_{i+1}, $i = 0, 1, 2, \ldots$ to solve a differential equation, it is termed a single-step method. The Taylor series method, Picard's method, Euler's method, Euler's modified formula and the Runge–Kutta method all fall under this category.

Multi-step Method If a method requires not only y_i but also at least one of the past values $y_{i-1}, y_{i-2}, \ldots$, etc., it is called a multi-step method. Finite difference method, the Adam–Bashforth predictor–corrector formula and Milne's predictor–corrector formula are all multi-step methods.

5.2 TAYLOR SERIES METHOD

Let us consider the ordinary differential equation

$$\frac{dy}{dx} = f(x, y)$$

with initial condition $$y_0 = y(x_0) \tag{5.1}$$

Let $y = f(x)$ be the exact solution of the above equation. Then, Taylor's series for $y = f(x)$ about $x = x_0$ is given by

$$y = f(x) = y_0 + \frac{(x - x_0)}{1!} y_0' + \frac{(x - x_0)^2}{2!} y_0'' + \frac{(x - x_0)^3}{3!} y_0''' + \cdots \tag{5.2}$$

Where $y_0', y_0'', \ldots$ are, respectively, the first, second, ... derivatives of y at $x = x_0$, provided they exist.

For $x = x_1 = x_0 + h$, equation (5.2) reduces to

$$y_1 = f(x_1) = y_0 + \frac{h}{1!}y_0' + \frac{h^2}{2!}y_0'' + \frac{h^3}{3!}y_0''' + \cdots$$

Thus, we obtain

$$y_{n+1} = y_n + \frac{h}{1!}y_n' + \frac{h^2}{2!}y_n'' + \frac{h^3}{3!}y_n''' + \cdots$$

Though this method is simple, the main disadvantage is that finding the derivatives of $f(x, y)$ is difficult if it is very complicated.

EXAMPLE 5.1 Using Taylor's method, solve $\frac{dy}{dx} = 1 + xy$ with $y(0) = 2$ at $x = 0.2$.

Solution: We have $\frac{dy}{dx} = 1 + xy$ with $y(0) = 2$.

Therefore $y' = 1 + xy$

and so $y_0' = 1 + 0 \times 2 = 1$

$y'' = y + xy'$

and so $y_0'' = 2 + 0 \times 1 = 2$

$y''' = 2y' + xy''$

and so $y_0''' = 2 \times 1 + 0 \times 2 = 2$

$y^{iv} = 3y'' + xy'''$

and so $y_0^{iv} = 3 \times 2 + 0 \times 2 = 6$ and so on.

Now, Taylor's series for $y = f(x)$ about $x = x_0$ is given by

$$y = f(x) = y_0 + \frac{(x - x_0)}{1!}y_0' + \frac{(x - x_0)^2}{2!}y_0'' + \frac{(x - x_0)^3}{3!}y_0''' + \cdots$$

Here $x = 0.2$ and $x_0 = 0$

and so $x - x_0 = 0.2 - 0 = 0.2$.

Therefore

$$y(0.2) = 2 + \frac{0.2}{1!} \times 1 + \frac{(0.2)^2}{2!} \times 2 + \frac{(0.2)^3}{3!} \times 2 + \frac{(0.2)^4}{4!} \times 6 + \cdots$$

$$= 2 + 0.2 + 0.04 + 0.0027 + 0.0004 + \cdots$$

$$\approx 2.203 \text{ (to 3 decimal places).}$$

EXAMPLE 5.2 Compute $y(1.1)$ and $y(1.2)$, correct to 5 decimal places using the Taylor series method, when $y(x)$ satisfies the equation $\frac{dy}{dx} = xy$ with $y(1.0) = 2$.

Solution: We have $\frac{dy}{dx} = xy$ with $y(1.0) = 2$.

Therefore $\qquad y' = xy$

and so $\qquad y_0' = 1 \times 2 = 2$

$$y'' = y + xy'$$

and so $\qquad y_0'' = 2 + 1 \times 2 = 4$

$$y''' = 2y' + xy''$$

and so $\qquad y_0''' = 2 \times 2 + 1 \times 4 = 8$

$$y^{iv} = 3y'' + xy'''$$

and so $\qquad y_0^{iv} = 3 \times 4 + 1 \times 8 = 20$ and so on.

Now, Taylor's series for $y = f(x)$ about $x = x_0$ is given by

$$y = f(x) = y_0 + \frac{(x - x_0)}{1!} y_0' + \frac{(x - x_0)^2}{2!} y_0'' + \frac{(x - x_0)^3}{3!} y_0''' + \cdots$$

(i) Here $\qquad x = 1.1$ and $x_0 = 1$

and so $\qquad x - x_0 = 1.1 - 1 = 0.1$

Therefore $\qquad y(1.1) = 2 + \frac{0.1}{1!} \times 2 + \frac{(0.1)^2}{2!} \times 4 + \frac{(0.1)^3}{3!} \times 8 + \frac{(0.1)^4}{4!} \times 20 + \cdots$

$$= 2 + 0.2 + 0.02 + 0.001333 + 0.000083 + \cdots$$

$$\approx 2.22142 \text{ (to 5 decimal places).}$$

(ii) Now $\qquad y_1' = 1.1 \times 2.22142 = 2.443562$

$$y_1'' = 2.22142 + 1.1 \times 2.443562 = 4.9093382$$

$$y_1''' = 2 \times 2.443562 + 1.1 \times 4.9093382 = 10.894622$$

$$y_0^{iv} = 3 \times 4.9093382 + 1.1 \times 10.894622 = 26.71412302 \text{ and so on}$$

Here $\qquad x = 1.2$ and $x_1 = 1.1$

and so $\qquad x - x_1 = 1.2 - 1.1 = 0.1$

Therefore $\qquad y(1.2) = 2.22142 + \frac{0.1}{1!} \times 2.443562 + \frac{(0.1)^2}{2!} \times 4.9093382$

$$+ \frac{(0.1)^3}{3!} \times 10.894622 + \frac{(0.1)^4}{4!} \times 26.71412302 + \cdots$$

$$= 2.22142 + 0.2443562 + 0.02454669$$

$$+ 0.001816077 + 0.000111309 + \cdots$$

$$\approx 2.49225 \text{ (to 5 decimal places).}$$

EXAMPLE 5.3 Solve by Taylor series method, $\frac{dy}{dx} = 2x + 3y^2$ with $y(0) = 0$ at $x = 0.2$.

Solution: We have $\frac{dy}{dx} = 2x + 3y^2$ with $y(0) = 0$.

Therefore $y' = 2x + 3y^2$

and so $y_0' = 2\times 0 + 3\times 0^2 = 0$

$y'' = 2 + 6yy'$

and so $y_0'' = 2 + 6\times 0\times 0 = 2$

$y''' = 6(y')^2 + 6yy''$

and so $y_0''' = 6\times 0^2 + 6\times 0\times 2 = 0$

$y^{iv} = 18y'y'' + 6yy'''$

and so $y_0^{iv} = 18\times 0\times 2 + 6\times 0\times 0 = 0$

$y^{v} = 18(y'')^2 + 24y'y''' + 6yy^{iv}$

and so $y_0^{v} = 18\times 2^2 + 0 + 0 = 72$ and so on.

Now, Taylor's series for $y = f(x)$ about $x = x_0$ is given by

$$y = f(x) = y_0 + \frac{(x - x_0)}{1!}y_0' + \frac{(x - x_0)^2}{2!}y_0'' + \frac{(x - x_0)^3}{3!}y_0''' + \cdots$$

Here $x = 0.2$ and $x_0 = 0$

and so $x - x_0 = 0.2 - 0 = 0.2$

Therefore

$$y(0.2) = 0 + \frac{0.2}{1!}\times 0 + \frac{(0.2)^2}{2!}\times 2 + \frac{(0.2)^3}{3!}\times 0 + \frac{(0.2)^4}{4!}\times 0 + \frac{(0.2)^5}{5!}\times 72 + \cdots$$

$$= 0 + 0 + 0.04 + 0 + 0 + 0.000192 + \cdots$$

$$\approx 0.04019 \text{ (to 5 decimal places).}$$

5.3 PICARD'S METHOD

Let us consider the ordinary differential equation

$$\frac{dy}{dx} = f(x, y)$$

with initial condition $y_0 = y(x_0)$ (5.3)

Integrating (5.3) between the limits x_0 and x, we get

$$\int_{y_0}^{y} dy = \int_{x_0}^{x} f(x, y)dx$$

or

$$y - y_0 = \int_{x_0}^{x} f(x, y)dx$$

or
$$y = y_0 + \int_{x_0}^{x} f(x, y)dx \tag{5.4}$$

For the first approximation of y, replace y with y_0 in $f(x, y)$ of (5.4) to get

$$y^{(1)} \approx y_0 + \int_{x_0}^{x} f(x, y_0)dx$$

For the second approximation of y, replace y with $y^{(1)}$ in $f(x, y)$ of (5.4) to get

$$y^{(2)} \approx y_0 + \int_{x_0}^{x} f(x, y^{(1)})dx$$

Similarly, for the $(n + 1)$th approximation of y, replace y with $y^{(n)}$ in $f(x, y)$ of (5.4) to get

$$y^{(n+1)} \approx y_0 + \int_{x_0}^{x} f\left(x, y^{(n)}\right)dx, \; n = 1, 2, 3, \ldots$$

EXAMPLE 5.4 Using Picard's method, find y at $x = 0.2$ correct to 3 decimal places, given that $\frac{dy}{dx} = x - y$ with $y(0) = 1$.

Solution: Here $f(x, y) = x - y$

and $x_0 = 0, y_0 = 1.$

Hence
$$y = y_0 + \int_{x_0}^{x} f(x, y)dx = 1 + \int_{x_0}^{x} (x - y)dx$$

Now
$$y^{(1)} \approx y_0 + \int_{x_0}^{x} f(x, y_0)dx = 1 + \int_{x_0}^{x} (x - 1)dx$$

$$= 1 + \frac{x^2}{2} - x$$

$\therefore$ at $x = 0.2$
$$y^{(1)} \approx 1 + \frac{(0.2)^2}{2} - 0.2 = 0.82$$

Again

$$y^{(2)} \approx y_0 + \int_{x_0}^{x} f\left(x, y^{(1)}\right)dx = 1 + \int_{x_0}^{x} \left(x - 1 - \frac{x^2}{2} + x\right)dx$$

$$= 1 + x^2 - \frac{x^3}{6} - x$$

$\therefore$ at $x = 0.2$
$$y^{(2)} \approx 1 + (0.2)^2 - \frac{(0.2)^3}{6} - 0.2 = 0.83867$$

The third approximation is

$$y^{(3)} \approx y_0 + \int_{x_0}^{x} f\left(x, y^{(2)}\right)dx = 1 + \int_{x_0}^{x} \left(x - 1 - x^2 + \frac{x^3}{6} + x\right)dx$$

$$= 1 + x^2 + \frac{x^4}{24} - \frac{x^3}{3} - x$$

$\therefore$ at $x = 0.2$ $\quad y^{(3)} \approx 1 + (0.2)^2 + \frac{(0.2)^4}{24} - \frac{(0.2)^3}{3} - 0.2 = 0.83740$

The fourth approximation is

$$y^{(4)} \approx y_0 + \int_{x_0}^{x} f\left(x, y^{(3)}\right) dx = 1 + \int_{x_0}^{x} \left(x - 1 - x^2 - \frac{x^4}{24} + \frac{x^3}{3} + x \right) dx$$

$$= 1 + x^2 - \frac{x^5}{120} + \frac{x^4}{12} - \frac{x^3}{3} - x$$

$\therefore$ at $x = 0.2$ $\quad y^{(3)} \approx 1 + (0.2)^2 - \frac{(0.2)^5}{120} + \frac{(0.2)^4}{12} - \frac{(0.2)^3}{3} - 0.2 = 0.83746$

Therefore, y at $\quad x = 0.2$ is 0.837.

EXAMPLE 5.5 Using Picard's method, obtain a solution up to the fourth approximation, given that $\frac{dy}{dx} = x + y$ with $y(0) = 1$.

Solution: Here $\quad f(x, y) = x - y$

and $\quad x_0 = 0, y_0 = 1.$

Hence

$$y = y_0 + \int_{x_0}^{x} f(x, y)\, dx = 1 + \int_{x_0}^{x} (x + y)\, dx$$

Therefore

$$y^{(1)} \approx y_0 + \int_{x_0}^{x} f(x, y_0)\, dx = 1 + \int_{x_0}^{x} (x + 1)\, dx$$

$$= 1 + x + \frac{x^2}{2}$$

$$y^{(2)} \approx y_0 + \int_{x_0}^{x} f\left(x, y^{(1)}\right) dx = 1 + \int_{x_0}^{x} \left(x + 1 + x + \frac{x^2}{2} \right) dx$$

$$= 1 + x + x^2 + \frac{x^3}{6}$$

$$y^{(3)} \approx y_0 + \int_{x_0}^{x} f\left(x, y^{(2)}\right) dx = 1 + \int_{x_0}^{x} \left(x + 1 + x + x^2 + \frac{x^3}{6} \right) dx$$

$$= 1 + x + x^2 + \frac{x^3}{3} + \frac{x^4}{24}$$

$$y^{(4)} \approx y_0 + \int_{x_0}^{x} f\left(x, y^{(3)}\right) dx = 1 + \int_{x_0}^{x} \left(x + 1 + x + x^2 + \frac{x^3}{3} + \frac{x^4}{24}\right) dx$$

$$= 1 + x + x^2 + \frac{x^3}{3} + \frac{x^4}{12} + \frac{x^5}{120}$$

5.4 EULER'S METHOD

Let us consider the differential equation

$$\frac{dy}{dx} = f(x, y)$$

with initial condition $y_0 = y(x_0)$ (5.5)

Let $y = \varnothing(x)$ be the solution of (5.5). (5.6)

Let (x_0, y_0), (x_1, y_1), ..., (x_n, y_n), (x_{n+1}, y_{n+1}), ... be the points on the curve of (5.6) where $x_0, x_1, \ldots, x_n$, x_{n+1}, ... are equidistant at equal interval h.

$\therefore$

$$y_{n+1} = \varnothing(x_{n+1}) = \varnothing(x_n + h)$$

$$= \varnothing(x_n) + h\varnothing'(x_n) + \frac{h^2}{2!}\varnothing''(x_n) + \cdots \quad (5.7)$$

$$\approx \varnothing(x_n) + h\varnothing'(x_n), \text{ assuming } h \text{ is a very small quantity}$$

$$= y_n + hy'(x_n)$$

$$= y_n + hf(x_n, y_n) \quad (5.8)$$

This iterative formula is known as Euler's formula.

Now, putting $n = 0, 1, 2, \ldots$, we get

$$y_1 \approx y_0 + hf(x_0, y_0)$$

$$y_2 \approx y_1 + hf(x_1, y_1)$$

$$\cdots$$

$$\cdots$$

This method is also called the **Runge–Kutta method of first order**.

Error in Euler's Method In equation (5.7), we neglected $\frac{h^2}{2!}\varnothing''(x_n)$ and the higher powers of h. Hence, the error (truncation error) committed in Euler's method is of the order of h^2, i.e., $\boldsymbol{O(h^2)}$.

Geometrical Interpretation of Euler's Method Let $y = \phi(x)$ be the solution of the equation $\frac{dy}{dx} = f(x, y)$ with initial condition $y_n = y(x_n)$. The graph of this solution is shown in Fig. 5.1 as the curve PQ. The ordinate of P, i.e., y_n is known.

Now, we want to find the ordinate y_{n+1}. As a very small portion of a smooth curve can be considered as a line segment, we can write

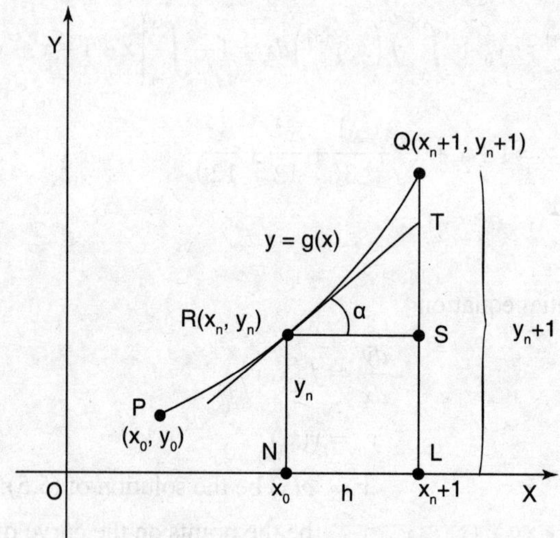

Fig. 5.1 *Euler's method*

$$\frac{y_{n+1} - y_n}{h} = \tan\alpha = \left(\frac{dy}{dx}\right)_{x_n, y_n}$$

$$= f(x_n, y_n)$$

Therefore, we get

$$y_{n+1} = y_n + hf(x_n, y_n)$$

which is the recursion formula for Euler's method. Clearly, here the error is TQ.

Note: Although Euler's method is very simple for implementation, it has some disadvantages:

(i) If h is not very small, the method yields erroneous results.

(ii) If h is very small, this becomes a very slow method to obtain a result to a desired degree of accuracy.

The following examples illustrate this method.

EXAMPLE 5.6 Given $\frac{dy}{dx} = \frac{y-x}{y+x}$ with initial condition $y = 1$ at $x = 0$, find y for $x = 0.1$ by Euler's method, correct to 4 decimal places, taking step length $h = 0.02$.

Solution: We are given $$\frac{dy}{dx} = \frac{y-x}{y+x},\ y(0) = 1$$

and $$h = 0.02.$$

Let $$f(x, y) = \frac{y-x}{y+x},\ x_0 = 0$$

so that $$y_0 = y(x_0) = 1$$

Euler's formula is given by $$y_{n+1} \approx y_n + hf(x_n, y_n),\ n = 0, 1, 2, \ldots$$

The calculations for $y(0.1)$ are given in the following table:

n	x_n	y_n	$f(x_n, y_n) = \dfrac{y_n - x_n}{y_n + x_n}$	$y_{n+1} \approx y_n + hf(x_n, y_n)$
0	0	1	$\dfrac{1-0}{1+0} = 1$	$1 + 0.02 \times 1 = 1.02$
1	0.02	1.02	$\dfrac{1.02-0.02}{1.02+0.02} = 0.961538$	$1.02 + 0.02 \times 0.961538 = 1.039231$
2	0.04	1.039231	$\dfrac{1.039231-0.04}{1.039231+0.04} = 0.925873$	$1.039231 + 0.02 \times 0.925873 = 1.057748$
3	0.06	1.057748	$\dfrac{1.057748-0.06}{1.057748+0.06} = 0.892641$	$1.057748 + 0.02 \times 0.892641 = 1.075601$
4	0.08	1.075601	$\dfrac{1.075601-0.08}{1.075601+0.08} = 0.861544$	$1.075601 + 0.02 \times 0.861544 = 1.092832$
5	1.0	**1.092832**	–	–

Therefore $\quad y(0.1) \approx 1.092832 \approx 1.0928$, correct to 4 decimal places.

EXAMPLE 5.7 Solve by using Euler's method, the following differential equation for $x = 1$, taking $h = 0.2$; $\dfrac{dy}{dx} = xy$, $y = 1$ when $x = 0$.

Solution: Here
$$\frac{dy}{dx} = xy,\ y(0) = 1$$
and
$$h = 0.2.$$
Let
$$f(x, y) = xy,\ x_0 = 0$$
so that
$$y_0 = y(x_0) = 1.$$
Euler's formula is given by
$$y_{n+1} \approx y_n + hf(x_n, y_n),\ n = 0, 1, 2, \ldots$$

The calculations for $y(1)$ are given in the following table:

n	x_n	y_n	$f(x_n, y_n) = x_n y_n$	$y_{n+1} \approx y_n + hf(x_n, y_n)$
0	0	1	$0 \times 1 = 0$	$1 + 0.2 \times 0 = 1$
1	0.2	1	$0.2 \times 1 = 0.2$	$1 + 0.2 \times 0.2 = 1.04$
2	0.4	1.04	$0.4 \times 1.04 = 0.416$	$1.04 + 0.2 \times 0.416 = 1.1232$
3	0.6	1.1232	$0.6 \times 1.1232 = 0.67392$	$1.1232 + 0.2 \times 0.67392 = 1.257984$
4	0.8	1.257984	$0.8 \times 1.257984 = 1.00639$	$1.257984 + 0.2 \times 1.00639 = 1.45926$
5	1.0	**1.45926**	–	–

Hence $\quad y(1) \approx 1.45926$ (approx.).

EXAMPLE 5.8 Using Euler's method, evaluate $y(2)$, correct to 3 decimal places from $\frac{dy}{dx}=\frac{1}{2}(x+y)$ with $y(0) = 2$, taking $h = 0.5$.

Solution: Given that $\frac{dy}{dx}=\frac{1}{2}(x+y), y(0)=2, h=0.5$

Let $f(x,y)=\frac{1}{2}(x+y),\ x_0=0$

so that $y_0 = y(x_0) = 2.$

Euler's formula is given by $y_{n+1} \approx y_n + hf(x_n, y_n),\ n = 0, 1, 2, \ldots$

The calculations for $y(2)$ are given in the following table:

n	x_n	y_n	$f(x_n, y_n)=\frac{1}{2}(x_n+y_n)$	$y_{n+1} \approx y_n + hf(x_n, y_n)$
0	0	2	$(0 + 2)/2 = 1$	$2 + 0.5 \times 1 = 2.5$
1	0.5	2.5	$(0.5 + 2.5)/2 = 1.5$	$2.5 + 0.5 \times 1.5 = 3.25$
2	1.0	3.25	$(1.0 + 3.25)/2 = 2.125$	$3.25 + 0.5 \times 2.125 = 4.3125$
3	1.5	4.3125	$(1.5 + 4.3125)/2 = 2.90625$	$4.3125 + 0.5 \times 2.90625 = 5.765625$
4	2.0	**5.765625**	–	–

So $y(2) \approx 5.765625 \approx 5.766$, correct to 3 decimal places.

5.5 EULER'S MODIFIED FORMULA

To minimise the error in Euler's first formula, he modified the formula as given below:

$$y_{n+1}=\varnothing(x_n)+h\varnothing'(x_n)+\frac{h^2}{2!}\varnothing''(x_n)+\frac{h^3}{3!}\varnothing'''(x_n)+\cdots \tag{5.9}$$

Differentiating (5.9) w.r.t. x, we get

$$\left(\frac{dy}{dx}\right)_{n+1}=\varnothing'(x_n)+h\varnothing''(x_n)+\frac{h^2}{2!}\varnothing'''(x_n)+\frac{h^3}{3!}\varnothing^{iv}(x_n)+\cdots$$

$$\Rightarrow \quad f(x_{n+1},y_{n+1})=\varnothing'(x_n)+h\varnothing''(x_n)+\frac{h^2}{2!}\varnothing'''(x_n)+\frac{h^3}{3!}\varnothing^{iv}(x_n)+\cdots$$

Multiplying both sides by $\frac{h}{2}$, we get

$$\frac{h}{2}f(x_{n+1},y_{n+1})=\frac{h}{2}\varnothing'(x_n)+\frac{h^2}{2}\varnothing''(x_n)+\frac{h^3}{2.2!}\varnothing'''(x_n)+\frac{h^4}{2.3!}\varnothing^{iv}(x_n)+\cdots \tag{5.10}$$

Now, subtracting (5.10) from (5.9), we get

$$y_{n+1}-\frac{h}{2}f\left(x_{n+1},y_{n+1}\right)=\varnothing\left(x_n\right)+\frac{h}{2}\varnothing'\left(x_n\right)+O\left(h^3\right)$$

$$\Rightarrow \qquad y_{n+1}\approx y_n+\frac{h}{2}\left[f\left(x_n,y_n\right)+f\left(x_{n+1},y_{n+1}\right)\right] \qquad (5.11)$$

Equation (5.11) is known as Euler's modified formula. However, to calculate y_{n+1}, we need the value of $f(x_{n+1}, y_{n+1})$, which cannot be calculated as y_{n+1} is unknown. So, calculate y_{n+1} using Euler's first formula

$$y_{n+1}\approx y_n+hf(x_n,y_n)$$

And denote this first approximated value by $y_{n+1}^{(1)}$

That is, for each stage, we need to use the following two formulae:

$$y_{n+1}^{(1)}\approx y_n+hf\left(x_n,y_n\right)$$

$$y_{n+1}^{(r)}\approx y_n+\frac{h}{2}\left[f\left(x_n,y_n\right)+f\left(x_{n+1},y_{n+1}^{(r-1)}\right)\right]$$

Where $r = 2, 3, \ldots$ for each $n = 0, 1, 2, \ldots$

This is also known as the **Runge–Kutta method of second order.**

Error in Euler's Modified Formula In equation (5.11), we ignored $-\frac{h^3}{12}\varnothing'''\left(x_n\right)$ and the higher powers of h. Hence, the error (truncation error) committed in Euler's method is of the order of h^3, i.e., $\boldsymbol{O(h^3)}$.

To illustrate the applications of this method, some examples are given below.

EXAMPLE 5.9 Given that $\frac{dy}{dx}+\frac{y}{x}=\frac{1}{x^2}$, $y(1)=1$, evaluate $y(1.2)$ by modified Euler's method.

Solution: Given $\qquad \frac{dy}{dx}+\frac{y}{x}=\frac{1}{x^2},\ y(1)=1$

Let $\qquad f(x,y)=-\frac{y}{x}+\frac{1}{x^2},\ x_0=1$

so that $\qquad y_0=y(x_0)=1.$

Let us consider the step length $\qquad h=0.2.$

So $\qquad f\left(x_0,y_0\right)=\left(-\frac{1}{1}+\frac{1}{1^2}\right)=0$

Therefore, by Euler's method, we get $y_1^{(1)}\approx y_0+hf\left(x_0,y_0\right)=1+0.2\times 0=1$

Now, Euler's modified formula for $n = 1$ is given by

$$y_1^{(r)}\approx y_0+\frac{h}{2}\left[f\left(x_0,y_0\right)+f\left(x_1,y_1^{(r-1)}\right)\right],\ r=2,3,\ldots$$

The procedure is shown in the following table:

r	$y_1^{(r-1)}$	$f\left(x_1, y_1^{(r-1)}\right)$	$y_1^{(r)} \approx y_0 + \frac{h}{2}\left[f(x_0, y_0) + f\left(x_1, y_1^{(r-1)}\right)\right]$
2	1	$-\frac{1}{1.2} + \frac{1}{(1.2)^2} = -0.13889$	1 + (0.2/2) × (0 – 0.13889) = 0.986111
3	0.986111	$-\frac{0.986111}{1.2} + \frac{1}{(1.2)^2} = -0.12731$	1 + (0.2/2) × (0 – 0.12731) = 0.987269
4	0.987269	$-\frac{0.987269}{1.2} + \frac{1}{(1.2)^2} = -0.12828$	1 + (0.2/2) × (0 – 0.12828) = **0.987**117
5	0.987117	$-\frac{0.987117}{1.2} + \frac{1}{(1.2)^2} = -0.128199$	1 + (0.2/2) × (0 – 0.128199) = **0.987**180

We see from the last two rows that the values of y at $x = 1.2$ are correct to 3 decimal places. Hence, we can write $y(1.2) \approx 0.987$

EXAMPLE 5.10 Solve by Euler's modified method, the differential equation $\frac{dy}{dx} = x^2 + y,\ y(0) = 1$ for $x = 0.02$, taking step length $h = 0.01$.

Solution: Here $\frac{dy}{dx} = x^2 + y,\ y(0) = 1,\ h = 0.01$

Let $f(x, y) = x^2 + y,\ x_0 = 0$

so that $y_0 = y(x_0) = 1.$

So $f(x_0, y_0) = 0^2 + 1 = 1.$

Therefore, by Euler's method, we get $y_1^{(1)} \approx y_0 + hf(x_0, y_0) = 1 + 0.01 \times 1$

$$= 1.01$$

Now, Euler's modified formula for $n = 1$ is given by

$$y_1^{(r)} \approx y_0 + \frac{h}{2}\left[f(x_0, y_0) + f\left(x_1, y_1^{(r-1)}\right)\right],\quad r = 2, 3, \ldots$$

The procedure is shown in the following table:

r	$y_1^{(r-1)}$	$f\left(x_1, y_1^{(r-1)}\right)$	$y_1^{(r)} \approx y_0 + \frac{h}{2}\left[f(x_0, y_0) + f\left(x_1, y_1^{(r-1)}\right)\right]$
2	1.01	$(0.01)^2 + 1.01 = 1.0101$	1 + (0.01/2) × (1 + 1.0101) = 1.00505
3	1.00505	$(0.01)^2 + 1.00505 = 1.00515$	1 + (0.01/2) × (1 + 1.00515) = 1.010026
4	1.010026	$(0.01)^2 + 1.010026 = 1.010126$	1 + (0.01/2) × (1 + 1.010126) = **1.01005**1
5	1.010051	$(0.01)^2 + 1.010051 = 1.010152$	1 + (0.01/2) × (1 + 1.010152) = **1.01005**2

We see from the last two rows that the values of y at $x = 0.01$ are correct to 3 decimal places. Hence, we can write $y(0.01) \approx 1.0101$, correct to 4 decimal places.

Therefore $$f(x_1, y_1) = (0.01)^2 + 1.0101 = 1.0102.$$

Using Euler's method, we get $$y_2^{(1)} \approx y_1 + hf(x_1, y_1) = 1.0101 + 0.01 \times 1.0102 = 1.020202$$

Now, Euler's modified formula for $n = 2$ is given by

$$y_2^{(r)} \approx y_1 + \frac{h}{2}\left[f(x_1, y_1) + f\left(x_2, y_2^{(r-1)}\right)\right], \; r = 2, 3, \ldots$$

The procedure is shown in the following table:

r	$y_2^{(r-1)}$	$f\left(x_2, y_2^{(r-1)}\right)$	$y_2^{(r)} \approx y_1 + \frac{h}{2}\left[f(x_1, y_1) + f\left(x_2, y_2^{(r-1)}\right)\right]$
2	1.020202	$(0.02)^2 + 1.020202 = 1.020602$	$1.0101 + (0.01/2) \times (1.0102 + 1.020602) =$ **1.020254**
3	1.020254	$(0.02)^2 + 1.020254 = 1.020654$	$1.0101 + (0.01/2) \times (1.0102 + 1.020654) =$ **1.020254**

We see from the last two rows that the values of y at $x = 0.02$ are correct to 5 decimal places. Hence, we can write $y(0.02) \approx 1.0203$, correct to 4 decimal places.

5.6 RUNGE–KUTTA METHOD

This is the most widely used method to solve first order, first degree differential equations. Here we discuss Runge–Kutta's second order, third order and fourth order methods.

Second Order Runge–Kutta Method Euler's modified formula gives

$$y_{n+1} \approx y_n + \frac{h}{2}\left[f(x_n, y_n) + f(x_{n+1}, y_{n+1})\right]$$

$$= y_n + \frac{h}{2}\left[f(x_n, y_n) + f(x_n + h, y_n + hf(x_n, y_n))\right]$$

Let $$k_1 = hf(x_n, y_n) \text{ and } k_2 = hf(x_n + h, y_n + hf(x_n, y_n)) = hf(x_n + h, y_n + k_1)$$

Therefore, using these, we get

$$y_{n+1} \approx y_n + \frac{1}{2}(k_1 + k_2), \; n = 0, 1, 2, \ldots \tag{5.12}$$

Equation (5.12) is known as the Runge–Kutta formula of second order.

Third Order Runge–Kutta Method This formula is given by

$$y_{n+1} \approx y_n + \frac{1}{6}(k_1 + 4k_2 + k_3), \; n = 0, 1, 2, \ldots$$

where $$k_1 = hf(x_n, y_n), \; k_2 = hf(x_n + h/2, y_n + k_1/2),$$

and $$k_3 = hf(x_n + h, y_n + k_2).$$

Fourth Order Runge–Kutta Method This formula is the most commonly used in practice and is given by

$$y_{n+1} \approx y_n + \frac{1}{6}\left(k_1 + 2k_2 + 2k_3 + k_4\right),\ n = 0, 1, 2, \ldots$$

where $k_1 = hf(x_n, y_n),\ k_2 = hf(x_n + h/2, y_n + k_1/2),$

$k_3 = hf(x_n + h/2, y_n + k_2/2)$

and $k_4 = hf(x_n + h, y_n + k_3).$

Error in Runge–Kutta Methods The truncation errors in Runge–Kutta's second order, third order and fourth order methods are, respectively, $\boldsymbol{O(h^3)}$, $\boldsymbol{O(h^4)}$ and $\boldsymbol{O(h^5)}$.

A few examples are provided below to illustrate the applications of the fourth order Runge–Kutta method.

EXAMPLE 5.11 Use the Runge–Kutta method of fourth order to calculate y(0.2) and y(0.4) for the equation $\dfrac{dy}{dx} = -xy, y(0) = 1.$

Solution: Given that $\dfrac{dy}{dx} = -xy, y(0) = 1.$

Let $f(x, y) = -xy, x_0 = 0$

so that $y_0 = y(x_0) = 1.$

Let us assume that $h = 0.2.$

Now, fourth order Runge–Kutta method is given by

$$y_{n+1} \approx y_n + \frac{1}{6}\left(k_1 + 2k_2 + 2k_3 + k_4\right),\ n = 0, 1, 2, \ldots$$

where $k_1 = hf(x_n, y_n),\ k_2 = hf(x_n + h/2, y_n + k_1/2),$

$k_3 = hf(x_n + h/2, y_n + k_2/2)$

and $k_4 = hf(x_n + h, y_n + k_3).$

The solution technique is as follows:

Calculation for n = 0:

$$k_1 = hf(x_0, y_0) = 0.2 \times (-0 \times 1)$$
$$= 0$$
$$k_2 = hf(x_0 + h/2, y_0 + k_1/2) = 0.2 \times [-(0 + 0.2/2) \times (1 + 0/2)]$$
$$= -0.02$$
$$k_3 = hf(x_0 + h/2, y_0 + k_2/2) = 0.2 \times [-(0 + 0.2/2) \times (1 - 0.02/2)]$$
$$= -0.0198$$
$$k_4 = hf(x_0 + h, y_0 + k_3) = 0.2 \times [-(0 + 0.2) \times (1 - 0.0198)]$$
$$= -0.039208$$

Therefore $$y_1 \approx y_0 + \frac{1}{6}\left(k_1 + 2k_2 + 2k_3 + k_4\right)$$

$$= 1 + \frac{1}{6}\left(0 + 2\times(-0.02) + 2\times(-0.0198) - 0.039208\right)$$

$$= 0.98019867 \approx 0.9802, \text{ correct to 4 decimal places.}$$

Therefore $\qquad y(0.2) \approx 0.9802.$

Calculation for $n = 1$:

$$k_1 = hf(x_1, y_1) = 0.2 \times (-0.2 \times 0.9802)$$
$$= -0.03921$$
$$k_2 = hf(x_1 + h/2, y_1 + k_1/2) = 0.2 \times [-(0.2 + 0.2/2) \times (0.9802 - 0.03921/2)]$$
$$= -0.05764$$
$$k_3 = hf(x_1 + h/2, y_1 + k_2/2) = 0.2 \times [-(0.2 + 0.2/2) \times (0.9802 - 0.05764/2)]$$
$$= -0.05708$$
$$k_4 = hf(x_1 + h, y_1 + k_3) = 0.2 \times [-(0.2 + 0.2) \times (0.9802 - 0.05708)]$$
$$= -0.07385$$

Therefore $\qquad y_2 \approx y_1 + \frac{1}{6}\left(k_1 + 2k_2 + 2k_3 + k_4\right)$

$$= 0.9802 + \frac{1}{6}\left(-0.03921 + 2\times(-0.05764) + 2\times(-0.05708) - 0.07385\right)$$

$$= 0.923114 \approx 0.9231, \text{ correct to 4 decimal places.}$$

Therefore $\qquad y(0.4) \approx 0.9231.$

EXAMPLE 5.12 Solve the initial value problem $10\frac{dy}{dx} = x^2 + y^2, y(0) = 1$ for $x = 0.1, 0.2$ using the Runge–Kutta fourth order method and find the solution correct to 4 decimal places.

Solution: Given that $10\frac{dy}{dx} = x^2 + y^2, y(0) = 1, h = 0.1$

Let $\qquad f(x, y) = (x^2 + y^2)/10, x_0 = 0$

so that $\qquad y_0 = y(x_0) = 1.$

Now, the fourth order Runge–Kutta method is given by

$$y_{n+1} \approx y_n + \frac{1}{6}\left(k_1 + 2k_2 + 2k_3 + k_4\right), n = 0, 1, 2, \ldots$$

where $\qquad k_1 = hf(x_n, y_n), k_2 = hf(x_n + h/2, y_n + k_1/2),$

$\qquad k_3 = hf(x_n + h/2, y_n + k_2/2)$

and $\qquad k_4 = hf(x_n + h, y_n + k_3).$

The solution technique is given below:

Calculation for $n = 0$:

$$k_1 = hf(x_0, y_0) = 0.1 \times [(0^2 + 1^2)/10]$$
$$= 0.01$$

$$k_2 = hf(x_0 + h/2, y_0 + k_1/2) = 0.1 \times [(0.05^2 + 1.005^2)/10]$$
$$= 0.010125$$
$$k_3 = hf(x_0 + h/2, y_0 + k_2/2) = 0.1 \times [(0.05^2 + 1.0050625^2)/10]$$
$$= 0.0101265$$
$$k_4 = hf(x_0 + h, y_0 + k_3) = 0.1 \times [(0.01^2 + 1.0101265^2)/10]$$
$$= 0.01030356$$

Therefore $$y_1 \approx y_0 + \frac{1}{6}(k_1 + 2k_2 + 2k_3 + k_4)$$

$$= 1 + \frac{1}{6}(0.01 + 2\times 0.010125 + 2\times 0.0101265 + 0.01030356)$$

$$= 1.0101344 \approx 1.0101, \text{ correct to 4 decimal places.}$$

Therefore $y(0.1) \approx 1.0101.$

Calculation for n = 1:

$$k_1 = hf(x_1, y_1) = 0.1 \times [(0.1^2 + 1.0101^2)/10]$$
$$= 0.01030372$$
$$k_2 = hf(x_1 + h/2, y_1 + k_1/2) = 0.1 \times [(0.15^2 + 1.015286^2)/10]$$
$$= 0.010533$$
$$k_3 = hf(x_1 + h/2, y_1 + k_2/2) = 0.1 \times [(0.15^2 + 1.0154009^2)/10]$$
$$= 0.010535$$
$$k_4 = hf(x_1 + h, y_1 + k_3) = 0.1 \times [(0.2^2 + 1.020669^2)/10]$$
$$= 0.010818$$

Therefore $$y_2 \approx y_1 + \frac{1}{6}(k_1 + 2k_2 + 2k_3 + k_4)$$

$$= 1.0101 + \frac{1}{6}(0.01030372 + 2\times 0.010533 + 2\times 0.010535 + 0.010818)$$

$$= 1.0206009 \approx 1.0206, \text{ correct to 4 decimal places.}$$

Therefore $y(0.2) \approx 1.0206.$

EXAMPLE 5.13 Find $y(1.1)$ using the Runge–Kutta method of the fourth order, given that $\frac{dy}{dx} = y^2 + xy, y(1) = 1$ and $h = 0.1$.

Solution: Given that $$\frac{dy}{dx} = y^2 + xy, y(1) = 1, h = 0.1.$$

Let $f(x, y) = y^2 + xy, x_0 = 1$

so that $y_0 = y(x_0) = 1.$

Now, the fourth order Runge–Kutta method is given by

$$y_{n+1} \approx y_n + \frac{1}{6}(k_1 + 2k_2 + 2k_3 + k_4),\ n = 0, 1, 2, \ldots$$

where $k_1 = hf(x_n, y_n),\ k_2 = hf(x_n + h/2, y_n + k_1/2)$

$k_3 = hf(x_n + h/2, y_n + k_2/2)$

and $k_4 = hf(x_n + h, y_n + k_3).$

Therefore, for $n = 0$,

$$k_1 = hf(x_0, y_0) = 0.1 \times [(1^2 + 1 \times 1)$$
$$= 0.2$$
$$k_2 = hf(x_0 + h/2, y_0 + k_1/2) = 0.1 \times (1.1^2 + 1.05 \times 1.1)$$
$$= 0.2365$$
$$k_3 = hf(x_0 + h/2, y_0 + k_2/2) = 0.1 \times (1.11825^2 + 1.05 \times 1.11825)$$
$$= 0.24246$$
$$k_4 = hf(x_0 + h, y_0 + k_3) = 0.1 \times (1.24246^2 + 1.1 \times 1.24246)$$
$$= 0.29104$$

Therefore
$$y_1 \approx y_0 + \frac{1}{6}(k_1 + 2k_2 + 2k_3 + k_4)$$
$$= 1 + \frac{1}{6}(0.2 + 2\times 0.2365 + 2\times 0.24246 + 0.29104)$$
$$= 1.241493 \approx 1.2415, \text{ correct to 4 decimal places.}$$

Therefore $y(1.1) \approx 1.2415.$

5.7 PREDICTOR–CORRECTOR METHOD

In this method, we require two formulae, namely, a predictor formula and a corrector formula. The predictor formula is used to predict the value y_{i+1} using a given set of values $y_i, y_{i-1}, y_{i-2}, \ldots$ Then we use the corrector formula to improve the value y_{i+1}, obtained by the predictor formula. In this context, three well-known formulae are discussed.

Euler's Predictor–Corrector Formula

Predictor formula: $$y_{n+1} \approx y_n + hy'_n$$

Corrector formula: $$y_{n+1} \approx y_n + \frac{1}{2}h(y'_n + y'_{n+1})$$

Milne's Predictor–Corrector Formula

Predictor formula: $$y_{n+1} \approx y_{n-3} + \frac{4}{3}h(2y'_{n-2} - y'_{n-1} + 2y'_n)$$

Corrector formula: $$y_{n+1} \approx y_{n-1} + \frac{1}{3}h(y'_{n+1} + 4y'_n + y'_{n-1})$$

Adam–Bashforth–Moulton Predictor–Corrector Formula

Adam–Bashforth predictor formula: $$y_{n+1} \approx y_n + \frac{1}{24}h(55y'_n - 59y'_{n-1} + 37y'_{n-2} - 9y'_{n-3})$$

Adam–Moulton corrector formula: $y_{n+1} \approx y_n + \frac{1}{24}h\left(9y'_{n+1} + 19y'_n - 5y'_{n-1} + y'_{n-2}\right)$

The above formulae can be explained with the following examples.

EXAMPLE 5.14 Given that $\frac{dy}{dx} = 1 + xy$, $y(0) = 2$, $y(0.1) = 2.1103$, $y(0.2) = 2.2430$, $y(0.3) = 2.4011$, find $y(0.4)$ by Milne's predictor–corrector method.

Solution: Given $\frac{dy}{dx} = 1 + xy$

i.e. $y' = 1 + xy$ (1)

Let $x_0 = 0, x_1 = 0.1$

$x_2 = 0.2, x_3 = 0.3$

so that $y_0 = 2, y_1 = 2.1103$

$y_2 = 2.2430, y_3 = 2.4011$

and let $h = 0.1$.

Now, Milne's predictor formula is given by

$$y_{n+1} \approx y_{n-3} + \frac{4}{3}h\left(2y'_{n-2} - y'_{n-1} + 2y'_n\right) \quad (2)$$

Putting $n = 3$ in (2), we get

$$y_4 \approx y_0 + \frac{4}{3}h\left(2y'_1 - y'_2 + 2y'_3\right) \quad (3)$$

Now, from (1) we get

$$y'_1 = 1 + x_1 y_1 = 1 + 0.1 \times 2.1103$$
$$= 1.21103$$
$$y'_2 = 1 + x_2 y_2 = 1 + 0.2 \times 2.2430$$
$$= 1.4486$$
$$y'_3 = 1 + x_3 y_3 = 1 + 0.3 \times 2.4011$$
$$= 1.72033$$

Therefore, using these in (3), the predicted value of y_4 is

$$y_4 \approx y_0 + \frac{4}{3}h\left(2y'_1 - y'_2 + 2y'_3\right)$$
$$= 2 + \frac{4 \times 0.1}{3}\left(2 \times 1.21103 - 1.4486 + 2 \times 1.72033\right)$$
$$= 2.5885$$

Now, Milne's corrector formula is

$$y_{n+1} \approx y_{n-1} + \frac{1}{3}h\left(y'_{n+1} + 4y'_n + y'_{n-1}\right) \quad (4)$$

Putting $n = 3$ in (4), we get

$$y_4 \approx y_2 + \frac{1}{3}h\left(y_4' + 4y_3' + y_2'\right) \tag{5}$$

From (1), we have

$$y_4' = 1 + x_4 y_4 = 1 + 0.4 \times 2.5885$$
$$= 2.0354$$

Therefore, from (5), we get the corrected value of y_4 as

$$y_4 \approx y_2 + \frac{1}{3}h\left(y_4' + 4y_3' + y_2'\right)$$
$$= 2.2430 + \frac{0.1}{3}\left(1.4486 + 4 \times 1.72033 + 2.0354\right)$$
$$= 2.5885$$

Hence, the required value of $y(0.4)$ is 2.5885.

EXAMPLE 5.15 Compute $y(0.4)$ by Milne's predictor–corrector method from the equation $\frac{dy}{dx} = y^2 + xy$, given that $y(0) = 1$, $y(0.1) = 1.1169$, $y(0.2) = 1.2773$, $y(0.3) = 1.5040$.

Solution: Given

$$\frac{dy}{dx} = y^2 + xy \tag{1}$$

i.e.

$$y' = y^2 + xy$$

Let

$$x_0 = 0,\ x_1 = 0.1$$
$$x_2 = 0.2,\ x_3 = 0.3$$

so that

$$y_0 = 1,\ y_1 = 1.1169$$
$$y_2 = 1.2773,\ y_3 = 1.5040$$

and let

$$h = 0.1.$$

Now, Milne's predictor formula is given by

$$y_{n+1} \approx y_{n-3} + \frac{4}{3}h\left(2y_{n-2}' - y_{n-1}' + 2y_n'\right) \tag{2}$$

Putting $n = 3$ in (2), we get

$$y_4 \approx y_0 + \frac{4}{3}h\left(2y_1' - y_2' + 2y_3'\right) \tag{3}$$

Now, from (1) we get

$$y_1' = y_1^2 + x_1 y_1 = (1.1169)^2 + 0.1 \times 1.1169$$
$$= 1.3591$$

$$y_2' = y_2^2 + x_2 y_2 = (1.2773)^2 + 0.2 \times 1.2773$$
$$= 1.8869$$

$$y_3' = y_3^2 + x_3 y_3 = (1.5040)^2 + 0.3 \times 1.5040$$
$$= 2.7132$$

Therefore, using these in (3), the predicted value of y_4 is

$$y_4 \approx y_0 + \frac{4}{3}h\left(2y_1' - y_2' + 2y_3'\right)$$

$$= 1 + \frac{4\times 0.1}{3}\left(2\times 1.3591 - 1.8869 + 2\times 2.7132\right) = 1.8344$$

Now, Milne's corrector formula is

$$y_{n+1} \approx y_{n-1} + \frac{1}{3}h\left(y_{n+1}' + 4y_n' + y_{n-1}'\right) \tag{4}$$

Putting $n = 3$ in (4), we get

$$y_4 \approx y_2 + \frac{1}{3}h\left(y_4' + 4y_3' + y_2'\right) \tag{5}$$

From (1), we have $\quad y_4' = y_4^{\,2} + x_4 y_4 = (1.8344)^2 + 0.4\times 2.5885$

$$= 4.0988$$

Therefore, from (5), we get the corrected value of y_4 as

$$y_4 \approx y_2 + \frac{1}{3}h\left(y_4' + 4y_3' + y_2'\right)$$

$$= 1.2773 + \frac{0.1}{3}\left(4.0988 + 4\times 2.7132 + 1.8869\right)$$

$$= 1.8386 \approx 1.839, \text{ correct to 3 decimal places.}$$

Hence, the required value of $y(0.4)$ is 1.839.

EXAMPLE 5.16 Compute $y(0.4)$ by Milne's predictor–corrector method from the equation $\frac{dy}{dx} = 4y$, given that $y(0) = 1$, $y(0.1) = 1.492$, $y(0.2) = 2.226$, $y(0.3) = 3.320$.

Solution: Given $\quad \frac{dy}{dx} = 4y \quad (1)$

i.e. $\quad y' = 4y$

Let $\quad x_0 = 0,\ x_1 = 0.1$

$\quad x_2 = 0.2,\ x_3 = 0.3$

so that $\quad y_0 = 1,\ y_1 = 1.492$

$\quad y_2 = 2.226,\ y_3 = 3.320$

and let $\quad h = 0.1.$

Now, Milne's predictor formula is given by

$$y_{n+1} \approx y_{n-3} + \frac{4}{3}h\left(2y_{n-2}' - y_{n-1}' + 2y_n'\right) \tag{2}$$

Putting $n = 3$ in (2), we get

$$y_4 \approx y_0 + \frac{4}{3}h\left(2y_1' - y_2' + 2y_3'\right) \tag{3}$$

Now, from (1) we get

$$y_1' = 4y_1 = 4\times 1.492$$

$$= 5.968$$

$$y_2' = 4y_2 = 4 \times 2.226$$
$$= 8.904$$
$$y_3' = 4y_3 = 4 \times 3.320$$
$$= 13.28$$

Therefore, using these in (3), the predicted value of y_4 is

$$y_4 \approx y_0 + \frac{4}{3}h\left(2y_1' - y_2' + 2y_3'\right)$$
$$= 1 + \frac{4 \times 0.1}{3}\left(2 \times 5.968 - 8.904 + 2 \times 13.28\right)$$
$$= 4.9456$$

Now, Milne's corrector formula is

$$y_{n+1} \approx y_{n-1} + \frac{1}{3}h\left(y_{n+1}' + 4y_n' + y_{n-1}'\right) \quad (4)$$

Putting $n = 3$ in (4), we get

$$y_4 \approx y_2 + \frac{1}{3}h\left(y_4' + 4y_3' + y_2'\right) \quad (5)$$

From (1), we have
$$y_4' = 4y_4 = 4 \times 4.9456$$
$$= 19.7824$$

Therefore, from (5), we get the corrected value of y_4 as

$$y_4 \approx y_2 + \frac{1}{3}h\left(y_4' + 4y_3' + y_2'\right)$$
$$= 2.226 + \frac{0.1}{3}\left(19.7824 + 4 \times 13.28 + 8.904\right)$$
$$= 4.95288 \approx 4.953, \text{ correct to 3 decimal places.}$$

Hence, the required value of $y(0.4)$ is 4.953.

EXAMPLE 5.17 Given that $\frac{dy}{dx} = 1 + xy$, $y(0) = 2$, $y(0.1) = 2.1103$, $y(0.2) = 2.2430$, $y(0.3) = 2.4011$, find $y(0.4)$ by the Adam–Bashforth–Moulton predictor–corrector method.

Solution: Given
$$\frac{dy}{dx} = 1 + xy$$

i.e.
$$y' = 1 + xy \quad (1)$$

Let
$$x_0 = 0,\ x_1 = 0.1$$
$$x_2 = 0.2,\ x_3 = 0.3$$

so that
$$y_0 = 2,\ y_1 = 2.1103$$
$$y_2 = 2.2430,\ y_3 = 2.4011$$

and let
$$h = 0.1.$$

Now, Adam–Bashforth's predictor formula is given by

$$y_{n+1} \approx y_n + \frac{1}{24}h\left(55y_n' - 59y_{n-1}' + 37y_{n-2}' - 9y_{n-3}'\right) \quad (2)$$

Putting $n = 3$ in (2), we get

$$y_4 \approx y_3 + \frac{1}{24}h\left(55y_3' - 59y_2' + 37y_1' - 9y_0'\right) \tag{3}$$

Now, from (1) we get

$$y_0' = 1 + x_0 y_0 = 1 + 0 \times 2$$
$$= 1$$
$$y_1' = 1 + x_1 y_1 = 1 + 0.1 \times 2.1103$$
$$= 1.21103$$
$$y_2' = 1 + x_2 y_2 = 1 + 0.2 \times 2.2430$$
$$= 1.4486$$
$$y_3' = 1 + x_3 y_3 = 1 + 0.3 \times 2.4011$$
$$= 1.72033$$

Therefore, using these in (3), the predicted value of y_4 is

$$y_4 \approx y_3 + \frac{1}{24}h\left(55y_3' - 59y_2' + 37y_1' - 9y_0'\right)$$
$$= 2.4011 + \frac{0.1}{24}(55 \times 1.72033 - 59 \times 1.4486 + 37 \times 1.21103 - 9 \times 1)$$
$$= 2.5884$$

Now, the Adam–Moulton corrector formula is

$$y_{n+1} \approx y_n + \frac{1}{24}h\left(9y_{n+1}' + 19y_n' - 5y_{n-1}' + y_{n-2}'\right) \tag{4}$$

Putting $n = 3$ in (4), we get

$$y_4 \approx y_3 + \frac{1}{24}h\left(9y_4' + 19y_3' - 5y_2' + y_1'\right) \tag{5}$$

From (1), we have $\quad y_4' = 1 + x_4 y_4 = 1 + 0.4 \times 2.5884$

$$= 2.03536$$

Therefore, from (5), we get the corrected value of y_4 as

$$y_4 \approx y_3 + \frac{1}{24}h\left(9y_4' + 19y_3' - 5y_2' + y_1'\right)$$
$$= 2.4011 + \frac{0.1}{24}(9 \times 2.03536 + 19 \times 1.72033 - 5 \times 1.4486 + 1.21103)$$
$$= 2.58848558 \approx 2.5885, \text{ correct to 4 decimal places.}$$

Hence, the required value of $y(0.4)$ is 2.5885.

EXAMPLE 5.18 Given that $\frac{dy}{dx} = \frac{1}{2}xy$, $y(0) = 1$, $y(0.1) = 1.01$, $y(0.2) = 1.022$, $y(0.3) = 1.023$, find $y(0.4)$ by the Adam–Bashforth–Moulton predictor–corrector method.

Solution: Given

$$\frac{dy}{dx} = \frac{1}{2}xy$$

i.e.

$$y' = \frac{1}{2}xy \tag{1}$$

Let $\quad x_0 = 0, x_1 = 0.1$

$\quad x_2 = 0.2, x_3 = 0.3$

so that $\quad y_0 = 1, y_1 = 1.01$

$\quad y_2 = 1.022, y_3 = 1.023$

and let $\quad h = 0.1.$

Now, the Adam–Bashforth's predictor formula is given by

$$y_{n+1} \approx y_n + \frac{1}{24} h\left(55y'_n - 59y'_{n-1} + 37y'_{n-2} - 9y'_{n-3}\right) \tag{2}$$

Putting $n = 3$ in (2), we get

$$y_4 \approx y_3 + \frac{1}{24} h\left(55y'_3 - 59y'_2 + 37y'_1 - 9y'_0\right) \tag{3}$$

Now, from (1) we get

$$y'_0 = \frac{1}{2} x_0 y_0 = \frac{1}{2} \times (0 \times 1)$$

$$= 0$$

$$y'_1 = \frac{1}{2} x_1 y_1 = \frac{1}{2} \times (0.1 \times 1.01)$$

$$= 0.0505$$

$$y'_2 = \frac{1}{2} x_2 y_2 = \frac{1}{2} \times (0.2 \times 1.022)$$

$$= 0.1022$$

$$y'_3 = \frac{1}{2} x_3 y_3 = \frac{1}{2} \times (0.3 \times 1.023)$$

$$= 0.1535$$

Therefore, using these in (3), the predicted value of y_4 is

$$y_4 \approx y_3 + \frac{1}{24} h\left(55y'_3 - 59y'_2 + 37y'_1 - 9y'_0\right)$$

$$= 1.023 + \frac{0.1}{24}(55 \times 0.1535 - 59 \times 0.1022 + 37 \times 0.0505 - 9 \times 0)$$

$$= 1.0408$$

Now, the Adam–Moulton corrector formula is

$$y_{n+1} \approx y_n + \frac{1}{24} h\left(9y'_{n+1} + 19y'_n - 5y'_{n-1} + y'_{n-2}\right) \tag{4}$$

Putting $n = 3$ in (4), we get

$$y_4 \approx y_3 + \frac{1}{24} h\left(9y'_4 + 19y'_3 - 5y'_2 + y'_1\right) \tag{5}$$

From (1), we have $\quad y'_4 = \frac{1}{2} x_4 y_4 = \frac{1}{2} \times (0.4 \times 1.0408)$

$$= 0.2082$$

Therefore, from (5), we get the corrected value of y_4 as

$$y_4 \approx y_3 + \frac{1}{24}h\left(9y_4' + 19y_3' - 5y_2' + y_1'\right)$$

$$= 1.023 + \frac{0.1}{24}(9\times 0.2082 + 19\times 0.1535 - 5\times 0.1022 + 0.0505)$$

$$= 1.04104083 \approx 1.041, \text{ correct to 3 decimal places.}$$

Hence, the required value of $y(0.4)$ is 1.041.

5.8 FINITE DIFFERENCE METHOD

This method is very useful for solving first order as well as second order ordinary differential equations with specified boundary values. In this method, the derivatives are replaced by the finite difference relations, as described below.

$$y(x_n + h) = y(x_n) + hy'(x_n) + \frac{h^2}{2!}y''(x_n) + \frac{h^3}{3!}y'''(x_n) + \cdots \tag{1}$$

and

$$y(x_n - h) = y(x_n) - hy'(x_n) + \frac{h^2}{2!}y''(x_n) - \frac{h^3}{3!}y'''(x_n) + \cdots \tag{2}$$

Adding (1) and (2), we get

$$y(x_n + h) + y(x_n - h) = 2y(x_n) + 2.\frac{h^2}{2!}y''(x_n) + O\left(h^4\right) \tag{3}$$

Assuming h as a small quantity and ignoring the terms involving h^4 and higher powers of h, from (3) we get

$$y(x_n + h) + y(x_n - h) \approx 2y(x_n) + h^2 y''(x_n)$$

$$\Rightarrow \qquad y''(x_n) \approx \frac{y(x_n + h) - 2y(x_n) + y(x_n - h)}{h^2}$$

i.e.

$$\boldsymbol{y_n'' \approx \frac{y_{n+1} - 2y_n + y_{n-1}}{h^2}} \tag{4}$$

Similarly, subtracting (2) from (1), we get

$$y(x_n + h) - y(x_n - h) = 2hy'(x_n) + O\left(h^3\right) \tag{5}$$

Neglecting the terms involving h^3 and higher powers of h, provided h is a small quantity, from (5) we get

$$y(x_n + h) - y(x_n - h) \approx 2y'(x_n)$$

$$\Rightarrow \qquad y'(x_n) \approx \frac{y(x_n + h) - y(x_n - h)}{2h}$$

i.e.

$$\boldsymbol{y_n' \approx \frac{y_{n+1} - y_{n-1}}{2h}} \tag{6}$$

The method is best explained by the examples given below.

EXAMPLE 5.19 Using the finite difference method, solve the boundary value problem $\frac{d^2y}{dx^2}+y=0$, $y(0)=0, y(1)=1$, taking $h=0.25$.

Solution: Given that $$\frac{d^2y}{dx^2}+y=0$$

i.e. $$y''=-y;\ y(0)=0,\ y(1)=1 \text{ and } h=0.25.$$

Let $$x_0=0$$

so that $$x_1=0.25,\ x_2=0.5$$

$$x_3=0.75,\ x_4=1.0$$

and $$y_0=0,\ y_4=1.$$

Now, we replace y'' with the relation $y_n''\approx\frac{y_{n+1}-2y_n+y_{n-1}}{h^2}$ and get

$$\frac{y_{n+1}-2y_n+y_{n-1}}{h^2}=-y_n$$

Putting $h=0.25$, we get

$$y_{n+1}-1.9375y_n+y_{n-1}=0 \tag{1}$$

Putting $n=1$, 2 and 3 we get

$$y_2-1.9375y_1+y_0=0$$
$$y_3-1.9375y_2+y_1=0$$
$$y_4-1.9375y_3+y_2=0$$

Using the boundary conditions $y_0=0$ and $y_4=1$, we have

$$y_2=1.9375y_1 \tag{2}$$
$$y_3-1.9375y_2+y_1=0 \tag{3}$$
$$1-1.9375y_3+y_2=0 \tag{4}$$

From (2) and (3), we get

$$y_3=2.75391y_1 \tag{5}$$

Using (2) and (5) in (4), we have

$$y_1=0.2943$$

From (2) and (5), we get $$y_2=0.5702 \text{ and } y_3=0.8105.$$

Therefore, $$y(0.25)=0.2943,\ y(0.5)=0.5702$$

and $$y(0.75)=0.8105.$$

EXAMPLE 5.20 Using the finite difference method, solve the boundary value problem $\frac{d^2y}{dx^2}=y$, $y(0)=0, y(2)=3.63$, taking $h=0.5$.

Solution: Given that $\frac{d^2y}{dx^2}=y$

i.e. $$y''=y;\ y(0)=0,\ y(2)=3.63$$

and $h = 0.5.$

Let $x_0 = 0$

so that $x_1 = 0.5, x_2 = 1.0$

$x_3 = 1.5, x_4 = 2.0$

and $y_0 = 0, y_4 = 3.63.$

Now, we replace y″ with the relation $y''_n \approx \dfrac{y_{n+1} - 2y_n + y_{n-1}}{h^2}$ and get

$$\frac{y_{n+1} - 2y_n + y_{n-1}}{h^2} = y_n$$

Putting $h = 0.5$, we get

$$y_{n+1} - 2.25y_n + y_{n-1} = 0 \quad (1)$$

Putting $n = 1$, 2 and 3 we get

$$y_2 - 2.25y_1 + y_0 = 0$$

$$y_3 - 2.25y_2 + y_1 = 0$$

$$y_4 - 2.25y_3 + y_2 = 0$$

Using the boundary conditions $y_0 = 0$ and $y_4 = 1$, we have

$$y_2 = 2.25y_1 \quad (2)$$

$$y_3 - 2.25y_2 + y_1 = 0 \quad (3)$$

$$3.63 - 2.25y_3 + y_2 = 0 \quad (4)$$

From (2) and (3), we get

$$y_3 = 4.0625y_1 \quad (5)$$

Using (2) and (5) in (4), we have

$$y_1 = 0.5268$$

From (2) and (5), we get $y_2 = 1.1853$ and $y_3 = 2.1401.$

Therefore, $y(0.5) = 0.5268$, $y(1.0) = 1.1853$ and $y(1.5) = 2.1401$

SOLVED EXAMPLES

EXAMPLE 5.21 Using Euler's method, obtain the solution of $\dfrac{dy}{dx} = x - y$ with $y(0) = 1$ and $h = 0.2$ at $x = 0.4$.

Solution: Here

$$\frac{dy}{dx} = x - y,\ y(0) = 1 \text{ and } h = 0.2.$$

Let $f(x, y) = x - y,\ x_0 = 0$

so that $y_0 = y(x_0) = 1.$

Euler's formula is given by $y_{n+1} \approx y_n + hf(x_n, y_n),\ n = 0, 1, 2, \ldots$

The calculations for $y(0.4)$ are given in the following table:

n	x_n	y_n	$f(x_n, y_n) = x_n - y_n$	$y_{n+1} \approx y_n + hf(x_n, y_n)$
0	0	1	$0 - 1 = -1$	$1 + 0.2 \times (-1) = 0.8$
1	0.2	0.8	$0.2 - 0.8 = -0.6$	$0.8 + 0.2 \times (-0.6) = 0.68$
2	0.4	**0.68**	—	—

Hence $y(0.4) \approx 0.68.$

EXAMPLE 5.22 Find the solution of the differential equation $\dfrac{dy}{dx} = x^2 - y,\ y(0) = 1$ for $x = 0.3$ taking $h = 0.1$ by Euler's method. Justify your result by putting $x = 0.3$ in its solution $y = x^2 - 2x + 2 - e^{-x}$.

Solution: Here $\dfrac{dy}{dx} = x^2 - y,\ y(0) = 1$ and $h = 0.1$.

Let $f(x, y) = x^2 - y, x_0 = 0$

so that $y_0 = y(x_0) = 1.$

Euler's formula is given by $y_{n+1} \approx y_n + hf(x_n, y_n),\ n = 0, 1, 2, \ldots$

The calculations for $y(0.3)$ are given in the following table:

n	x_n	y_n	$f(x_n, y_n) = x_n^2 - y_n$	$y_{n+1} \approx y_n + hf(x_n, y_n)$
0	0	1	$(0)^2 - 1 = -1$	$1 + 0.1 \times (-1) = 0.9$
1	0.1	0.9	$(0.1)^2 - 0.9 = -0.89$	$0.9 + 0.1 \times (-0.89) = 0.811$
2	0.2	0.811	$(0.2)^2 - 0.811 = -0.771$	$0.811 + 0.1 \times (-0.771) = 0.7339$
3	0.3	**0.7339**	—	—

Hence $y(0.3) \approx 0.7339.$

Now, putting $x = 0.3$ in the given solution $y = x^2 - 2x + 2 - e^{-x}$, we get

$$y(0.3) = (0.3)^2 - 2 \times 0.3 + 2 - e^{-0.3} = 0.7492.$$

Hence, the relative error is $\dfrac{|0.7492 - 0.7339|}{0.7492} = \mathbf{0.0204}.$

This shows that the approximate solution obtained by the numerical method is correct to 1 significant figure.

EXAMPLE 5.23 Solve the following initial value problem using Euler's method: $\dfrac{dy}{dx} = x^2 + y$ with $y(0) = 1$. Compute the first steps of the solution with $h = 0.1$. Compare the result (% relative error) with those obtained from the exact solution $y = 3e^x - x^2 - 2x - 2$.

Solution: We are given $\dfrac{dy}{dx} = x^2 + y,\ y(0) = 1$

and $h = 0.1.$

Let $f(x, y) = x^2 + y, x_0 = 0$

so that $y_0 = y(x_0) = 1.$

Euler's formula is given by $y_{n+1} \approx y_n + hf(x_n, y_n),\ n = 0, 1, 2, \ldots$

The calculations for five steps are given in the following table:

n	x_n	y_n	$f(x_n, y_n) = x_n^2 + y_n$	$y_{n+1} \approx y_n + hf(x_n, y_n)$
0	0	1	$(0)^2 + 1 = 1$	$1 + 0.1 \times 1 = 1.1$
1	0.1	1.1	$(0.1)^2 + 1.1 = 1.11$	$1.1 + 0.1 \times 1.11 = 1.211$
2	0.2	1.211	$(0.2)^2 + 1.211 = 1.251$	$1.211 + 0.1 \times 1.251 = 1.3361$
3	0.3	1.3361	$(0.3)^2 + 1.3361 = 1.4261$	$1.3361 + 0.1 \times 1.4261 = 1.47871$
4	0.4	1.47871	$(0.4)^2 + 1.47871 = 1.63871$	$1.47871 + 0.1 \times 1.63871 = 1.642581$

Hence, using Euler's method, we obtain $y(0.1) \approx 1.1$, $y(0.2) \approx 1.211$, $y(0.3) \approx 1.3361$, $y(0.4) \approx 1.47871$ and $y(0.5) \approx 1.642581$.

Now, putting $x = 0.1, 0.2, 0.3$ and 0.4 in the given solution $y = 3e^x - x^2 - 2x - 2$, we get

$$y(0.1) = 3 \times e^{0.1} - (0.1)^2 - 2 \times 0.1 - 2 = 1.105513$$

$$y(0.2) = 3 \times e^{0.2} - (0.2)^2 - 2 \times 0.2 - 2 = 1.224208$$

$$y(0.3) = 3 \times e^{0.3} - (0.3)^2 - 2 \times 0.3 - 2 = 1.359576$$

$$y(0.4) = 3 \times e^{0.4} - (0.4)^2 - 2 \times 0.4 - 2 = 1.515474$$

$$y(0.5) = 3 \times e^{0.5} - (0.5)^2 - 2 \times 0.5 - 2 = 1.696164$$

Hence, the % error in

$$y_1 = y(0.1) \text{ is } \frac{|1.105513 - 1.1|}{1.105513} \times 100\% = 0.499\%$$

$$y_2 = y(0.2) \text{ is } \frac{|1.224208 - 1.211|}{1.224208} \times 100\% = 1.079\%$$

$$y_3 = y(0.3) \text{ is } \frac{|1.359576 - 1.3361|}{1.359576} \times 100\% = 1.727\%$$

$$y_4 = y(0.4) \text{ is } \frac{|1.515474 - 1.47871|}{1.515474} \times 100\% = 2.426\%$$

$$y_5 = y(0.5) \text{ is } \frac{|1.696164 - 1.642581|}{1.696164} \times 100\% = 3.163\%$$

EXAMPLE 5.24 Solve the differential equation $\frac{dy}{dx} = 1 - y,\ y(0) = 0$ by Euler's method at $x = 0.1, 0.2$.

Solution: Here $\frac{dy}{dx} = 1 - y,\ y(0) = 0.$

Let $$f(x, y) = 1 - y, x_0 = 0$$

so that $$y_0 = y(x_0) = 1$$

and let $$h = 0.1.$$

Euler's formula is given by $$y_{n+1} \approx y_n + hf(x_n, y_n), n = 0, 1, 2, \ldots$$

The calculations for $y(0.4)$ are given in the following table:

n	x_n	y_n	$f(x_n, y_n) = 1 - y_n$	$y_{n+1} \approx y_n + hf(x_n, y_n)$
0	0	0	$1 - 0 = 1$	$0 + 0.1 \times 1 = 0.1$
1	0.1	**0.1**	$1 - 0.1 = 0.9$	$0.1 + 0.1 \times 0.9 = 0.19$
2	0.2	**0.19**	–	–

Hence $$y(0.1) \approx 0.1 \text{ and } y(0.2) \approx 0.19.$$

EXAMPLE 5.25 Find the solution of the following differential equation by Euler's method for $x = 1$ by taking $h = 0.2$: $\frac{dy}{dx} = x + y$ with $y = 1$ when $x = 0$.

Solution: Given $$\frac{dy}{dx} = x + y,\ y(0) = 1$$

and $$h = 0.2.$$

Let $$f(x, y) = x + y, x_0 = 0$$

so that $$y_0 = y(x_0) = 1.$$

Euler's formula is given by $$y_{n+1} \approx y_n + hf(x_n, y_n), n = 0, 1, 2, \ldots$$

The calculations for $y(1)$ are given in the following table:

n	x_n	y_n	$f(x_n, y_n) = x_n + y_n$	$y_{n+1} \approx y_n + hf(x_n, y_n)$
0	0	1	$0 + 1 = 1$	$1 + 0.2 \times 1 = 1.2$
1	0.2	1.2	$0.2 + 1.2 = 1.4$	$1.2 + 0.2 \times 1.4 = 1.48$
2	0.4	1.48	$0.4 + 1.48 = 1.88$	$1.48 + 0.2 \times 1.88 = 1.856$
3	0.6	1.856	$0.6 + 1.856 = 2.456$	$1.856 + 0.2 \times 2.456 = 2.3472$
4	0.8	2.3472	$0.8 + 2.3472 = 3.1472$	$2.3472 + 0.2 \times 3.1472 = 2.97664$
5	1.0	**2.97664**	–	–

Hence $$y(1) \approx 2.97664$$

EXAMPLE 5.26 Given $\frac{dy}{dx} = \log_{10}(x + y)$ with the initial condition that $y = 1$ when $x = 0$, find y for $x = 0.2$ and $x = 0.5$ using Euler's modified formula.

Solution: Here $$\frac{dy}{dx} = \log_{10}(x + y),\ y(0) = 1.$$

Let $$f(x, y) = \log_{10}(x + y), x_0 = 0$$

so that $$y_0 = y(x_0) = 1.$$

So $$f(x_0, y_0) = \log_{10}(0 + 1) = 0.$$

To find $y(0.2)$, let

$$h = 0.2.$$

Therefore, by Euler's method, we get $y_1^{(1)} \approx y_0 + hf(x_0, y_0) = 1 + 0.2 \times 0 = 1$

Now, Euler's modified formula for $n = 1$ is given by

$$y_1^{(r)} \approx y_0 + \frac{h}{2}\left[f(x_0, y_0) + f\left(x_1, y_1^{(r-1)}\right)\right], \quad r = 2, 3, \ldots$$

The procedure is shown in the following table:

r	$y_1^{(r-1)}$	$f\left(x_1, y_1^{(r-1)}\right)$	$y_1^{(r)} \approx y_0 + \frac{h}{2}\left[f(x_0, y_0) + f\left(x_1, y_1^{(r-1)}\right)\right]$
2	1	$\log_{10}(0.2 + 1) = 0.0792$	$1 + (0.2/2) \times (1 + 0.0792) = 1.10792$
3	1.10792	$\log_{10}(0.2 + 1.10792) = 0.1166$	$1 + (0.2/2) \times (1 + 0.1166) = 1.11166$
4	1.11166	$\log_{10}(0.2 + 1.11166) = 0.1178$	$1 + (0.2/2) \times (1 + 0.1178) =$ **1.11178**
5	1.11178	$\log_{10}(0.2 + 1.11178) – 0.1179$	$1 + (0.2/2) \times (1 + 0.1179) =$ **1.11179**

We see from the last two rows that the values of y at $x = 0.2$ are correct to 3 decimal places. Hence, we can write $y(0.2) \approx 1.1118$, correct to 4 decimal places.

Again, to find $y(0.5)$, let

$$h = 0.5.$$

Therefore, by Euler's method, we get $y_1^{(1)} \approx y_0 + hf(x_0, y_0) = 1 + 0.5 \times 0 = 1$

Now, Euler's modified formula for $n = 1$ is given by

$$y_1^{(r)} \approx y_0 + \frac{h}{2}\left[f(x_0, y_0) + f\left(x_1, y_1^{(r-1)}\right)\right], \quad r = 2, 3, \ldots$$

The procedure is shown in the following table:

r	$y_1^{(r-1)}$	$f\left(x_1, y_1^{(r-1)}\right)$	$y_1^{(r)} \approx y_0 + \frac{h}{2}\left[f(x_0, y_0) + f\left(x_1, y_1^{(r-1)}\right)\right]$
2	1	$\log_{10}(0.5 + 1) = 0.1761$	$1 + (0.2/2) \times (1 + 0.1761) = 1.11761$
3	1.11761	$\log_{10}(0.5 + 1.11761) = 0.2089$	$1 + (0.2/2) \times (1 + 0.2089) = 1.12089$
4	1.12089	$\log_{10}(0.5 + 1.12089) = 0.2098$	$1 + (0.2/2) \times (1 + 0.2098) =$ **1.12098**
5	1.12098	$\log_{10}(0.5 + 1.12098) = 0.2098$	$1 + (0.2/2) \times (1 + 0.\ 2098) =$ **1.12098**

We see from the last two rows that the values of y at $x = 0.5$ are correct to 5 decimal places. Hence, we can write $y(0.5) \approx 1.12098$, correct to 5 decimal places.

EXAMPLE 5.27 Find the values of $y(0.1)$, $y(0.2)$ and $y(0.3)$ using the Runge–Kutta method of the fourth order, given that $\frac{dy}{dx} = xy + y^2$, $y(0) = 1$.

Solution: Given that $$\frac{dy}{dx} = y^2 + xy,\ y(1) = 1.$$

Let $$f(x, y) = y^2 + xy,\ x_0 = 1$$

so that $$y_0 = y(x_0) = 1$$

and let $$h = 0.1.$$

Now, fourth order Runge–Kutta method is given by

$$y_{n+1} \approx y_n + \frac{1}{6}(k_1 + 2k_2 + 2k_3 + k_4),\ n = 0, 1, 2, \ldots$$

where
$$k_1 = hf(x_n, y_n)$$
$$k_2 = hf(x_n + h/2, y_n + k_1/2)$$
$$k_3 = hf(x_n + h/2, y_n + k_2/2)$$

and
$$k_4 = hf(x_n + h, y_n + k_3).$$

Calculation for $n = 0$:

$$k_1 = hf(x_0, y_0) = 0.1\times(1^2 + 0\times 1)$$
$$= 0.1$$
$$k_2 = hf(x_0 + h/2, y_0 + k_1/2) = 0.1\times(1.05^2 + 0.05\times 1.05)$$
$$= 0.1155$$
$$k_3 = hf(x_0 + h/2, y_0 + k_2/2) = 0.1\times(1.05775^2 + 0.05\times 1.05775)$$
$$= 0.1172$$
$$k_4 = hf(x_0 + h, y_0 + k_3) = 0.1\times(1.1172^2 + 0.1\times 1.1172)$$
$$= 0.13598$$

Therefore $$y_1 \approx y_0 + \frac{1}{6}(k_1 + 2k_2 + 2k_3 + k_4)$$
$$= 1 + \frac{1}{6}(0.1 + 2\times 0.1155 + 2\times 0.1172 + 0.13598)$$
$$= 1.116896667 \approx 1.1169, \text{ correct to 4 decimal places.}$$

Therefore $y(0.1) \approx 1.1169$.

Calculation for $n = 1$:

$$k_1 = hf(x_1, y_1) = 0.1\times(1.1169^2 + 0.1\times 1.1169)$$
$$= 0.1359$$

$$k_2 = hf\left(x_1 + h/2, y_1 + k_1/2\right) = 0.1\times\left(1.1849^2 + 0.15\times 1.1849\right)$$
$$= 0.1582$$
$$k_3 = hf\left(x_1 + h/2, y_1 + k_2/2\right) = 0.1\times\left(1.196^2 + 0.15\times 1.196\right)$$
$$= 0.1610$$
$$k_4 = hf\left(x_1 + h, y_1 + k_3\right) = 0.1\times\left(1.2779^2 + 0.2\times 1.2779\right)$$
$$= 0.1889$$

Therefore $$y_2 \approx y_1 + \frac{1}{6}(k_1 + 2k_2 + 2k_3 + k_4)$$
$$= 1.1169 + \frac{1}{6}(0.1359 + 2\times 0.1582 + 2\times 0.1610 + 0.1889)$$
$$= 1.277433 \approx 1.2774, \text{ correct to 4 decimal places.}$$

Therefore $y(0.2) \approx 1.2774.$

Calculation for n = 2:

$$k_1 = hf\left(x_2, y_2\right) = 0.1\times\left(1.2774^2 + 0.2\times 1.2774\right)$$
$$= 0.1887$$
$$k_2 = hf\left(x_2 + h/2, y_2 + k_1/2\right) = 0.1\times\left(1.3718^2 + 0.25\times 1.3718\right)$$
$$= 0.2225$$
$$k_3 = hf\left(x_2 + h/2, y_2 + k_2/2\right) = 0.1\times\left(1.3886^2 + 0.25\times 1.3886\right)$$
$$= 0.2275$$
$$k_4 = hf\left(x_2 + h, y_2 + k_3\right) = 0.1\times\left(1.5049^2 + 0.3\times 1.5049\right)$$
$$= 0.2716$$

Therefore $$y_3 \approx y_2 + \frac{1}{6}(k_1 + 2k_2 + 2k_3 + k_4)$$
$$= 1.2774 + \frac{1}{6}(0.1887 + 2\times 0.2225 + 2\times 0.2275 + 0.2716)$$
$$= 1.504116667 \approx 1.5041, \text{ correct to 4 decimal places.}$$

Therefore $y(0.3) \approx 1.1.5041.$

EXAMPLE 5.28 Using the fourth order Runge–Kutta method, evaluate the value of y when $x = 1.2$, given $\frac{dy}{dx} + \frac{y}{x} = \frac{1}{x^2}, y(1) = 1.$

Solution: Given that $$\frac{dy}{dx} + \frac{y}{x} = \frac{1}{x^2}, y(1) = 1.$$

Let $$f(x,y)=-\frac{y}{x}+\frac{1}{x^2}, x_0=1$$

so that $$y_0 = y(x_0) = 1$$

and let $$h = 0.2.$$

Now, the fourth order Runge–Kutta method is given by

$$y_{n+1} \approx y_n + \frac{1}{6}(k_1 + 2k_2 + 2k_3 + k_4),\ n = 0, 1, 2, \ldots$$

where
$$k_1 = hf(x_n, y_n)$$
$$k_2 = hf(x_n + h/2, y_n + k_1/2)$$
$$k_3 = hf(x_n + h/2, y_n + k_2/2)$$

and $$k_4 = hf(x_n + h, y_n + k_3).$$

Therefore, for $$n = 0$$

$$k_1 = hf(x_0, y_0) = 0.2\times\left[-\frac{1}{1}+\frac{1}{1^2}\right]$$

$$= 0$$

$$k_2 = hf(x_0 + h/2, y_0 + k_1/2) = 0.2\times\left[-\frac{1}{1.1}+\frac{1}{(1.1)^2}\right]$$

$$= -0.016529$$

$$k_3 = hf(x_0 + h/2, y_0 + k_2/2) = 0.2\times\left[-\frac{0.9917355}{1.1}+\frac{1}{(1.1)^2}\right]$$

$$= -0.015026$$

$$k_4 = hf(x_0 + h, y_0 + k_3) = 0.2\times\left[-\frac{0.98497}{1.2}+\frac{1}{(1.2)^2}\right]$$

$$= -0.025273$$

Therefore $$y_1 \approx y_0 + \frac{1}{6}(k_1 + 2k_2 + 2k_3 + k_4)$$

$$= 1 + \frac{1}{6}(0 + 2\times(-0.016529) + 2\times(-0.015026) - 0.025273)$$

$$= 0.9852695 \approx 0.9853, \text{ correct to 4 decimal places.}$$

Therefore $$y(1.2) \approx 0.9853.$$

EXAMPLE 5.29 Use the fourth order Runge–Kutta method to find an approximate value of $y(0.2)$, given that $y(0) = 0$ and $\frac{dy}{dx} = 1 + y^2$.

Solution: Given that $$\frac{dy}{dx} = 1 + y^2, y(0) = 0$$

Let $$f(x, y) = 1 + y^2, x_0 = 0$$
so that $$y_0 = y(x_0) = 1$$
and let $$h = 0.2.$$

Now, the fourth order Runge–Kutta method is given by

$$y_{n+1} \approx y_n + \frac{1}{6}(k_1 + 2k_2 + 2k_3 + k_4),\ n = 0, 1, 2, \ldots$$

where
$$k_1 = hf(x_n, y_n)$$
$$k_2 = hf(x_n + h/2, y_n + k_1/2)$$
$$k_3 = hf(x_n + h/2, y_n + k_2/2)$$
and
$$k_4 = hf(x_n + h, y_n + k_3).$$

Therefore, for $$n = 0$$

$$k_1 = hf(x_0, y_0) = 0.2 \times \left[1 + 0^2\right]$$
$$= 0.2$$
$$k_2 = hf(x_0 + h/2, y_0 + k_1/2) = 0.2 \times \left[1 + (0.1)^2\right]$$
$$= 0.202$$
$$k_3 = hf(x_0 + h/2, y_0 + k_2/2) = 0.2 \times \left[1 + (0.101)^2\right]$$
$$= 0.2020402$$
$$k_4 = hf(x_0 + h, y_0 + k_3) = 0.2 \times \left[1 + (0.1010201)^2\right]$$
$$= 0.2020401012$$

Therefore
$$y_1 \approx y_0 + \frac{1}{6}(k_1 + 2k_2 + 2k_3 + k_4)$$
$$= 0 + \frac{1}{6}(0.2 + 2 \times 0.202 + 2 \times 0.2020402 + 0.2020401012)$$
$$= 0.20168675 \approx 0.2017, \text{ correct to 4 decimal places.}$$

Therefore $$y(0.2) \approx 0.2017.$$

EXAMPLE 5.30 Given $\dfrac{dy}{dx} = \dfrac{1}{2}\left(1 + x^2\right)y^2$, $y(0) = 1$, $y(0.1) = 1.06$, $y(0.2) = 1.12$, $y(0.3) = 1.21$, calculate $y(0.4)$ by Milne's predictor–corrector method.

Solution: Given
$$\frac{dy}{dx} = \frac{1}{2}\left(1 + x^2\right)y^2$$

i.e.
$$y' = \frac{1}{2}\left(1 + x^2\right)y^2 \tag{1}$$

Let
$$x_0 = 0, x_1 = 0.1$$
$$x_2 = 0.2, x_3 = 0.3$$
so that
$$y_0 = 1, y_1 = 1.06$$
$$y_2 = 1.12, y_3 = 1.21$$
and let
$$h = 0.1.$$

Now, Milne's predictor formula is given by

$$y_{n+1} \approx y_{n-3} + \frac{4}{3}h\left(2y'_{n-2} - y'_{n-1} + 2y'_n\right) \tag{2}$$

Putting $n = 3$ in (2), we get

$$y_4 \approx y_0 + \frac{4}{3}h\left(2y'_1 - y'_2 + 2y'_3\right) \tag{3}$$

Now, from (1) we get

$$y'_1 = \frac{1}{2}\left(1 + x_1^{\,2}\right)y_1^{\,2} = \frac{1}{2}\left[1 + (0.1)^2\right] \times (1.06)^2$$

$$= 0.5674$$

$$y'_2 = \frac{1}{2}\left(1 + x_2^{\,2}\right)y_2^{\,2} = \frac{1}{2}\left[1 + (0.2)^2\right] \times (1.12)^2$$

$$= 0.6522$$

$$y'_3 = \frac{1}{2}\left(1 + x_3^{\,2}\right)y_3^{\,2} = \frac{1}{2}\left[1 + (0.3)^2\right] \times (1.21)^2$$

$$= 0.7980$$

Therefore, using these in (3), the predicted value of y_4 is

$$y_4 \approx y_0 + \frac{4}{3}h\left(2y'_1 - y'_2 + 2y'_3\right)$$

$$= 1 + \frac{4 \times 0.1}{3}\left(2 \times 0.5674 - 0.6522 + 2 \times 0.7980\right)$$

$$= 1.27715 \approx 1.2772$$

Now, Milne's corrector formula is

$$y_{n+1} \approx y_{n-1} + \frac{1}{3}h\left(y'_{n+1} + 4y'_n + y'_{n-1}\right) \tag{4}$$

Putting $n = 3$ in (4), we get

$$y_4 \approx y_2 + \frac{1}{3}h\left(y'_4 + 4y'_3 + y'_2\right) \tag{5}$$

From (1), we have $$y'_4 = \frac{1}{2}\left(1 + x_4^{\,2}\right)y_4^{\,2} = \frac{1}{2}\left[1 + (0.4)^2\right] \times (1.2772)^2$$

$$= 0.9458$$

Therefore, from (5), we get the corrected value of y_4 as

$$y_4 \approx y_2 + \frac{1}{3}h\left(y'_4 + 4y'_3 + y'_2\right)$$

$$= 1.12 + \frac{0.1}{3}\left(0.9458 + 4 \times 0.7980 + 0.6522\right)$$

$= 1.2796667 \approx 1.2797$, correct to 4 decimal places.

Hence, the required value of $y(0.4)$ is 1.2797.

EXAMPLE 5.31 Compute $y(0.5)$ by Milne's predictor–corrector method from $\frac{dy}{dx} = 2e^x - y$ given that $y(0.1) = 2.0100$, $y(0.2) = 2.0401$, $y(0.3) = 2.0907$, $y(0.4) = 2.1621$.

Solution: Given $$\frac{dy}{dx} = 2e^x - y$$

i.e. $$y' = 2e^x - y \tag{1}$$

Let $$x_0 = 0.1, x_1 = 0.2$$
$$x_2 = 0.3, x_3 = 0.4$$

so that $$y_0 = 2.0100, y_1 = 2.0101$$
$$y_2 = 2.0907, y_3 = 2.1621$$

and let $$h = 0.1.$$

Now, Milne's predictor formula is given by

$$y_{n+1} \approx y_{n-3} + \frac{4}{3}h\left(2y'_{n-2} - y'_{n-1} + 2y'_n\right) \tag{2}$$

Putting $n = 3$ in (2), we get

$$y_4 \approx y_0 + \frac{4}{3}h\left(2y'_1 - y'_2 + 2y'_3\right) \tag{3}$$

Now, from (1) we get

$$y'_1 = 2e^{x_1} - y_1 = 2e^{0.2} - 2.0101$$
$$= 0.43271$$
$$y'_2 = 2e^{x_2} - y_2 = 2e^{0.3} - 2.0907$$
$$= 0.60902$$
$$y'_3 = 2e^{x_3} - y_3 = 2e^{0.4} - 2.1621$$
$$= 0.82155$$

Therefore, using these in (3), the predicted value of y_4 is

$$y_4 \approx y_0 + \frac{4}{3}h\left(2y'_1 - y'_2 + 2y'_3\right)$$
$$= 2.0100 + \frac{4 \times 0.1}{3}\left(2 \times 0.43271 - 0.60902 + 2 \times 0.82155\right)$$
$$\approx 2.26327$$

Now, Milne's corrector formula is

$$y_{n+1} \approx y_{n-1} + \frac{1}{3}h\left(y'_{n+1} + 4y'_n + y'_{n-1}\right) \tag{4}$$

Putting $n = 3$ in (4), we get

$$y_4 \approx y_2 + \frac{1}{3}h\left(y'_4 + 4y'_3 + y'_2\right) \tag{5}$$

From (1), we have $$y'_4 = 2e^{x_4} - y_4 = 2e^{0.5} - 2.26327$$
$$= 1.03417$$

Therefore, from (5), we get the corrected value of y_4 as

$$y_4 \approx y_2 + \frac{1}{3}h\left(y_4' + 4y_3' + y_2'\right)$$

$$= 2.0907 + \frac{0.1}{3}(1.03417 + 4\times 0.82155 + 0.60902)$$

$$\approx 2.2550, \text{ correct to 4 decimal places.}$$

Hence, the required value of $y(0.5)$ is 2.2550.

EXAMPLE 5.32 Apply Milne's method to find y(0.8) for the equation $\frac{dy}{dx} = x + y^2$, given that $y(0) = 0$, $y(0.2) = 0.02$, $y(0.4) = 0.0805$, $y(0.6) = 0.1839$.

Solution: Given $\quad \frac{dy}{dx} = x + y^2$

i.e. $\quad y' = x + y^2 \quad (1)$

Let $\quad x_0 = 0, x_1 = 0.2$

$x_2 = 0.4, x_3 = 0.6$

so that $\quad y_0 = 0, y_1 = 0.02$

$y_2 = 0.0805, y_3 = 0.1839$

and let $\quad h = 0.2.$

Now, Milne's predictor formula is given by

$$y_{n+1} \approx y_{n-3} + \frac{4}{3}h\left(2y_{n-2}' - y_{n-1}' + 2y_n'\right) \quad (2)$$

Putting $n = 3$ in (2), we get

$$y_4 \approx y_0 + \frac{4}{3}h\left(2y_1' - y_2' + 2y_3'\right) \quad (3)$$

Now, from (1) we get

$$y_1' = x_1 + y_1^2 = 0.2 + (0.02)^2$$

$$= 0.2004$$

$$y_2' = x_2 + y_2^2 = 0.4 + (0.0805)^2$$

$$= 0.40648$$

$$y_3' = x_3 + y_3^2 = 0.6 + (0.1839)^2$$

$$= 0.63382$$

Therefore, using these in (3), the predicted value of y_4 is

$$y_4 \approx y_0 + \frac{4}{3}h\left(2y_1' - y_2' + 2y_3'\right)$$

$$= 0 + \frac{4\times 0.2}{3}(2\times 0.2004 - 0.40648 + 2\times 0.63382)$$

$$\approx 0.3365$$

Now, Milne's corrector formula is

$$y_{n+1} \approx y_{n-1} + \frac{1}{3}h\left(y'_{n+1} + 4y'_n + y'_{n-1}\right) \tag{4}$$

Putting $n = 3$ in (4), we get

$$y_4 \approx y_2 + \frac{1}{3}h\left(y'_4 + 4y'_3 + y'_2\right) \tag{5}$$

From (1), we have $\quad y'_4 = x_4 + {y_4}^2 = 0.8 + (0.3365)^2$

$$= 0.91323$$

Therefore, from (5), we get the corrected value of y_4 as

$$y_4 \approx y_2 + \frac{1}{3}h\left(y'_4 + 4y'_3 + y'_2\right)$$

$$= 0.0805 + \frac{0.2}{3}(0.91323 + 4 \times 0.63382 + 0.40648)$$

$$\approx 0.3375, \text{ correct to 4 decimal places.}$$

Hence, the required value of $y(0.8)$ is 0.3375.

EXAMPLE 5.33 Using the finite difference method, solve the boundary value problem:

$$\frac{d^2y}{dx^2} + y + 1 = 0 \text{ with } y(0) = 0, y(1) = 0.$$

Solution: Given that $\quad \dfrac{d^2y}{dx^2} + y + 1 = 0$

i.e. $\quad y'' = -y - 1; y(0) = 0, y(1) = 0.$

Let $\quad x_0 = 0$

and $\quad h = 0.25$

so that $\quad x_1 = 0.25, x_2 = 0.5$

$\quad x_3 = 0.75, x_4 = 1.0$

and $\quad y_0 = 0, y_4 = 0.$

Now, we replace y'' with the relation $\; y''_n \approx \dfrac{y_{n+1} - 2y_n + y_{n-1}}{h^2}$ and get

$$\frac{y_{n+1} - 2y_n + y_{n-1}}{h^2} = -y_n - 1$$

Putting $h = 0.25$, we get

$$y_{n+1} - 1.9375y_n + y_{n-1} + 0.0625 = 0 \tag{1}$$

Putting $n = 1, 2$ and 3 we get

$$y_2 - 1.9375y_1 + y_0 + 0.0625 = 0$$

$$y_3 - 1.9375y_2 + y_1 + 0.0625 = 0$$

$$y_4 - 1.9375y_3 + y_2 + 0.0625 = 0$$

Using the boundary conditions $y_0 = 0$ and $y_4 = 0$, we have

$$y_2 = 1.9375y_1 - 0.0625 \quad (2)$$

$$y_3 - 1.9375y_2 + y_1 + 0.0625 = 0 \quad (3)$$

$$y_2 = 1.9375y_3 - 0.0625 \quad (4)$$

From (2) and (4), we get

$$y_3 = y_1 \quad (5)$$

Using (2) and (5) in (3), we have

$$y_1 - 1.9375(1.9375y_1 - 0.0625) + y_1 + 0.0625 = 0$$

$$\Rightarrow \quad y_1 = 0.10467706 \approx 0.1047$$

From (2) and (5), we get $y_2 = 0.1404$.

Therefore, $y(0.25) = 0.1047, y(0.5) = 0.1404$
$y(0.75) = 0.1047.$

EXAMPLE 5.34 Using the Adam–Bashforth–Moulton method, find $y(1.4)$, given $\frac{dy}{dx} = (1+y)x^2$, $y(1) = 1, y(1.1) = 1.233, y(1.2) = 1.548, y(1.3) = 1.979.$

Solution: Given $\frac{dy}{dx} = (1+y)x^2$

i.e. $y' = (1+y)x^2 \quad (1)$

Let $x_0 = 1, x_1 = 1.1$
$x_2 = 1.2, x_3 = 1.3$

so that $y_0 = 1, y_1 = 1.233$
$y_2 = 1.548, y_3 = 1.979$

and let $h = 0.1.$

Now, Adam–Bashforth's predictor formula is given by

$$y_{n+1} \approx y_n + \frac{1}{24}h(55y'_n - 59y'_{n-1} + 37y'_{n-2} - 9y'_{n-3}) \quad (2)$$

Putting $n = 3$ in (2), we get

$$y_4 \approx y_3 + \frac{1}{24}h(55y'_3 - 59y'_2 + 37y'_1 - 9y'_0) \quad (3)$$

Now, from (1) we get

$$y'_0 = (1+y_0)x_0^2 = (1+1)\times(1)^2$$
$$= 2$$

$$y'_1 = (1+y_1)x_1^2 = (1+1.233)\times(1.1)^2$$
$$= 2.70193$$

$$y'_2 = (1+y_2)x_2^2 = (1+1.548)\times(1.2)^2$$
$$= 3.66912$$

$$y'_3 = (1+y_3)x_3^2 = (1+1.979)\times(1.3)^2$$
$$= 5.03451$$

Therefore, using these in (3), the predicted value of y_4 is

$$y_4 \approx y_3 + \frac{1}{24}h(55y_3' - 59y_2' + 37y_1' - 9y_0')$$

$$= 1.979 + \frac{0.1}{24}(55\times5.03451 - 59\times3.66912 + 37\times2.70193 - 9\times2)$$

$$= 2.5722974$$

Now, the Adam–Moulton's corrector formula is

$$y_{n+1} \approx y_n + \frac{1}{24}h(9y_{n+1}' + 19y_n' - 5y_{n-1}' + y_{n-2}') \quad (4)$$

Putting $n = 3$ in (4), we get

$$y_4 \approx y_3 + \frac{1}{24}h(9y_4' + 19y_3' - 5y_2' + y_1') \quad (5)$$

From (1), we have $y_4' = (1+y_4)x_4^2 = (1+2.5722974)\times(1.4)^2$

$$= 7.0017029$$

Therefore, from (5), we get the corrected value of y_4 as

$$y_4 \approx y_3 + \frac{1}{24}h(9y_4' + 19y_3' - 5y_2' + y_1')$$

$$= 1.979 + \frac{0.1}{24}(9\times7.0017029 + 19\times5.03451 - 5\times3.66912 + 2.70193)$$

$= 2.5749473 \approx 2.5749$, correct to 4 decimal places.

Hence, the required value of $y(0.4)$ is 2.5749.

MULTIPLE CHOICE QUESTIONS

1. The Runge–Kutta formula has a truncation error, which is of the order of
   (a) h^2 (b) h^4 (c) h^5 (d) none of these
2. Error of the fourth order Runge–Kutta method is of the order of
   (a) h^4 (b) h^5 (c) h^6 (d) none of these
3. If $\frac{dy}{dx} = x + y$ and $y(1) = 0$, then $y(1.1)$ according to Euler's method $[h = 0.1]$ is
   (a) 0.1 (b) 0.3 (c) 0.5 (d) 0.9
4. The error involved in fourth order R–K method is given by
   (a) $O(h^2)$ (b) $O(h^4)$ (c) $O(h^3)$ (d) $O(h^5)$
5. The truncation error in modified Euler's formula is of the order of
   (a) h^2 (b) h^3 (c) h^4 (d) h^5
6. The finite difference method is used to solve
   (a) a system of simultaneous linear equations
   (b) a system of simultaneous non-linear equations

(c) partial differential equations
(d) non-linear equations

7. Condition of convergence for Euler's method

(a) $|1+hf'(x_i,y_i)|<1$ (b) $|1+hf'(x_i,y_i)|\le 1$

(c) $|1+hf'(x_i,y_i)|>1$ (d) $|1+hf'(x_i,y_i)|\ge 1$

8. Milne's corrector formula is

(a) $y_{n+1} \approx y_n + \frac{h}{3}(y'_{n-1} + 4y'_n + 4y'_{n+1})$ (b) $y_{n+1} \approx y_{n-1} + \frac{h}{3}(y'_{n-1} + 4y'_n + 4y'_{n+1})$

(c) $y_{n+1} \approx y_n + \frac{4h}{3}(y'_{n-1} + 4y'_n + 4y'_{n+1}$ (d) none of these

9. If $\frac{dy}{dx} = x^2 + y$ and $y(0) = 1$, then $y(0.02)$ according to Euler's method $[h = 0.01]$ is
(a) 1.02 (b) 1.04 (c) 1.0 (d) 0.99

10. The finite difference method is used to solve
(a) a system of ordinary differential equations
(b) a boundary value problem
(c) a partial differential equation
(d) a system of transcendental equations

11. The truncation error in Euler's method for the solution of ODE of first order is

(a) $O(h^2)$ (b) $\frac{h^2}{2}y''(x_m + \theta h)$

(c) $h^2 y''(x_m + \theta h)$ (d) none of these

12. In finite difference method, $\frac{d^2y}{dx^2}$ is replaced by

(a) $\frac{y_{n+1} - 2y_{n-1} + y_n}{2h^2}$ (b) $\frac{y_{n+1} - 2y_n + y_{n-1}}{h^2}$

(c) $\frac{y_{n+1} - 2y_{n-1} + y_n}{2h}$ (d) $\frac{y_{n+1} - 2y_{n-1} + y_n}{4h^2}$

13. For $\frac{dy}{dx} = xy$ and $y(0) = 2$, the value of k_2 according to the Runge–Kutta method of fourth order $[h = 0.2]$ is
(a) 0.1 (b) 0.01 (c) 0.4 (d) 0.04

14. The ordinary differential equation is solved by
(a) bisection method (b) Gauss–Seidel method
(c) Euler's method (d) Simpson's 1/3 rule

15. For solving differential equations, Euler's formula is

(a) $y_{n+1} \approx y_n + hf(x_n, y_{n-1})$ (b) $y_{n+1} \approx y_n + hf(x_n, y_{n+1})$

(c) $y_{n+1} \approx y_n + hf(x_n, y_n)$ (d) $y_{n+1} \approx hf(x_n, y_n)$

16. Error in second order Runge–Kutta method is

(a) $O(h^2)$ (b) $O(h^3)$ (c) $O(h^4)$ (d) $O(h^5)$

17. Milne's predictor formula is

(a) $y_{n+1} \approx y_{n-3} + \frac{4h}{3}\left(2y'_{n-2} - y'_{n-1} + 2y'_n\right)$ (b) $y_{n-1} \approx y_{n-3} + \frac{4h}{3}\left(2y'_{n-2} - y'_{n-1} + 2y'_n\right)$

(c) $y_{n+1} \approx y_{n-3} + \frac{4h}{3}\left(2y'_{n-2} - y'_{n-1} + 2y'_{n+1}\right)$ (d) $y_{n-1} \approx y_{n-3} + \frac{4h}{3}(2y'_{n-2} - y'_{n-1} + 2y'_n$

18. The Adam–Bashforth predictor formula is given by

1. $y_{n+1} \approx y_n + \frac{1}{24}h\left(55y'_{n+1} - 59y'_{n-1} + 37y'_{n-2} - 9y'_{n-3}\right)$

2. $y_{n+1} \approx y_n + \frac{1}{24}h\left(55y'_n - 59y'_{n-1} + 37y'_{n-2} - 9y'_{n-3}\right)$

3. $y_{n-1} \approx y_n + \frac{1}{24}h\left(55y'_n - 59y'_{n-1} + 37y'_{n-2} - 9y'_{n-3}\right)$

4. $y_{n+1} \approx y_{n-1} + \frac{1}{24}h\left(55y'_n - 59y'_{n-1} + 37y'_{n-2} - 9y'_{n-3}\right)$

19. The Adam–Moulton corrector formula is

(a) $y_{n+1} \approx y_{n-1} + \frac{1}{24}h\left(9y'_{n+1} + 19y'_n - 5y'_{n-1} + y'_{n-2}\right)$

(b) $y_{n-1} \approx y_n + \frac{1}{24}h\left(9y'_{n+1} + 19y'_n - 5y'_{n-1} + y'_{n-2}\right)$

(c) $y_n \approx y_{n-1} + \frac{1}{24}h\left(9y'_{n+1} + 19y'_n - 5y'_{n-1} + y'_{n-2}\right)$

(d) $y_{n+1} \approx y_n + \frac{1}{24}h\left(9y'_{n+1} + 19y'_n - 5y'_{n-1} + y'_{n-2}\right)$

20. In the finite difference method, $\frac{dy}{dx}$ is replaced with

(a) $\frac{y_{n+1} - y_{n-1}}{2h}$ (b) $\frac{y_{n-1} - y_{n+1}}{2h}$ (c) $\frac{y_{n+1} - y_n}{2h}$ (d) $\frac{y_n - y_{n-1}}{2h}$

Answers

1. (d)	2. (b)	3. (a)	4. (d)	5. (b)	6. (c)	7. (b)	8. (d)
9. (a)	10. (b) and (c)	11. (a) and (b)	12. (b)	13. (d)	14. (c)	15. (c)	
16. (b)	17. (a)	18. (b)	19. (d)	20. (a)			

SHORT ANSWER QUESTIONS

1. Describe Euler's method to find the solution of the differential equation $\frac{dy}{dx} = f(x, y)$ with $y(x_0) = y_0$. Interpret the error in this method geometrically.
2. Establish the second order Runge–Kutta method.
3. Use Euler's method to find $y(0.4)$, where $\frac{dy}{dx} = \log_{10}(x+y)$, $y(0) = 1$ taking $h = 0.1$.
4. Using Euler's modified formula, find $y(0.1)$, correct to 3 decimal places, given that $\frac{dy}{dx} = x^2 + y$, $y(0) = 0.94$.
5. Evaluate $y(0.4)$ by Euler's method, given that $\frac{dy}{dx} = y^2 + 3x$, $y(0) = 1$, $h = 0.1$.
6. Obtain the recursion formula for solving a second order boundary value problem by the finite difference method.
7. Find $y(0.10)$ and $y(0.5)$ by Euler's method from the differential equation $\frac{dy}{dx} = y^2 + x^2$ with $y(0) = 0$ correct to 4 decimal places, taking step length $h = 0.05$.
8. Solve using Euler's method the differential equation $\frac{dy}{dx} = xy$, $y = 1$ when $x = 0$, for $x = 1$ by taking $h = 0.2$.
9. Solve the differential equation $\frac{dy}{dx} = 1 - y$, $y(0) = 0$ by Euler's method at $x = 0.1, 0.2$.
10. Find the solution of the differential equation $\frac{dy}{dx} = x + y$ with $y = 1$ when $x = 0$, by Euler's method for $x = 1$, taking $h = 0.2$.
11. Use the fourth order Runge–Kutta method to find an approximate value of $y(0.2)$, given that $y(0) = 0$ and $\frac{dy}{dx} = 1 + y^2$
12. Using the Runge–Kutta method of fourth order, solve $\frac{dy}{dx} = \frac{y^2 - x^2}{y^2 + x^2}$ with $y(0) = 1$ at $x = 0.2$.
13. Given that $\frac{dy}{dx} = y + x$, $y(0) = 1$, find $y(0.1)$ by Euler's method, taking $h = 0.05$.
14. Use Euler's method to find $y(0.1)$ when $\frac{dy}{dx} = \frac{y - x}{y + x}$, $y(0) = 1$.
15. Using Euler's method, obtain the solution of $\frac{dy}{dx} = x - y$ with $y(0) = 1$ and $h = 0.2$ at $x = 0.4$.
16. Find $y(1.1)$ using the Runge–Kutta method of the fourth order, given that $\frac{dy}{dx} = y^2 + xy$, $y(1) = 1$ and $h = 0.1$.
17. Evaluate $y(1.1)$ using the Runge–Kutta method of fourth order for the problem $\frac{dy}{dx} = x^2 + y^2$, $y(1) = 0$.

Answers

4. $y(0.1) = 1.039$	5. $y(0.4) = 1.7729$	7. $y(0.10) = 0.0001$; $y(0.15) = 0.0006$	
8. $y(1) = 1.4593$	9. $y(0.1) = 0.1$; $y(0.2) = 0.19$	10. $y(1) = 2.9766$	11. $y(0.2) = 0.2027$
12. $y(0.2) = 1.1960$	13. $y(0.1) = 1.1050$	14. $y(0.1) = 1.1$	15. $y(0.4) = 0.68$
16. $y(1.1) = 1.2415$	17. $y(1.1) = 0.1107$		

LONG ANSWER QUESTIONS

1. Given $\frac{dy}{dx} = \frac{y-x}{y+x}$ with initial condition $y = 1$ at $x = 0$, find y for $x = 0.1$ by Euler's method, correct to 4 decimal places, taking step length $h = 0.02$.

2. Solve by using Euler's method the differential equation $\frac{dy}{dx} = xy$, $y = 1$ when $x = 0$ for $x = 1$, taking $h = 0.2$.

3. Using Euler's method evaluate $y(2)$, correct to 3 decimal places from $\frac{dy}{dx} = \frac{1}{2}(x+y)$ with $y(0) = 2$ taking $h = 0.5$.

4. Use Euler's method to find the solution of $\frac{dy}{dx} = x - y$ with $y(0) = 1$, $h = 0.2$ at $x = 0.4$.

5. Solve the following initial value problem using Euler's method: $\frac{dy}{dx} = x^2 + y$ with $y(0) = 1$. Compute the first steps of the solution with $h = 0.1$. Compare the result (% relative error) with those obtained from the exact solution $y = 3e^x - x^2 - 2x - 2$.

6. Use Euler's method to find the numerical solution of the following differential equation: $f'(x) = 1 + x - x^2$ with $y(0) = 1$, $h = 0.02$, find $y(0.1)$.

7. Apply Euler's method to find y at $x = 0.01$ and 0.02 from $\frac{dy}{dx} = y + e^x$ with $y(0) = 0$.

8. Given that $\frac{dy}{dx} = \frac{x-y}{x+y}$, $y(1) = 1$, compute $y(1.9)$ by Euler's modified method.

9. Given that $\frac{dy}{dx} + \frac{y}{x} = \frac{1}{x^2}$, $y(1) = 1$, evaluate $y(1.2)$ by Euler's modified method correct up to 4 decimal places.

10. Solve by Euler's modified method, the differential equation $\frac{dy}{dx} = x^2 + y$, $y(0) = 1$ for $x = 0.02$ by taking step length $h = 0.01$.

11. Given that $\frac{dy}{dx} = \log_{10}(x+y)$ with the initial condition that $y = 1$ when $x = 0$, find y for $x = 0.2$ and $x = 0.5$ using Euler's modified formula.

12. Use the Runge–Kutta method of fourth order to calculate y(0.2) and y(0.4) for the equation $\frac{dy}{dx} = -xy$, $y(0) = 1$.

13. Use the Runge–Kutta method to compute $y(0.4)$ from $\frac{dy}{dx} = x^2 y$ with $y(0) = 2$, taking $h = 0.1$.

14. Solve the initial value problem $10\frac{dy}{dx} = x^2 + y^2$, $y(0) = 1$ for $x = 0.1, 0.2$ using the Runge–Kutta fourth order method and find the solution correct to 4 decimal places.

15. Solve the equation $\frac{dy}{dx} = x^2 + y^2$, $y(0) = 1$ for $x = 0.1$ using the Runge–Kutta fourth order method and find the solution correct to 4 decimal places($h = 0.05$).

16. Find the values of $y(0.1)$, $y(0.2)$ and $y(0.3)$ using the Runge–Kutta method of fourth order, given that $\frac{dy}{dx} = xy + y^2$, $y(0) = 1$.

17. Solve the equation $\frac{dy}{dx} = \frac{1}{x + y}$, $y(0) = 1$ for $y(0.1)$ and $y(0.2)$, using the Runge–Kutta method of fourth order.

18. Solve the initial value problem $10\frac{dy}{dx} = x^2 + y^2$, $y(0) = 1$ for $x = 0.1, 0.2$ using the Runge–Kutta fourth order method and find the solution correct to 2 decimal places.

19. Using the Runge–Kutta method of fourth order, find $y(0.2)$, given that $\frac{dy}{dx} = 3e^x + 2y$, $y(0) = 1$, taking $h = 0.1$.

20. Use the fourth order Runge–Kutta method to find an approximate value of $y(0.2)$, given that $y(0) = 0$ and $\frac{dy}{dx} = 1 + y^2$.

21. Compute $y(0.5)$ by Milne's predictor–corrector method from $\frac{dy}{dx} = 2e^x - y$, given that $y(0.1) = 2.0100$, $y(0.2) = 2.0401$, $y(0.3) = 2.0907$, $y(0.4) = 2.1621$.

22. Apply Milne's method to find y(0.8) for the equation $\frac{dy}{dx} = x + y^2$, given that $y(0) = 0$, $y(0.2) = 0.02$, $y(0.4) = 0.0805$, $y(0.6) = 0.1839$.

23. Given $\frac{dy}{dx} = \frac{1}{2}\left(1 + x^2\right)y^2$, $y(0) = 1$, $y(0.1) = 1.06$, $y(0.2) = 1.12$, $y(0.3) = 1.21$, calculate $y(0.4)$ by Milne's predictor–corrector method.

24. Use the finite difference method to solve the boundary value problem $\frac{d^2y}{dx^2} + y + 1 = 0$ with $y(0) = 0$, $y(1) = 0$.

25. Using the Adam–Bashforth–Moulton method, find $y(1.4)$, given $\frac{dy}{dx} = (1 + y)x^2$, $y(1) = 1$, $y(1.1) = 1.233$, $y(1.2) = 1.548$, $y(1.3) = 1.979$.

Answers

1. $y(0.1) = 1.1095$
2. $y(1) = 1.4593$
3. $y(2) = 5.7656$
4. $y(0.4) = 0.68$
5. $y(0.1) = 1.100$; $y(0.1) = 1.1055$; 0.004975
6. $y(0.1) = 1.1256$
7. $y(0.01) = 0.01$; $y(0.02) = 0.0202$
8. $y(1.9) = 1.1261$
9. $y(1.2) = 0.98718$
10. $y(0.02) = 1.0203$
11. $y(0.2) = 1.1118$; $y(0.5) = 1.12098$
12. $y(0.2) = 0.9802$; $y(0.4) = 0.9231$
13. $y(0.4) = 2.0431$
14. $y(0.1) = 1.0101$; $y(0.2) = 1.0207$
15. $y(0.1) = 1.1115$
16. $y(0.1) = 1.1169$; $y(0.2) = 1.2774$; $y(0.3) = 1.5041$
17. $y(0.1) = 1.0914$; $y(0.2) = 1.1696$
18. $y(0.1) = 1.01$; $y(0.2) = 1.02$
19. $y(0.2) = 2.3031$
20. $y(0.2) = 0.2027$
21. $y(0.5) = 2.255$
22. $y(0.8) = 0.3375$
23. $y(0.4) = 1.2797$
24. $y(0.25) = 0.1047$; $y(0.5) = 0.1404$; $y(0.75) = 0.1047$
25. $y(1.4) = 2.5749$

6

Numerical Solutions of Algebraic and Transcendental Equations

6.1 INTRODUCTION

This chapter deals with some numerical solution techniques for solving algebraic equations. The equation $f(x) = 0$ is called algebraic if $f(x)$ is a polynomial of x. For example, $x^3 - 3x - 5 = 0, x^4 - 2x - 10 = 0$, are algebraic equations. This type of equation arises frequently in applied mathematics and engineering applications. We need to determine one or more roots for these equations. However, this may not always be possible using analytical methods. For instance, examine the following equations:

$$x^2 - 5x + 4 = 0,$$
$$x^3 - 3x + 2 = 0,$$
$$x^4 - 2x + 7 = 0,$$
$$x^5 + 5x + 2 = 0, \ldots,$$
$$x^{95} + 3x - 2 = 0.$$

Though these equations are algebraic, analytical methods may fail to find their real root. It is easy to state that $x = 1$ is a real root of $x^2 - 5x + 4 = 0$. However, this may not be applicable for $x^{95} + 3x - 2 = 0$. In such situations, we require approximate solutions of these equations up to a specified degree of accuracy. This can be achieved by applying numerical techniques.

6.2 MATHEMATICAL PRELIMINARIES

We must first make some mathematical assumptions:

(i) The function $y = f(x)$, defining the equation $f(x) = 0$, is continuous and continuously differentiable up to a sufficient number of times.

(ii) The equation $f(x) = 0$ should not have multiple roots, i.e., if α is a real root of $f(x) = 0$, then $f(\alpha) = 0$ and $f'(\alpha) \neq 0$.

Also, we apply a searching process to determine a sufficiently small interval $[a, b]$ containing one and only one real root of the equation $f(x) = 0$. The theorem given below implements this.

Theorem If $f(x)$ is a real-valued continuous function defined on $[a, b]$ and $f(a)$, $f(b)$ have opposite signs, then there exists at least one real root between a and b.

Different numerical techniques can be used to solve algebraic equations. In this context, we include (i) method of iteration, (ii) bisection method, (iii) regula falsi method, (iv) secant method and (v) Newton–Raphson method. All of these are **iterative** methods.

6.3 ORDER OF CONVERGENCE

Let x_n and x_{n+1} be two approximations in the nth and $(n+1)$th iterations, respectively, to the root α of the equation $f(x) = 0$. Let ε_n and ε_{n+1} be the corresponding errors. If the iterative method converges to the root α, then we can find two constants $k \geq 1$ and $A > 0$ so that $\lim_{n\to\infty} \left|\frac{\varepsilon_{n+1}}{\varepsilon_n^k}\right| = A.$

The constant k is known as the **order of convergence** and A is called the **asymptotic error constant**.

6.4 METHOD OF ITERATION

Let $f(x) = 0$ be an equation to be solved by successive approximations. We can re-write it as

$$x = \phi(x) \tag{6.1}$$

Let x_0 be an initial approximation of the actual root α of the equation $f(x) = 0$. Then, the first approximation x_1 is given by

$$x_1 = \phi(x_0)$$

Similarly, the second approximation x_2 is given by

$$x_2 = \phi(x_1)$$

Proceeding in this way, the $(n+1)$th approximation is given by

$$x_{n+1} = \phi(x_n)\ n = 0,1,2, \ldots \tag{6.2}$$

Thus, we get a sequence of successive approximations $x_1, x_2, \ldots, x_n$. Equation (6.2) is known as the iterative or recursion formula and the method of solving it is called the method of iteration.

Sufficient Conditions for the Convergence of the Iteration Method

If

(i) α be a root of $f(x) = 0$ which is equivalent to $x = \phi(x)$
(ii) I be any interval containing the root α
(iii) $|\phi'(x)| < 1$ for all x in I

then the sequence of approximations $x_1, x_2, \ldots, x_n$ will converge to the root α, if the initial approximation x_0 is chosen in I.

EXAMPLE 6.1 Find a root between 5 and 6 of the equation $x^2 - 6x + 2 = 0$ by the method of iteration.

Solution: We have $\quad x^2 - 6x + 2 = 0$

$\Rightarrow \quad x^2 = 6x - 2$, i.e., $x = 6 - \dfrac{2}{x} = \phi(x)$, say.

$\therefore \quad \phi'(x) = \dfrac{2}{x^2}$

Clearly, $|\phi'(x)| = \dfrac{2}{x^2} < 1$ for all x in the interval [5, 6].

Hence, the method is convergent.
Let $x_0 = 5$. The iterative process is shown in the following table:

n	x_n	$x_{n+1} = \phi(x_n) = 6 - \dfrac{2}{x_n}$
0	5	5.6
1	5.6	5.64286
2	5.64286	5.64574
3	5.64574	5.64575
4	5.64575	5.64575

Therefore, the desired root is 5.646 (correct to 3 decimal places).

EXAMPLE 6.2 Find the real root of the equation $x + \log x - 2 = 0$, correct to 4 significant digits, by the method of iteration.

Solution: Let $f(x) = x + \log x - 2$. Now, $f(1) = -1 < 0$ and $f(2) = \log 2 > 0$. Therefore, a real root lies between 1 and 2.

Now $$x + \log x - 2 = 0$$

$\Rightarrow$ $$x = 2 - \log x = \phi(x), \text{ say.}$$

$\therefore$ $$\phi'(x) = -\frac{1}{x}$$

Clearly, $|\phi'(x)| = \frac{1}{x} < 1$ for all x in the interval [1, 2].

Hence, the method is convergent.
Let $x_0 = 2$. The iterative process is shown in the following table:

n	x_n	$x_{n+1} = \phi(x_n) = 2 - \log x_n$
0	2	1.6999
1	1.6999	1.7698
2	1.7698	1.7521
3	1.7521	1.7564
4	1.7564	1.7554
5	1.7554	1.7556
6	1.7556	1.75556

Therefore, the desired root is 1.756 (correct to 4 significant digits).

6.5 BISECTION METHOD

One of the simplest iterative methods is the bisection method, also known as the **half-interval** or **Bolzano** method. This is based on the above theorem. In this method, first a sufficiently small interval $[a_0, b_0]$ is determined so that $f(a_0).f(b_0) < 0$ assuming that there exists a real root α such that $a_0 < \alpha < b_0$. Then, we calculate $x_1 = \frac{1}{2}(a_0 + b_0)$ (Fig. 6.1). Now, if $f(a_0).f(x_1) < 0$, we set $a_1 = a_0$ and $b_1 = x_1$. Otherwise, $a_1 = x_1$ and $b_1 = a_0$. Thus, we obtain a new interval $[a_1, b_1]$, containing the root α, the length of which is half that of the previous interval $[a_0, b_0]$. Next, we calculate $x_2 = \frac{1}{2}(a_1 + b_1)$. If $f(a_1).f(x_2) < 0$, we set $a_2 = a_1$ and $b_2 = x_2$. Otherwise, $a_2 = x_2$ and $b_2 = a_1$.

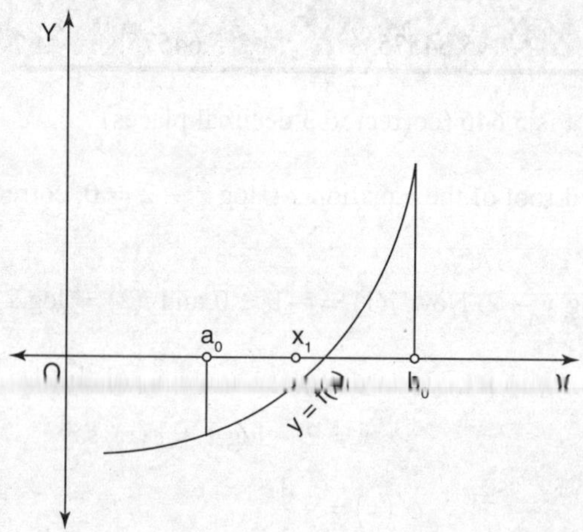

Fig. 6.1 *Bisection method*

This will give another smaller interval $[a_2, b_2]$ containing the root α. Continuing this procedure, we obtain the sequence $\{x_1, x_2, x_3, \ldots\}$ of iterative values. This leads to the iterative formula

$$\boldsymbol{x_{n+1} = \frac{1}{2}(a_n + b_n), \; n = 0, 1, 2, \ldots}$$

We stop if $|x_{n+1} - x_n| \le a_c$, where a_c is the desired accuracy.

Convergence of the Bisection Method Let x_n and x_{n+1} be two approximations in the nth and $(n + 1)$th iterations, respectively, to the root α of the equation $f(x) = 0$ using the bisection method. Let ε_n and ε_{n+1} be the corresponding errors. Then, in the $(n + 1)$th step, we have

$$\varepsilon_{n+1} = |\alpha - x_{n+1}| < |b_n - a_n| \le \frac{1}{2}|b_{n-1} - a_{n-1}| \le \cdots \le \frac{1}{2^n}|b_0 - a_0|$$

Therefore, as $n \to \infty$, $\varepsilon_{n+1} \to 0$.
This shows that the **bisection method is always convergent**.

Again, we see that the error in any step is half that of the previous step. Thus, we can write $\varepsilon_{n+1} = \frac{1}{2}\varepsilon_n$, i.e., $\frac{\varepsilon_{n+1}}{\varepsilon_n^1} = \frac{1}{2}$. Clearly, the **order of convergence of the bisection method is 1**. That is, it is **linearly** convergent.

Important Guidelines for Computations The following instructions are very useful for controlling computations for the bisection method:

(i) Choose the interval $[a, b]$ so that $f(a).f(b) < 0$.

(ii) Compute $x = \frac{1}{2}(a+b)$.

(iii) If $f(a)$ and $f(x)$ have same sign, shift the point a to x. Otherwise, shift b to x. That is, if you see the **same sign, shift the point.**

To illustrate, some examples are given below.

EXAMPLE 6.3 Find the positive root of the equation $x^3 - 3x + 1.06 = 0$ by the bisection method, correct to 3 significant figures.

Solution: Let $f(x) = x^3 - 3x + 1.06$. We see that $f(1) = -0.94 < 0$ and $f(2) = 3.06 > 0$. So, there exists at least one positive root between 1 and 2. Hence, we set $[a_0, b_0] = [1, 2]$. The computations for finding the approximate value of the root by the bisection method are given in the following table:

n	a_n	b_n	$x_{n+1} = \frac{a_n + b_n}{2}$	$f(a_n)$	$f(b_n)$	$f(x_{n+1})$
0	1	2	1.5	**−ve**	+ve	$\mathbf{-0.065} < 0$
1	1.5	2	1.75	−ve	**+ve**	$\mathbf{1.1694} > 0$
2	1.5	1.75	1.625	−ve	+ve	$0.4760 > 0$
3	1.5	1.625	1.5625	−ve	+ve	$0.1872 > 0$
4	1.5	1.5625	1.53125	−ve	+ve	$0.05661 > 0$
5	1.5	1.53125	1.51562	**−ve**	+ve	$\mathbf{-0.0053} < 0$
6	1.51562	1.53125	1.5234	−ve	**+ve**	$\mathbf{0.02537} > 0$
7	1.51562	1.5234	1.5195	−ve	+ve	$0.00988 > 0$
8	1.51562	1.5195	1.5176	–	–	–

We see that in column x_{n+1}, two values 1.5195 and 1.5176 are correct to 3 significant figures. Hence, the required positive root is **1.52** (correct to 3 significant figures).

Note: In the first row, $f(a)$ and $f(x)$ have the same sign. That is why in the next row, a is shifted from 1 (value of a) to 1.5 (value of x). Similar techniques are applied to the next row.

EXAMPLE 6.4 Find the real root of the equation $x^3 - 4x - 9 = 0$ by the bisection method, correct to 4 significant figures.

Solution: Let $f(x) = x^3 - 4x - 9$. We see that $f(2) = -9 < 0$ and $f(3) = 6 > 0$. So, there exists at least one positive root between 2 and 3. Hence, we set $[a_0, b_0] = [2,3]$. The computations for finding the approximate value of the root by the bisection method are given in the following table:

n	a_n	b_n	$x_{n+1} = \dfrac{a_n + b_n}{2}$	$f(a_n)$	$f(b_n)$	$f(x_{n+1})$
0	2	3	2.5	$-ve$	$+ve$	$-3.375 < 0$
1	2.5	3	2.75	$-ve$	$+ve$	$0.797 > 0$
2	2.5	2.75	2.625	$-ve$	$+ve$	$-1.412 < 0$
3	2.625	2.75	2.6875	$-ve$	$+ve$	$-0.339 < 0$
4	2.6875	2.75	2.71875	$-ve$	$+ve$	$0.221 > 0$
5	2.6875	2.71875	2.702875	$-ve$	$+ve$	$-0.0656 < 0$
6	2.702875	2.71875	2.7108125	$-ve$	$+ve$	$0.0772 > 0$
7	2.702875	2.7108125	2.70684375	$-ve$	$+ve$	$0.057 > 0$
8	2.702875	2.70684375	2.7048594	$-ve$	$+ve$	$-0.030 < 0$
9	2.7048594	2.70684375	2.7058516	$-ve$	$+ve$	$-0.012 < 0$
10	2.7058516	2.70684375	2.7063477	$-ve$	$+ve$	$-0.032 < 0$
11	2.7063477	2.70684375	2.7065957	$-ve$	$+ve$	$0.012 > 0$
12	2.7063477	2.7065957	2.7064717	–	–	–

We see that in column x_{n+1}, two values 2.7065957 and 2.7064717 are correct to 4 significant figures. Hence, the required positive root is **2.707** (correct to 4 significant figures).

EXAMPLE 6.5 Find the root of the equation $x^3 - x - 4 = 0$ between 1 and 2 by the bisection method, correct to 3 decimal places.

Solution: Let $f(x) = x^3 - x - 4$. We see that $f(1) = -4 < 0$ and $f(2) = 2 > 0$. So, there exists at least one positive root between 1 and 2. Hence, we set $[a_0, b_0] = [1, 2]$. The computations for finding the approximate value of the root by the bisection method are given in the following table:

n	a_n	b_n	$x_{n+1} = \dfrac{a_n + b_n}{2}$	$f(a_n)$	$f(b_n)$	$f(x_{n+1})$
0	1	2	1.5	$-ve$	$+ve$	$-2.125 < 0$
1	1.5	2	1.75	$-ve$	$+ve$	$-0.391 < 0$
2	1.75	2	1.875	$-ve$	$+ve$	$0.717 > 0$
3	1.75	1.875	1.8125	$-ve$	$+ve$	$0.142 > 0$
4	1.75	1.8125	1.78125	$-ve$	$+ve$	$-0.130 < 0$
5	1.78125	1.8125	1.796875	$-ve$	$+ve$	$0.048 > 0$
6	1.78125	1.796875	1.7890625	$-ve$	$+ve$	$-0.063 < 0$

Cont'd

Table cont'd

n	a_n	b_n	$x_{n+1} = \frac{a_n + b_n}{2}$	$f(a_n)$	$f(b_n)$	$f(x_{n+1})$
7	1.7890625	1.796875	1.792969	$-ve$	$+ve$	$-0.029 < 0$
8	1.792969	1.796875	1.794922	$-ve$	$+ve$	$-0.012 < 0$
9	1.794922	1.796875	1.795899	$-ve$	$+ve$	$-0.037 < 0$
10	1.795899	1.796875	1.796387	$-ve$	$+ve$	$0.057 > 0$
11	1.795899	1.796387	1.796143	–	–	–

We see that in column x_{n+1}, two values 1.796387 and 1.796143 are correct to 3 decimal places. Hence, the required root is **1.796** (correct to 3 decimal places).

Advantages and Disadvantages of the Bisection Method

Advantages

(i) The method is very simple and easy to implement.

(ii) It is **unconditionally convergent**, i.e., there is certainty of finding the root.

Disadvantages

(i) The method is linearly convergent, so it is very slow.

(ii) A large number of iterations are required to obtain a moderately accurate result.

6.6 REGULA FALSI METHOD (METHOD OF FALSE POSITION)

The basics of this method are similar to those of the bisection method. Let $f(x) = 0$ be the equation and α be a root of it, α being real. First, find a sufficiently small interval $[a_0, b_0]$ so that $f(a_0).f(b_0) < 0$, which implies that the root α lies between a_0 and b_0. Next, draw two perpendiculars at $x = a_0$ and $x = b_0$ on the x-axis. Let them cut the function $y = f(x)$ at $A(a_0, f(a_0))$ and $B(b_0, f(b_0))$, respectively. Then, the equation of the straight line joining A and B can be given by

$$\frac{y - f(a_0)}{x - a_0} = \frac{f(b_0) - f(a_0)}{b_0 - a_0} \tag{6.3}$$

As $f(a_0)$ and $f(b_0)$ are of opposite signs, the straight line AB must cut the x-axis. This point of intersection of AB and the x-axis will be considered as the next approximation to the root α (Fig. 6.2).

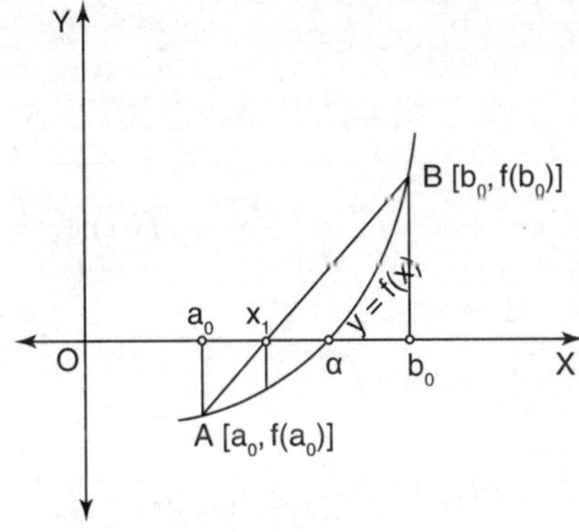

Fig. 6.2 *Regula falsi method*

To obtain this approximation, say x_1, put $y = 0$ in (6.3)

$$\frac{0-f(a_0)}{x_1-a_0}=\frac{f(b_0)-f(a_0)}{b_0-a_0}$$

$$\Rightarrow \qquad x_1=a_0-f(a_0).\frac{(b_0-a_0)}{f(b_0)-f(a_0)}$$

$$\Rightarrow \qquad x_1=\frac{a_0 f(b_0)-b_0 f(a_0)}{f(b_0)-f(a_0)}$$

Now, if $f(a_0).f(x_1) < 0$, then $a_1 = a_0$ and $b_1 = x_1$. Otherwise, $a_1 = x_1$ and $b_1 = b_0$. Thus, a new interval $[a_1, b_1]$ containing α is obtained and $[a_1, b_1] \subset [a_0, b_0]$. Now, drawing a perpendicular at $x = x_1$ and continuing the same, a sequence $\{x_1, x_2, x_3, \ldots\}$ is obtained. Thus the iterative formula for the regula falsi method is given by

$$x_{n+1}=a_n-f(a_n).\frac{(b_n-a_n)}{f(b_n)-f(a_n)}$$

or

$$x_{n+1}=\frac{a_n f(b_n)-b_n f(a_n)}{f(b_n)-f(a_n)}, n=0,1,2,\ldots$$

Note: In the regula falsi method, we use straight lines to determine the iterative values, and all these values are within the initial estimated interval $[a_0, b_0]$. Therefore, this method is an interpolation. As the equation of the interpolant (straight line) is linear, this method is sometimes called **linear interpolation**.

Convergence of the Regula Falsi Method To show this, let x_{n+1} be the $(n + 1)$th approximation to the root α of the equation $f(x) = 0$. Then clearly, either $x_n = a_n$ or $x_n = b_n$. Let $x_n = a_n$. Then we can denote $b_n = x_{n-1.}$ Let ε_{n+1} be the error in x_{n+1} so that $x_{n+1} = \alpha + \varepsilon_{n+1}$. Then, from the iterative formula, we get

$$x_{n+1}=x_n-f(x_n).\frac{(x_{n-1}-x_n)}{f(x_{n-1})-f(x_n)}$$

$$\Rightarrow \quad \alpha+\varepsilon_{n+1}=\alpha+\varepsilon_n-f(\alpha+\varepsilon_n).\frac{(\alpha+\varepsilon_{n-1}-\alpha-\varepsilon_n)}{f(\alpha+\varepsilon_{n-1})-f(\alpha+\varepsilon_n)}$$

$$\Rightarrow \quad \varepsilon_{n+1}=\varepsilon_n-f(\alpha+\varepsilon_n).\frac{(\varepsilon_{n-1}-\varepsilon_n)}{f(\alpha+\varepsilon_{n-1})-f(\alpha+\varepsilon_n)}$$

$$=\varepsilon_n-\frac{(\varepsilon_n-\varepsilon_{n-1})\left[f(\alpha)+\varepsilon_n f'(\alpha)+\frac{\varepsilon_n^{\ 2}}{2}f''(\alpha)+\cdots\right]}{\left[f(\alpha)+\varepsilon_n f'(\alpha)+\frac{\varepsilon_n^{\ 2}}{2}f''(\alpha)+\cdots\right]-\left[f(\alpha)+\varepsilon_{n-1} f'(\alpha)+\frac{\varepsilon_{n-1}^{\ 2}}{2}f''(\alpha)+\cdots\right]}$$

$$=\varepsilon_n-\frac{(\varepsilon_n-\varepsilon_{n-1})\left[\varepsilon_n f'(\alpha)+\frac{\varepsilon_n^{\ 2}}{2}f''(\alpha)+\cdots\right]}{\left[\varepsilon_n f'(\alpha)+\frac{\varepsilon_n^{\ 2}}{2}f''(\alpha)+\cdots\right]-\left[\varepsilon_{n-1} f'(\alpha)+\frac{\varepsilon_{n-1}^{\ 2}}{2}f''(\alpha)+\cdots\right]}, \text{ since } f(\alpha)=0.$$

$$= \varepsilon_n - \frac{(\varepsilon_n - \varepsilon_{n-1})\left[\varepsilon_n f'(\alpha) + \frac{\varepsilon_n^2}{2} f''(\alpha) + \cdots\right]}{(\varepsilon_n - \varepsilon_{n-1})\left[f'(\alpha) + \frac{(\varepsilon_n + \varepsilon_{n-1})}{2} f''(\alpha) + \cdots\right]}$$

$$= \varepsilon_n - \varepsilon_n\left[1 + \frac{\varepsilon_n}{2}\frac{f''(\alpha)}{f'(\alpha)} + \cdots\right] \times \left[1 + \frac{(\varepsilon_n + \varepsilon_{n-1})}{2}\frac{f''(\alpha)}{f'(\alpha)} + \cdots\right]^{-1}$$

$$= \frac{1}{2}\frac{f''(\alpha)}{f'(\alpha)}\varepsilon_n\varepsilon_{n-1} + O\left(\varepsilon_n^2\varepsilon_{n-1} + \varepsilon_n\varepsilon_{n-1}^2\right)$$

Therefore, neglecting the higher powers of ε_n, we get

$$\varepsilon_{n+1} = \frac{1}{2}\frac{f''(\alpha)}{f'(\alpha)}\varepsilon_n\varepsilon_{n-1} = C\varepsilon_n\varepsilon_{n-1}$$

where
$$C = \frac{1}{2}\frac{f''(\alpha)}{f'(\alpha)}$$

Now, to find the order of convergence, let $\varepsilon_{n+1} = A\varepsilon_n^{\ k}$.
Using this, we get

$$A\varepsilon_n^{\ k} = C.\varepsilon_n.A^{-1/k}\varepsilon_n^{\ 1/k} = CA^{-1/k}\varepsilon_n^{\ 1+1/k}$$

Equating the power of ε_n, we get

$$k = 1 + 1/k$$

or
$$k^2 - k - 1 = 0$$

$\Rightarrow$
$$k = \frac{1}{2}\left(1 \pm \sqrt{5}\right) \approx 1.618 \text{ (taking the positive sign).}$$

Therefore
$$\varepsilon_{n+1} = A\varepsilon_n^{\ 1.618}$$

Hence, the regula falsi method has the **order of convergence 1.618**.

Important Guidelines for Computations The following instructions are very useful for controlling the computations for the regula falsi method:

(i) Choose the interval $[a, b]$ so that $f(a).f(b) < 0$.

(ii) Compute $x = \dfrac{af(b) - bf(a)}{f(b) - f(a)}$.

(iii) If $f(a)$ and $f(x)$ have the same sign, shift the point a to x. Otherwise, shift b to x. That is, if you see the **same sign, shift the point**.

To illustrate, some examples are given below.

EXAMPLE 6.6 Find the real root of the equation $x^3 - 3x - 5 = 0$ by the method of false position, correct to 3 decimal places.

Solution: Let $f(x) = x^3 - 3x - 5$. We see that $f(2) = -3 < 0$ and that $f(3) = 13 > 0$. So, there exists at least one positive root between 2 and 3. Hence, we set $[a_0, b_0] = [2,3]$. The computations for finding the approximate value of the root by the method of false position are given in the following table:

n	a_n	b_n	$x_{n+1} = \dfrac{a_n f(b_n) - b_n f(a_n)}{f(b_n) - f(a_n)}$	$f(a_n)$	$f(b_n)$	$f(x_{n+1})$
0	2	3	2.1875	$-ve$	$+ve$	$-1.095 < 0$
1	2.1875	3	2.2506	$-ve$	$+ve$	$-0.352 < 0$
2	2.2506	3	2.27034	$-ve$	$+ve$	$-0.108 < 0$
3	2.27034	3	2.27636	$-ve$	$+ve$	$-0.033 < 0$
4	2.27636	3	2.2782	$-ve$	$+ve$	$-0.0102 < 0$
5	2.2782	3	2.2788	$-ve$	$+ve$	$-0.003 < 0$
6	2.2788	3	2.2789	–	–	–

We see that in column x_{n+1}, two values 2.2788 and 2.2789 are correct to 3 decimal places. Hence, the required root is **2.279** (correct to 3 decimal places).

EXAMPLE 6.7 Find out the root of the equation $x^3 - 5x - 7 = 0$ between 2 and 3 by the regula falsi method, correct to 4 decimal places.

Solution: Let $f(x) = x^3 - 5x - 7$. We see that $f(2) = -9 < 0$ and $f(3) = 5 > 0$. So, there exists at least one positive root between 2 and 3. Hence, we set $[a_0, b_0] = [2, 3]$. The computations for finding the approximate value of the root by the regula falsi method are given in the following table:

n	a_n	b_n	$x_{n+1} = \dfrac{a_n f(b_n) - b_n f(a_n)}{f(b_n) - f(a_n)}$	$f(a_n)$	$f(b_n)$	$f(x_{n+1})$
0	2	3	2.642857	$-ve$	$+ve$	$-1.7547 < 0$
1	2.642857	3	2.735635	$-ve$	$+ve$	$-0.2055 < 0$
2	2.735635	3	2.746072	$-ve$	$+ve$	$-0.0225 < 0$
3	2.746072	3	2.747208	$-ve$	$+ve$	$-0.0024 < 0$
4	2.747208	3	2.747332	$-ve$	$+ve$	$-0.00026 < 0$
5	2.747332	3	2.747345	$-ve$	$+ve$	$-0.00003 < 0$
6	2.747345	3	2.747346	–	–	–

We see that in column x_{n+1}, two values 2.747345 and 2.747346 are correct to 4 decimal places. Hence, the required root is **2.7473** (correct to 4 decimal places).

EXAMPLE 6.8 Find the root of the equation $x^3 + x^2 + x + 7 = 0$ using the regula falsi method.

Solution: Let $f(x) = x^3 + x^2 + x + 7$. We see that $f(-3) = -14 < 0$ and $f(-2) = 1 > 0$. So, there exists at least one real root between -3 and -2. Hence, we set $[a_0, b_0] = [-3, -2]$. The computations for finding the approximate value of the root by the regula falsi method are given in the following table:

n	a_n	b_n	$x_{n+1} = \dfrac{a_n f(b_n) - b_n f(a_n)}{f(b_n) - f(a_n)}$	$f(a_n)$	$f(b_n)$	$f(x_{n+1})$
0	−3	−2	−2.86667	−ve	+ve	$-11.2065 < 0$
1	−2.86667	−2	−2.07100	−ve	+ve	$0.33543 > 0$
2	−2.86667	−2.07100	−2.09412	−ve	+ve	$0.10776 > 0$
3	−2.86667	−2.09412	−2.10148	−ve	+ve	$0.03417 > 0$
4	−2.86667	−2.10148	−2.10379	−ve	+ve	$0.01085 > 0$
5	−2.86667	−2.10379	−2.10268	−ve	+ve	$0.0221 > 0$
6	−2.86667	−2.10268	−2.10418	−ve	+ve	$0.00698 > 0$
7	−2.86667	−2.10418	−2.10465	−ve	+ve	$0.00225 > 0$
8	−2.86667	−2.10465	−2.10480	–	–	–

We see that in column x_{n+1}, two values −2.10465 and −2.10480 are correct to 3 decimal places. Hence, the required root is **−2.105** (correct to 3 decimal places).

Advantages and Disadvantages of the Regula Falsi Method

Advantages

(i) The method is very simple and easy to implement.
(ii) It is always **convergent**, i.e., there is certainty in finding the root.

Disadvantages

(i) The method is quite slow in the sense of convergence.
(ii) For practical computation, it is not a good tool.

6.7 THE SECANT METHOD

This method is quite similar to the regula falsi method except for the condition $f(a)\,f(x) < 0$. In this case, the interval at each iteration may not contain the root. The rationale is that if the root lies between $x = a_0$ and $x = b_0$, then either a_0 or b_0 is kept fixed throughout the iterative process. This is shown below.

Let us consider a sufficiently small interval $[a_0, b_0]$ so that condition $f(a_0).f(b_0) < 0$, which implies that the root α lies between a_0 and b_0. Then, the first approximation x_1 is given by

$$x_1 = \frac{a_0 f(b_0) - b_0 f(a_0)}{f(b_0) - f(a_0)}$$

Now, we have a new interval $[a_1, b_1]$ where $a_1 = a_0$ and $b_1 = x_1$. Note that in this case we fix the point a_0. Therefore, the second approximation x_2 is given by

$$x_2 = \frac{a_1 f(b_1) - b_1 f(a_1)}{f(b_1) - f(a_1)}$$

Similarly, the $(n + 1)$th approximation is given by

$$x_{n+1} = \frac{a_n f(b_n) - b_n f(a_n)}{f(b_n) - f(a_n)}, n = 0,1,2,\ldots$$

The order of convergence of this method is that of the regula falsi method, i.e., **1.618**.

EXAMPLE 6.9 Find the real root of the equation $x^3 - 3x - 5 = 0$ by the secant method, correct to 3 decimal places.

Solution: Let $f(x) = x^3 - 3x - 5$. We see that $f(2) = -3 < 0$ and $f(3) = 13 > 0$. So, there exists at least one positive root between 2 and 3. Hence, we set $[a_0, b_0] = [2, 3]$. The computations for finding the approximate value of the root by the secant method are given in the following table:

n	a_n	b_n	$x_{n+1} = \frac{a_n f(b_n) - b_n f(a_n)}{f(b_n) - f(a_n)}$	$f(x_{n+1})$
0	2	3	2.1875	−1.095
1	2.1875	3	2.2506	−0.352
2	2.2506	3	2.27034	−0.108
3	2.27034	3	2.27636	−0.033
4	2.27636	3	2.2782	0.0102
5	2.2782	3	2.2788	−0.003
6	2.2788	3	2.2789	–

We see that in column x_{n+1}, two values 2.2788 and 2.2789 are correct to 3 decimal places. Hence, the required root is **2.279** (correct to 3 decimal places).

EXAMPLE 6.10 Find the positive root of the equation $x^3 - 4x - 9 = 0$, correct to 4 decimal places, by the secant method.

Solution: Let $f(x) = x^3 - 4x - 9$. We see that $f(2) = -9 < 0$ and $f(3) = 6 > 0$. So, there exists at least one positive root between 2 and 3. Hence, we set $[a_0, b_0] = [2,3]$. The computations for finding the approximate value of the root by the secant method are given in the following table:

n	a_n	b_n	$x_{n+1} = \frac{a_n f(b_n) - b_n f(a_n)}{f(b_n) - f(a_n)}$	$f(x_{n+1})$
0	2	3	2.6	−1.824
1	2.6	3	2.69325	−0.237
2	2. 69325	3	2.7049	−0.029
3	2.7049	3	2.7063	−0.0041
4	2.7063	3	2.7064957	−0.0058
5	2.7064957	3	2.706520157	–

We see that in column x_{n+1}, two values 2.7064957 and 2.706520157 are correct to 4 significant digits. Hence, the required root is **2.707** (correct to 4 decimal places).

EXAMPLE 6.11 Find the root of the equation $3x - \cos x - 1 = 0$ using the secant method, correct to 4 significant digits.

Solution: Let $f(x) = 3x - \cos x - 1$. We see that $f(0) = -1 < 0$ and $f(1) = 1.46 > 0$. So, there exists at least one real root between 0 and 1. Hence, we set $[a_0, b_0] = [0,1]$. The computations for finding the approximate value of the root by the secant method are given in the following table:

n	a_n	b_n	$x_{n+1} = \dfrac{a_n f(b_n) - b_n f(a_n)}{f(b_n) - f(a_n)}$	$f(x_{n+1})$
0	0	1	0.41	−0.67
1	0.41	1	0.59	−0.061
2	0.59	1	0.6064	−0.0025
3	0.6064	1	0.60707	−0.000113
4	0.60707	1	0.6071004	−0.0000045
5	0.6071004	1	0.6071016	–

We see that in column x_{n+1}, two values 0.6071004 and 0.6071016 are correct to 5 significant digits. Hence, the required root is **0.6071** (correct to 4 significant digits).

6.8 NEWTON–RAPHSON METHOD

This is the most satisfactory and widely used iterative method. In this method, let x_0 be the initial approximation for the root α of the equation $f(x) = 0$. The assumption should such that x_0 is **very close** to α. Let h be the correction for x_0 so that $x_0 + h = \alpha$. As α is a root of $f(x) = 0$, it implies

$$f(\alpha) = 0$$

$$\Rightarrow \quad f(x_0 + h) = 0$$

Expanding in Taylor's series and assuming $f(x)$ possesses derivatives, we get

$$f(x_0) + hf'(x_0) + \frac{h^2}{2} f''(x_0) + \cdots = 0$$

As h is a very small quantity, neglecting the terms containing h^2 and higher powers of h, we get

$$f(x_0) + hf'(x_0) \approx 0$$

$$\Rightarrow \quad h \approx -\frac{f(x_0)}{f'(x_0)}, f'(x_0) \neq 0$$

Thus, a better approximation to the root α is

$$x_1 = x_0 + \text{approximate value of } h.$$

$$= x_0 + \left\{-\frac{f(x_0)}{f'(x_0)}\right\} = x_0 - \frac{f(x_0)}{f'(x_0)}$$

Similarly, the next approximation is

$$x_2 = x_1 - \frac{f(x_1)}{f'(x_1)}, f'(x_1) \neq 0$$

Thus, using this approach, we obtain the iterative formula in the $(n + 1)$th iteration as

$$x_{n+1} = x_n - \frac{f(x_n)}{f'(x_n)}, f'(x_n) \neq 0, n = 0,1,2,\ldots$$

Geometrical Interpretation Let x_0 be an initial approximation which is very close to the root α of the equation $f(x) = 0$. Therefore, the equation of the tangent at the point $P_0(x_0, f(x_0))$ on the curve $y = f(x)$ is

$$y - f(x_0) = f'(x_0)(x - x_0) \tag{6.4}$$

Suppose this tangent cuts the x-axis at x_1, which can be obtained by putting $y = 0$ and $x = x_1$ in equation (6.4). Thus, we get

$$x_1 = x_0 - \frac{f(x_0)}{f'(x_0)}$$

This is the first approximation for α.

Again, the tangent, drawn at $P_1(x_1, f(x_1))$ on the curve $y = f(x)$, will cut the x-axis at x_2, where

$$x_2 = x_1 - \frac{f(x_1)}{f'(x_1)}$$

This is the second approximation for α.

Thus, geometrically, the approximations of the root, i.e., $x_1, x_2, \ldots, x_{n+1}, \ldots$ are obtained, respectively, by the points at which the tangents at $(x_0, f(x_0))$, $(x_1, f(x_1))$, … ,$(x_n, f(x_n))$, … to the curve $y = f(x)$ meet the x-axis (Fig. 6.3). This is why this method is often called the **tangent method**.

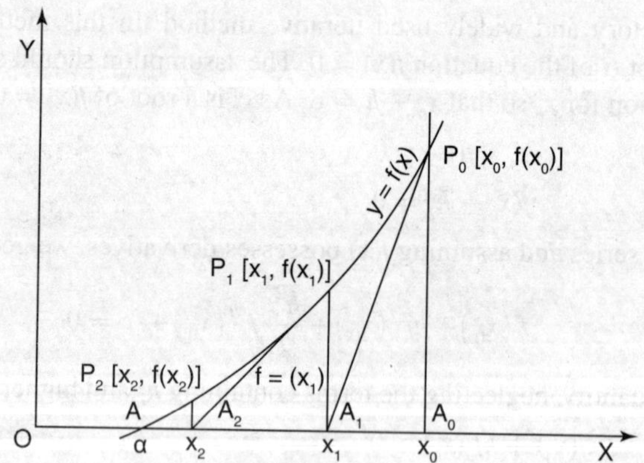

Fig. 6.3 *Newton–Raphson method*

Order of Convergence Let x_n and x_{n+1} be two approximations in the nth and $(n + 1)$th iterations, respectively, to the root α of the equation $f(x) = 0$. Let ε_n and ε_{n+1} be the corresponding errors. Then, we have $x_n = \alpha + \varepsilon_n$ and $x_{n+1} = \alpha + \varepsilon_{n+1}$. Using these in Newton's iterative formula, we get

$$\alpha + \varepsilon_{n+1} = \alpha + \varepsilon_n - \frac{f(\alpha + \varepsilon_n)}{f'(\alpha + \varepsilon_n)}$$

$$\Rightarrow \quad \varepsilon_{n+1} = \varepsilon_n - \frac{\left[f(\alpha)+\varepsilon_n f'(\alpha)+\frac{\varepsilon_n^{\,2}}{2}f''(\alpha)+\cdots\right]}{\left[f'(\alpha)+\varepsilon_n f''(\alpha)+\frac{\varepsilon_n^{\,2}}{2}f'''(\alpha)+\cdots\right]} = \varepsilon_n - \frac{\left[\varepsilon_n f'(\alpha)+\frac{\varepsilon_n^{\,2}}{2}f''(\alpha)+\cdots\right]}{\left[f'(\alpha)+\varepsilon_n f''(\alpha)+\frac{\varepsilon_n^{\,2}}{2}f'''(\alpha)+\cdots\right]},$$

as $f(\alpha) = 0$

$$= \varepsilon_n - \frac{\varepsilon_n\left[f'(\alpha)+\frac{\varepsilon_n}{2}f''(\alpha)+\cdots\right]}{\left[f'(\alpha)+\varepsilon_n f''(\alpha)+\frac{\varepsilon_n^{\,2}}{2}f'''(\alpha)+\cdots\right]}$$

$$= \varepsilon_n - \varepsilon_n\left[1+\frac{\varepsilon_n}{2}\frac{f''(\alpha)}{f'(\alpha)}+\cdots\right]\times\left[1+\varepsilon_n\frac{f''(\alpha)}{f'(\alpha)}+\frac{\varepsilon_n^{\,2}}{2}\frac{f'''(\alpha)}{f'(\alpha)}\cdots\right]^{-1}$$

$$= \frac{\varepsilon_n^{\,2}}{2}\frac{f'''(\alpha)}{f'(\alpha)}+O\left(\varepsilon_n^{\,3}\right)$$

Therefore

$$\lim_{n\to\infty}\left|\frac{\varepsilon_{n+1}}{\varepsilon_n^{\,2}}\right| = \frac{1}{2}\left|\frac{f'''(\alpha)}{f'(\alpha)}\right|, \text{ as } \varepsilon_n \to 0 \text{ as } n \to \infty$$

This shows that the Newton–Raphson method has the **order of convergence 2**. That is, it is **quadratic convergent**.

Some Important Applications This method has a special application in evaluating an approximate result of the ***m*th root** and **reciprocal** of a **positive real number**.

- *Evaluation of mth root of a positive real number*

  Let us consider the problem $\sqrt[m]{R}$, where m is any positive number and $R > 0$. Let $x = \sqrt[m]{R}$ which gives $x^m - R = 0$. Let $f(x) = x^m - R$. Therefore, $f'(x) = mx^{m-1}$. Now, the Newton–Raphson iterative formula gives

$$x_{n+1} = x_n - \frac{f(x_n)}{f'(x_n)} = x_n - \frac{x_n^{\,m} - R}{mx_n^{\,m-1}}$$

$$= \frac{(m-1)x_n^{\,m} + R}{mx_n^{\,m-1}}, n = 0, 1, 2, \ldots$$

Now, we consider the following cases:

(i) $m - 2$:

$$x_{n+1} = \frac{1}{2}\left[x_n + \frac{R}{x_n}\right]$$

(ii) $m = 3$:

$$x_{n+1} = \frac{1}{3}\left[x_n + \frac{R}{x_n^{\,2}}\right]$$

and so on.

- *Evaluation of the reciprocal of a positive real number*

Let $x = \frac{1}{R}$ where R is a positive real number. Then, we have $\frac{1}{x} - R = 0$. Let $f(x) = \frac{1}{x} - R$. Therefore, $f'(x) = -\frac{1}{x^2}$. Then, the Newton–Raphson iterative formula gives

$$x_{n+1} = x_n - \frac{f(x_n)}{f'(x_n)}$$

$$= x_n - \frac{\frac{1}{x_n} - R}{-\frac{1}{x_n^2}}$$

$$= x_n + x_n(1 - Rx_n) = x_n(2 - Rx_n), n = 0,1,2,\ldots$$

Some examples are given below to better illustrate this method.

EXAMPLE 6.12 Find the smallest positive root of the equation $3x^3 - 9x^2 + 8 = 0$, correct to 3 decimal places, using the Newton–Raphson method.

Solution: Let $f(x) = 3x^3 - 9x^2 + 8$ and so $f'(x) = 9x^2 - 18x$. We see that $f(1) = 2 > 0$ and $f(2) = -4 < 0$. So, there exists at least one positive root between 1 and 2. To set the initial approximation, we make use of the bisection method, i.e., $x_0 = \frac{1+2}{2} = 1.5$. The computations for finding the approximate value of the root by the Newton–Raphson method are given in the following table:

n	x_n	$f(x_n)$	$f'(x_n)$	$x_{n+1} = x_n - \frac{f(x_n)}{f'(x_n)}$
0	0.5	−0.75	3	0.75
1	0.75	0.0625	3.5	0.732143
2	0.732143	0.00032	3.46429	0.732051
3	0.732051	0.0000007	3.464102	0.732051

As the last two consecutive iterative values of x_{n+1} are correct to 6 decimal places, the required root is **0.732** (correct to 3 significant figures).

EXAMPLE 6.13 Find the positive root of $x^2 + 2x - 2 = 0$ using the Newton–Raphson method, correct to 5 significant figures.

Solution: Let $f(x) = x^2 + 2x - 2$ and so $f'(x) = 2x + 2$. We see that $f(0) = -2 < 0$ and $f(1) = 1 > 0$. So, there exists at least one positive root between 0 and 1. To set the initial approximation, we make use of the bisection method, i.e., $x_0 = \frac{0+1}{2} = 0.5$. The computations for finding the approximate value of the root by the Newton–Raphson method are given in the following table:

n	x_n	$f(x_n)$	$f'(x_n)$	$x_{n+1} = x_n - \dfrac{f(x_n)}{f'(x_n)}$
0	0.5	−0.75	3	0.75
1	0.75	0.0625	3.5	0.732143
2	0.732143	0.00032	3.46429	0.732051
3	0.732051	0.0000007	3.464102	0.732051

As the last two consecutive iterative values of x_{n+1} are correct to 6 decimal places, the required root is **0.732** (correct to 3 significant figures).

EXAMPLE 6.14 Find the approximate value of $\sqrt{5}$ using the Newton–Raphson method.

Solution: Let $x = \sqrt{5}$. Then we have $x^2 - 5 = 0$. Let $f(x) = x^2 - 5$. We see that $f(2) = -1 < 0$ and $f(3) = 4 > 0$. So, the value of $\sqrt{5}$ lies between 0 and 1. To set the initial approximation, we make use of the bisection method, i.e., $x_0 = \dfrac{2+3}{2} = 2.5$. Now, the iterative formula for finding the square root of 5 is given by

$$x_{n+1} = \frac{1}{2}\left[x_n + \frac{5}{x_n}\right]$$

The computations for finding the approximate value of the root by the Newton–Raphson method are given in the following table:

n	x_n	$x_{n+1} = \dfrac{1}{2}\left[x_n + \dfrac{5}{x_n}\right]$
0	2.5	2.25
1	2.25	2.236111
2	2.236111	2.236068

As the last two consecutive iterative values of x_{n+1} are correct to 3 decimal places, the approximate value of $\sqrt{5}$ is **2.2361** (correct to 4 decimal places).

EXAMPLE 6.15 Determine the Newton–Raphson iterative formula to find the pth root of a positive number N and hence find the cube root of 17.

Solution: Let $x = \sqrt[p]{N}$ which gives $x^P - \mathrm{N} = 0$. Let $f(x) = x^P - N$. Therefore, $f'(x) = px^{p-1}$. Now, the Newton–Raphson iterative formula gives

$$x_{n+1} = x_n - \frac{f(x_n)}{f'(x_n)} = x_n - \frac{x_n^{\,p} - N}{p x_n^{\,p-1}}$$

$$= \frac{(p-1)x_n^{\,p} + N}{p x_n^{\,p-1}} = \frac{1}{p}\left[(p-1)x_n + \frac{N}{x_n^{\,p-1}}\right] \qquad (6.5)$$

which is the required iterative formula.

Now, putting $p = 3$ and $n = 17$ in (6.5), we get

$$x_{n+1} = \frac{1}{3}\left[2x_n + \frac{17}{x_n^2}\right]$$

As $8 < 17 < 27$, i.e., $2 < \sqrt[3]{17} < 3$, we consider the initial approximation to be $x_0 = \frac{2+3}{2} = 2.5$.

The computations are shown in the following table:

n	x_n	$x_{n+1} = \frac{1}{3}\left[2x_n + \frac{17}{x_n^2}\right]$
0	2.5	2.573333
1	2.573333	2.571283
2	2.571283	2.571282

As the last two consecutive iterative values of x_{n+1} are correct to 5 decimal places, the approximate value of $\sqrt[3]{17}$ is **2.57128** (correct to 5 decimal places).

Advantages and Disadvantages of the Newton–Raphson Method

Advantages

(i) The method is very fast as the order of convergence is 2.

(ii) It is the most satisfactory method and easy for computer programming.

Disadvantages

(i) The method may diverge if the initial approximation is not very close to the root.

(ii) The iterative formula requires derivatives of $f(x)$ at every iterative value. But if $f(x)$ is of a complicated form, its derivative may be difficult to obtain. Moreover, if $f'(x)$ becomes zero or tends to zero or is very small in the neighbourhood of the root, the Newton–Raphson method will fail.

SOLVED EXAMPLES

EXAMPLE 6.16 Using the bisection method, find the real root of the equation $x^3 - 5x + 1 = 0$, correct to 3 decimal places.

Solution: Let $f(x) = x^3 - 5x + 1$. We see that $f(2) = -1 < 0$ and $f(3) = 13 > 0$. So, there exists at least one positive root between 2 and 3. Hence, we set $[a_0, b_0] = [2, 3]$. The computations for finding the approximate value of the root by the bisection method are given in the following table:

n	a_n	b_n	$x_{n+1} = \frac{a_n + b_n}{2}$	$f(a_n)$	$f(b_n)$	$f(x_{n+1})$
0	2	3	2.5	$-ve$	$+ve$	$4.125 > 0$
1	2	2.5	2.25	$-ve$	$+ve$	$1.141 > 0$
2	2	2.25	2.125	$-ve$	$+ve$	$-0.029 < 0$
3	2.125	2.25	2.1875	$-ve$	$+ve$	$0.530 > 0$
4	2.125	2.1875	2.15625	$-ve$	$+ve$	$0.244 > 0$
5	2.125	2.15625	2.140625	$-ve$	$+ve$	$0.106 > 0$
6	2.125	2.140625	2.132812	$-ve$	$+ve$	$0.038 > 0$
7	2.125	2.132812	2.128906	$-ve$	$+ve$	$0.0042 > 0$
8	2.125	2.128906	2.126953	$-ve$	$+ve$	$-0.026 < 0$
9	2.126953	2.128906	2.12793	$-ve$	$+ve$	$-0.0042 < 0$
10	2.12793	2.128906	2.12842	$-ve$	$+ve$	$0.000008 > 0$
11	2.12793	2.12842	2.12817	–	–	–

We see that in column x_{n+1}, two values 2.12842 and 2.12817 are correct to 3 decimal places. Hence, the required root is **2.128** (correct to 3 decimal places).

EXAMPLE 6.17 Use the bisection method to find the real root of the equation $x^3 - 9x + 1 = 0$, correct to 2 decimal places.

Solution: Let $f(x) = x^3 - 9x + 1$. We see that $f(2) = -9 < 0$ and $f(3) = 1 > 0$. So, there exists at least one root between 2 and 3. Hence, we set $[a_0, b_0] = [2, 3]$. The computations for finding the approximate value of the root by the bisection method are given in the following table:

n	a_n	b_n	$x_{n+1} = \frac{a_n + b_n}{2}$	$f(a_n)$	$f(b_n)$	$f(x_{n+1})$
0	2	3	2.5	$-ve$	$+ve$	$-5.8 < 0$
1	2.5	3	2.75	$-ve$	$+ve$	$-2.9 < 0$
2	2.75	3	2.875	$-ve$	$+ve$	$-1.03 < 0$
3	2.875	3	2.9375	$-ve$	$+ve$	$-0.05 < 0$
4	2.9375	3	2.96875	$-ve$	$+ve$	$0.47 > 0$
5	2.9375	2.96875	2.95312	$-ve$	$+ve$	$0.18 > 0$
6	2.9375	2.95312	2.94531	$-ve$	$+ve$	$0.042 > 0$

Cont'd

Table cont'd

n	a_n	b_n	$x_{n+1} = \dfrac{a_n + b_n}{2}$	$f(a_n)$	$f(b_n)$	$f(x_{n+1})$
7	2.9375	2.94531	2.941405	$-ve$	$+ve$	$-0.024 < 0$
8	2.941405	2.94531	2.94336	–	–	–

We see that in column x_{n+1}, two values 2.941405 and 2.94336 are correct to 2 decimal places. Hence, the required root is **2.94** (correct to 2 decimal places).

EXAMPLE 6.18 Evaluate $\sqrt{12}$, correct to 3 decimal places, using the Newton–Raphson method.

Solution: Let $x = \sqrt{12}$. Then, we have $x^2 - 12 = 0$. Let $f(x) = x^2 - 12$. We see that $f(3) = -3 < 0$ and $f(4) = 4 > 0$. So, the value of $\sqrt{12}$ lies between 3 and 4. To set the initial approximation, we make use of the bisection method, i.e., $x_0 = \dfrac{3+4}{2} = 3.5$. Now, the iterative formula for finding the square root of 12 is given by

$$x_{n+1} = \frac{1}{2}\left[x_n + \frac{12}{x_n}\right]$$

The computations for finding the approximate value of the root by the Newton–Raphson method are given in the following table:

n	x_n	$x_{n+1} = \dfrac{1}{2}\left[x_n + \dfrac{12}{x_n}\right]$
0	3.5	3.46429
1	3.46429	3.464102
2	3.464102	3.464102

As the last two consecutive iterative values of x_{n+1} are correct to 6 decimal places, the approximate value of $\sqrt{12}$ is **3.464** (correct to 3 decimal places).

EXAMPLE 6.19 Evaluate 1/31, correct to 3 places of decimal, using the Newton–Raphson method.

Solution: Let $x = 1/31$. Then, we have $\dfrac{1}{x} - 31 = 0$. Let $f(x) = \dfrac{1}{x} - 31$. We see that $f(x)$ is $-$ve for $\dfrac{1}{x} = 30$, i.e., for $x = 0.03333$ and +ve for $\dfrac{1}{x} = 32$, i.e., for $x = 0.03125$. So, the value of 1/31 lies between 0.03333 and 0.03125. To set the initial approximation, we make use of the bisection method, i.e., $x_0 = \dfrac{0.03333 + 0.03125}{2} = 0.03229$. Now, the iterative formula for finding the reciprocal of 31 is given by

$$x_{n+1} = x_n(2 - 31x_n)$$

The computations for finding the approximate value by the Newton–Raphson method are given in the following table:

n	x_n	$x_{n+1} = x_n(2 - 31x_n)$
0	0.03229	0.032258
1	0.032258	0.032258

As the last two consecutive iterative values of x_{n+1} are correct to 6 decimal places, the approximate value of 1/31 is **0.032** (correct to 3 decimal places).

EXAMPLE 6.20 Find the positive real root of $x^3 = 18$ using the bisection method of four iterations.

Solution: Let $f(x) = x^3 - 18$. We see that $f(2) = -10 < 0$ and $f(3) = 9 > 0$. So, there exists at least one root between 2 and 3. Hence, we set $[a_0, b_0] = [2, 3]$. The computations for finding the approximate value of the root by the bisection method are given in the following table:

n	a_n	b_n	$x_{n+1} = \dfrac{a_n + b_n}{2}$	$f(a_n)$	$f(b_n)$	$f(x_{n+1})$
0	2	3	2.5	$-ve$	$+ve$	$-2.375 < 0$
1	2.5	3	2.75	$-ve$	$+ve$	$2.8 > 0$
2	2.5	2.75	2.625	$-ve$	$+ve$	$0.09 > 0$
3	2.5	2.625	2.5625	$-ve$	$+ve$	$-1.17 < 0$
4	2.5625	2.625	2.59375	–	–	–

Hence, the required root is **2.6** (correct to 2 significant digits).

EXAMPLE 6.21 Find the root of the equation $x^3 - 2x - 5 = 0$ by the regula falsi method, correct to 4 decimal places.

Solution: Let $f(x) = x^3 - 2x - 5$. We see that $f(2) = -1 < 0$ and $f(3) = 16 > 0$. So, there exists at least one positive root between 2 and 3. Hence, we set $[a_0, b_0] = [2, 3]$. The computations for finding the approximate value of the root by the regula falsi method are given in the following table:

n	a_n	b_n	$x_{n+1} = \dfrac{a_n f(b_n) - b_n f(a_n)}{f(b_n) - f(a_n)}$	$f(a_n)$	$f(b_n)$	$f(x_{n+1})$
0	2	3	2.0588	$-ve$	$+ve$	$-0.3908 < 0$
1	2.0588	3	2.08126	$-ve$	$+ve$	$-0.1472 < 0$
2	2.08126	3	2.08964	$-ve$	$+ve$	$-0.0547 < 0$
3	2.08964	3	2.09274	$-ve$	$+ve$	$-0.0202 < 0$

Cont'd

Table cont'd

n	a_n	b_n	$x_{n+1} = \dfrac{a_n f(b_n) - b_n f(a_n)}{f(b_n) - f(a_n)}$	$f(a_n)$	$f(b_n)$	$f(x_{n+1})$
4	2.09274	3	2.09388	$-ve$	$+ve$	$-0.0075 < 0$
5	2.09388	3	2.09430	$-ve$	$+ve$	$-0.0028 < 0$
6	2.09430	3	2.09445	$-ve$	$+ve$	$-0.00105 < 0$
7	2.09445	3	2.09451	$-ve$	$+ve$	$-0.00047 < 0$
8	2.09451	3	2.09454	$-ve$	$+ve$	$-0.00013 < 0$
9	2.09454	3	2.094547	–	–	–

We see that in column x_{n+1}, two values 2.09454 and 2.094547 are correct to 4 decimal places. Hence, the required root is **2.0945** (correct to 4 decimal places).

EXAMPLE 6.22 Find the value of 1/23 by the Newton–Raphson method. The result must be corrected to 4 decimal places.

Solution: Let $x = 1/23$. Then, we have $\dfrac{1}{x} - 23 = 0$. Let $f(x) = \dfrac{1}{x} - 23$. We see that $f(x)$ is $-$ve for $\dfrac{1}{x} = 20$, i.e., for $x = 0.05$ and +ve for $\dfrac{1}{x} = 25$, i.e., for $x = 0.04$. So, the value of 1/31 lies between 0.04 and 0.05. To set the initial approximation, we make use of the bisection method, i.e., $x_0 = \dfrac{0.04 + 0.05}{2} = 0.045$. Now, the iterative formula for finding the reciprocal of 23 is given by

$$x_{n+1} = x_n(2 - 23x_n)$$

The computations for finding the approximate value by the Newton–Raphson method are given in the following table:

n	x_n	$x_{n+1} = x_n(2 - 23x_n)$
0	0.045	0.043425
1	0.043425	0.043478
2	0.043478	0.043478

As the last two consecutive iterative values of x_{n+1} are correct to 6 decimal places, the approximate value of 1/23 is **0.0435** (correct to 4 decimal places).

EXAMPLE 6.23 Find the root of the equation $x \tan x = 1.28$ that lies in the interval (0, 1), correct to 3 decimal places, using the bisection method.

Solution: Let $f(x) = x \tan x - 1.28$. We see that $f(0) = -1.28 < 0$ and $f(1) = 0.2774 > 0$. So, there exists at least one positive root between 0 and 1. Hence, we set $[a_0, b_0] = [0, 1]$. The computations for finding the approximate value of the root by the bisection method are given in the following table:

n	a_n	b_n	$x_{n+1} = \frac{a_n + b_n}{2}$	$f(a_n)$	$f(b_n)$	$f(x_{n+1})$
0	0	1	0.5	$-ve$	$+ve$	$-1.007 < 0$
1	0.5	1	0.75	$-ve$	$+ve$	$-0.581 < 0$
2	0.75	1	0.875	$-ve$	$+ve$	$-0.232 < 0$
3	0.875	1	0.9375	$-ve$	$+ve$	$-0.003 < 0$
4	0.9375	1	0.96875	$-ve$	$+ve$	$0.130 > 0$
5	0.9375	0.96875	0.953125	$-ve$	$+ve$	$0.062 > 0$
6	0.9375	0.953125	0.945312	$-ve$	$+ve$	$0.029 > 0$
7	0.9375	0.945312	0.941406	$-ve$	$+ve$	$0.013 > 0$
8	0.9375	0.941406	0.939453	$-ve$	$+ve$	$0.005 > 0$
9	0.9375	0.939453	0.938477	$-ve$	$+ve$	$0.001 > 0$
10	0.9375	0.938477	0.937989	$-ve$	$+ve$	$-0.001 < 0$
11	0.937989	0.938477	0.938232	–	–	–

We see that in column x_{n+1}, two values 2.12842 and 2.12817 are correct to 3 decimal places. Hence, the required root is **2.128** (correct to 3 decimal places).

MULTIPLE CHOICE QUESTIONS

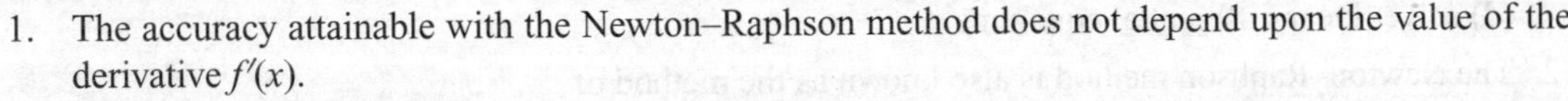

1. The accuracy attainable with the Newton–Raphson method does not depend upon the value of the derivative $f'(x)$.
(a) true (b) false

2. The regula falsi method, used for finding the real root of a numerical equation, is an/a
(a) analytical method (b) graphical method
(c) iterative method (d) none of these

3. The convergence of the bisection method is
(a) linear (b) quadratic (c) cubic (d) none of these

4. If $f(x)$ is continuous in the interval (a, b) and if $f(a)$ and $f(b)$ are of opposite signs, then there is
(a) at least one root of $f(x) = 0$ between a and b
(b) at most one root of $f(x) = 0$ between a and b
(c) no real root of $f(x) = 0$ between a and b
(d) none of these

5. Which of the following is a disadvantage of the regula falsi method?
(a) it is very difficult (b) it is very complex
(c) it is very slow (d) none of these

6. The method of bisection is
   (a) conditionally convergent (b) always convergent
   (c) non-convergent (d) none of these
7. The condition for convergence of the Newton–Raphson method is
   (a) $|f(x)f'(x)| < \{f''(x)\}^2$
   (b) $|f(x)f''(x)| < \{f'(x)\}^2$
   (c) $|f(x)f'(x)| > \{f''(x)\}^2$
   (d) $|f(x)f''(x)| > \{f'(x)\}^2$
8. Which of the following does not always guarantee convergence?
   (a) secant method (b) bisection method
   (c) regula falsi method (d) Newton–Raphson method
9. In the bisection method, if $[a_0, b_0]$ be the initial interval, then the condition satisfied by a_0 and b_0 for the existence of a root for the equation $f(x) = 0$ is
   (a) $f(a_0).f(b_0) < 1$ (b) $f(a_0).f(b_0) > 1$
   (c) $f(a_0).f(b_0) < 0$ (d) $f(a_0).f(b_0) > 0$
10. According to the Newton–Raphson method, the algorithm for iteration x_{n+1} $(n = 0,1,2,\ldots)$ is
    (a) $x_{n+1} = x_n + \dfrac{f(x_n)}{f'(x_n)}$ (b) $x_{n+1} = x_n + \dfrac{f'(x_n)}{f(x_n)}$
    (c) $x_{n+1} = x_n - \dfrac{f(x_n)}{f'(x_n)}$ (d) $x_{n+1} = x_n - \dfrac{f'(x_n)}{f(x_n)}$
11. The regula falsi method is used to
    (a) find the root of a system of linear simultaneous equations
    (b) differentiate
    (c) find the root of an algebraic or transcendental equation
    (d) solve linear differential equations
12. The Newton–Raphson method is also known as the method of
    (a) straight line (b) tangent (c) normal (d) cyclic
13. The bisection method is ____________ but ____________.
    (a) fast, fails (b) fast, sometimes fails (c) fast, never fails (d) slow, never fails
14. Under the condition that $f(a), f(b)$ have opposite signs and $a < b$, the first approximation of one of the roots of $f(x) = 0$ by the regula falsi method is given by
    (a) $\dfrac{af(b)+bf(a)}{f(b)+f(a)}$ (b) $\dfrac{af(a)-bf(b)}{f(a)-f(b)}$
    (c) $\dfrac{af(b)-bf(a)}{f(b)-f(a)}$ (d) $\dfrac{af(a)+bf(b)}{f(b)+f(a)}$
15. The regula falsi method is
    (a) conditionally convergent (b) linearly convergent
    (c) divergent (d) none of these
16. The value of x that satisfies $f(x) = 0$ is called the
    (a) root of an equation $f(x) = 0$ (b) root of a function $f(x)$
    (c) zero of an equation $f(x) = 0$ (d) none of these

17. The Newton–Raphson method fails when
(a) $f'(x) = 1$ (b) $f'(x) = 0$ (c) $f'(x) = -1$ (d) $f''(x) = 0$

18. The regula falsi method has a convergence rate of the order of
(a) 2 (b) 1.62 (c) 1 (d) none of these

19. One root of the equation $x^2 + 2x - 2 = 0$ lies between
(a) 1 and 2 (b) 0 and 0.5 (c) 0.5 and 1 (d) none of these

20. Which of the following statements applies to the bisection method, used for finding roots of functions?
(a) convergence within a few iterations
(b) guaranteed to work for all continuous functions
(c) is faster than the Newton–Raphson method
(d) requires that there be no error in determining the sign of the function

21. The order of convergence of the bisection method is
(a) 1 (b) between 0 and 1 (c) –1 (d) between –1 and 0

22. The Newton–Raphson method has an order of convergence of
(a) 1 (b) 1.62 (c) 2 (d) 3

23. The Newton–Raphson iterative formula for finding the square root of a positive number R is

(a) $x_{n+1} = \frac{1}{2}\left[x_n + \frac{R}{x_n}\right]$ (b) $x_{n+1} = \frac{1}{2}\left[x_n - \frac{R}{x_n}\right]$

(c) $x_{n+1} = x_n + \frac{R}{x_n}$ (d) $x_{n+1} = x_n - \frac{R}{x_n}$

24. The Newton–Raphson iterative formula for $\frac{1}{N}$ (where N is a positive number) is given by

(a) $x_{n+1} = Nx_n$ (b) $x_{n+1} = x_n(2 - Nx_n)$ (c) $x_{n+1} = x_n(2 + Nx_n)$ (d) $x_{n+1} = Nx_{n-1}$

Answers

1. (b) 2. (c) 3. (a) 4. (a) 5. (c) 6. (b) 7. (b) 8. (d) 9. (c) 10. (c) 11. (c)
12. (b) 13. (d) 14. (c) 15. (a) 16. (a) 17. (b) 18. (b) 19. (c) 20. (b) 21. (a) 22. (c)
23. (a) 24. (b)

SHORT ANSWER QUESTIONS

1. Show that the Newton–Raphson method has a quadratic rate of convergence.
2. Find an expression of the inherent error in the Newton–Raphson method.
3. Explain the geometrical significance of the Newton–Raphson method.
4. State the condition of convergence of the Newton–Raphson method.
5. Explain the Newton–Raphson method for computing the real root of an equation $f(x) = 0$.
6. Discuss the advantages and disadvantages of the regula falsi method.
7. Show that the bisection method is linearly convergent.

8. Interpret the regula falsi method geometrically. State the advantages and disadvantages of this method.
9. Evaluate $\sqrt{12}$ to 3 decimal places by the Newton–Raphson method.
10. Find the approximate value of $\sqrt{5}$ using the Newton–Raphson method.
11. Find the cube root of 10 using the Newton–Raphson method.
12. Find the value of 1/23 using the Newton–Raphson method. The result must be corrected up to 4 decimal places.
13. Using the Newton–Raphson method, find $\sqrt[4]{32}$ correct to 4 decimal places.
14. Find the value of $(19)^{\frac{1}{3}}$ correct to 4 decimal points by the Newton–Raphson method.
15. Derive the order of convergence for the Newton–Raphson method.
16. Find the positive real root of $x^3 = 18$ using the bisection method of 4 iterations.
17. Using the Newton–Raphson method, show that the iteration formula for finding the mth root of a positive real number R in $x_{n+1} = \dfrac{(m-1)x_n^{\ m} + R}{mx_n^{\ m-1}}$.
18. Derive the iterative formula for finding the reciprocal of a positive real number R by applying the Newton–Raphson method.
19. Obtain the order of convergence of the regula falsi method.
20. Find the positive real root of $x^3 - x^2 - 1 = 0$ using the bisection method of 4 iterations.
21. Using the method of bisection, compute a root of $x^3 - x - 1 = 0$ correct to 2 significant digits.

Answers

9. 3.464 10. 2.236 11. 2.154 12. 0.0435 13. 2.3784 14. 2.6684 16. 2.6207 20. 1.466
21. 1.3

LONG ANSWER QUESTIONS

1. Find the real root of the following algebraic equations, correct to 3 decimal places by the bisection method:
   (i) $x^3 - 2x - 5 = 0$
   (ii) $x^3 - 3x - 5 = 0$
   (iii) $x^3 - 5x - 7 = 0$
   (iv) $x^4 - x - 10 = 0$
   (v) $x^3 - 9x + 1 = 0$
2. Find the real root of the following algebraic equations, correct to 4 significant figures by the bisection method:
   (i) $x^3 - 2x - 9 = 0$
   (ii) $x^4 - 5x - 17 = 0$
   (iii) $x^3 - 5x^2 - 7 = 0$
   (iv) $x^3 - 3x - 6.5 = 0$
   (v) $x^2 + 4.5x - 21 = 0$

3. Find the positive root of the equation $x^3 - 3x + 1.06 = 0$ by the method of bisection, correct to 3 significant figures.
4. Find the smallest real roots of the following equations using the regula falsi method, correct to 3 decimal places:
   (i) $x^3 - 2x - 5 = 0$
   (ii) $x^3 - 4x + 1 = 0$
   (iii) $x^3 - 3x - 6 = 0$
   (iv) $x^3 + x^2 - 1 = 0$
   (v) $x^3 - x - 0.1 = 0$
5. Find the smallest real roots of the following equations using the regula falsi method, correct to 4 significant digits:
   (i) $x^3 - 2.1x - 9 = 0$
   (ii) $x^2 + 7x - 25 = 0$
   (iii) $x^3 - x - 3 = 0$
   (iv) $x^3 - 3x + 1 = 0$
   (v) $x^4 + x - 35.4 = 0$
6. Find the real root of the equation $x^3 - 3x - 5 = 0$ by the method of false position, correct to 3 decimal places.
7. Find the root of the equation $x^3 - 2x - 5 = 0$ by the regula falsi method, correct to 4 decimal places
8. Find the root of the equation $x^3 - 5x - 7 = 0$ between 2 and 3 by the regula falsi method, correct to 4 decimal places.
9. Find the root of the equation $x^3 + x^2 + x + 7 = 0$ using the regula falsi method.
10. Find the real root of the equation $x^3 - 2x - 5 = 0$ using the regula falsi method.
11. Find the real root of each of the following equations, correct to 4 decimal places by the Newton–Raphson method:
    (i) $x^3 + 2x - 6 = 0$
    (ii) $3x^3 + 5x - 7 = 0$
    (iii) $2x^3 + x - 5 = 0$
    (iv) $x^4 - x - 3 = 0$
    (v) $3x^3 - 2x - 12.7 = 0$
12. Using the Newton–Raphson method, find the real root, correct to 5 significant figures, for the following equations:
    (i) $x^3 - 3x - 5 = 0$
    (ii) $x^4 - x - 9 = 0$
    (iii) $x^3 - 10x - 2.1 = 0$
    (iv) $x^3 + 2x - 8 = 0$
    (v) $x^3 - 2.1x + 5.6 = 0$
13. Find the smallest positive root of the equation $3x^3 - 9x^2 + 8 = 0$, correct to 4 decimal places using the Newton–Raphson method.
14. Find the positive root of $x^2 + 2x - 2 = 0$ by the Newton–Raphson method, correct to 3 significant figures.
15. Evaluate $\sqrt{132}$ to 3 decimal places by the Newton–Raphson method.

16. Construct the Newton–Raphson iterative formula to evaluate the mth root of a positive real number R and hence calculate
(i) $\sqrt{21}$ (ii) $\sqrt[3]{7}$ (iii) $\sqrt[4]{2}$ (iv) $\sqrt[3]{5}$ (v) $\sqrt{17}$ (vi) $\sqrt{31}$
17. Find the iterative method based on the Newton–Raphson method for finding $\sqrt{N}$, N is positive integer. Apply this method to $N = 18$ to obtain the result correct to 2 decimal places.
18. Find the Newton–Raphson iterative formula to find the pth root of positive number N and hence find the cube root of 17.
19. Use the Newton–Raphson iterative formula to evaluate the following. The result must be correct to 5 decimal places.
(i) $\sqrt[7]{21}$ (ii) $\sqrt[6]{17}$ (iii) $\sqrt[9]{2}$ (iv) $\sqrt[4]{53}$ (v) $\sqrt[5]{19}$ (vi) $\sqrt[3]{153}$
20. Use the Newton–Raphson method to evaluate $\sqrt[7]{111}$ correct to 4 significant figures.
21. Deduce the Newton–Raphson iteration for finding the reciprocal of a positive real number R and calculate
(i) 1/21 (ii) 1/31 (iii) 1/13 (iv) 1/19 (v) 1/15
22. Find the value of 1/23 by the Newton–Raphson method. The result must be corrected up to 4 decimal places.
23. Find a root of the equation $x^4 - x - 10 = 0$ that lies between 1 and 2 using the Newton–Raphson method correct to 3 decimal places.

Answers

1. (i) 2.095 (ii) 2.279 (iii) 2.747 (iv) 2.309 (v) 0.111
2. (i) 2.398 (ii) −1.706 (iii) 5.254 (iv) 2.391 (v) 2.855
3. 0.3703 4. (i) 2.095 (ii) −2.115 (iii) 2.355 (iv) 0.755 (v) −0.946
5. (i) 2.4141 (ii) −9.6033 (iii) 1.6717 (iv) −1.8794 (v) −2.4809
6. 2.279 7. 2.0946 8. 2.7473 9. −2.1049 10. 2.0946
11. (i) 1.4562 (ii) 0.9251 (iii) 1.2348 (iv) 1.4526 (v) 2.4177
12. (i) 2.2790 (ii) 1.8134 (iii) 3.2625 (iv) 1.6702 (v) −2.1649
13. 1.2261 14. 0.732 15. 11.4891
16. (i) 4.5826 (ii) 1.9129 (iii) 1.1892 (iv) 1.710 (v) 4.1231 (vi) 5.5678
17. 4.24 18. 2.5713
19. (i) 1.54486 (ii) 1.60352 (iii) 1.08006 (iv) 2.69817 (v) 1.80198 (vi) 5.34848
20. 1.9597 21. (i) 0.0476 (ii) 0.03226 (iii) 0.0769 (iv) 0.0526 (v) 0.0667
22. 0.0435 23. 1.856

7

Numerical Solution of a System of Linear Equations

7.1 INTRODUCTION

In modelling physical situations such as determining the potential in certain electrical networks, stress in a building frame, flow rates in a hydraulic system, etc., we may come across several equations instead of one equation. Also, in optimisation theory, we may have to solve a system of linear equations simultaneously. If each of these equations is linear, we call it a system of simultaneous linear equations.

There are a huge number of solution techniques in both analytical and numerical methods, provided the system has a solution. In this chapter, we discuss some well-known numerical techniques, some of which are known as **direct methods** while others are **iterative** in nature.

7.2 DIRECT METHODS

This type of method yields the exact solution after a finite number of steps in the absence of rounding errors. Direct methods are also called **elimination** or **reduction methods**. In this case, the number of computations can be specified in advance and the **desired accuracy is irrelevant**. In this section, we discuss some commonly used direct methods.

7.2.1 Gauss Elimination Method

This method is a systematic process of eliminating the unknowns so that the system is reduced to an equivalent upper triangular system which could be solved by back substitution.

Let us consider the following system of n linear equations with n unknowns

$$\begin{aligned} a_{11}x_1 + a_{12}x_2 + \cdots + a_{1n}x_n &= b_1 \\ a_{21}x_1 + a_{22}x_2 + \cdots + a_{2n}x_n &= b_2 \\ &\cdots \\ a_{n1}x_1 + a_{n2}x_2 + \cdots + a_{nn}x_n &= b_n \end{aligned}$$

where the coefficients a_{ij} $(i, j = 1, 2, 3, \ldots, n)$ and the constants b_i $(i = 1, 2, 3, \ldots, n)$ are known. The system can also be written in matrix form as

$$AX = B$$

where $=\left[a_{ij}\right]_{n\times n}=\begin{bmatrix} a_{11} & a_{12} & a_{13} & \dots & a_{1n} \\ a_{21} & a_{22} & a_{23} & \dots & a_{2n} \\ a_{31} & a_{32} & a_{33} & \dots & a_{3n} \\ \dots & \dots & \dots & \dots & \dots \\ a_{n1} & a_{n1} & a_{n1} & \dots & a_{n1} \end{bmatrix}$, known as the coefficient matrix

$$B=\begin{bmatrix} b_1 \\ b_2 \\ b_3 \\ \vdots \\ b_n \end{bmatrix},$$ known as the right-hand side vector is given, while

$$X=\begin{bmatrix} x_1 \\ x_2 \\ x_3 \\ \vdots \\ x_n \end{bmatrix},$$ known as the solution vector is to be determined.

The Gauss elimination method consists of two steps:

Forward Elimination of Unknowns In this step

(i) x_1 is eliminated from each equation except the first equation. After that,
(ii) x_2 is eliminated from each equation except the first and second equations. Next
(iii) x_3 is eliminated from each equation except the first, second and third equations, *and so on*.

The process is executed in a systematic manner. First, a matrix is constructed with the coefficient matrix A, appended with the constant matrix B. This matrix is known as the **augmented matrix** and is given by

$$\left[a_{ij}\,|b_i\right]_{n\times(n+1)}=\left[\begin{array}{ccccc|c} a_{11} & a_{12} & a_{13} & \dots & a_{1n} & b_1 \\ a_{21} & a_{22} & a_{23} & \dots & a_{2n} & b_2 \\ a_{31} & a_{32} & a_{33} & \dots & a_{3n} & b_3 \\ \dots & \dots & \dots & \dots & \dots & \dots \\ a_{n1} & a_{n1} & a_{n1} & \dots & anl & b_n \end{array}\right] \tag{1}$$

The aim is to reduce this augmented matrix $[a_{ij}:b_i]_{n\times(n+1)}$ to an upper triangular matrix.

Step 1: Assume $a_{11}\neq 0$. Now, multiply the first row of (1) by $\dfrac{a_{i1}}{a_{11}}$ and subtract from the ith row ($i=2,3,\dots,n$). By this, all elements in the first column except a_{11} are made zero. Thus, (1) takes the form

$$\left[\begin{array}{ccccc|c} a_{11} & a_{12} & a_{13} & \dots & a_{1n} & b_1 \\ 0 & a_{22}^{(1)} & a_{23}^{(1)} & \dots & a_{2n}^{(1)} & b_2^{(1)} \\ 0 & a_{32}^{(1)} & a_{33}^{(1)} & \dots & a_{3n}^{(1)} & b_3^{(1)} \\ \dots & \dots & \dots & \dots & \dots & \dots \\ 0 & a_{n2}^{(1)} & a_{n3}^{(1)} & \dots & a_{nn}^{(1)} & b_n^{(1)} \end{array}\right] \tag{2}$$

where
$$a_{ij}^{(1)} = a_{ij} - a_{1j}\frac{a_{i1}}{a_{11}}, \quad i, j = 2, 3, \ldots, n$$

and
$$b_i^{(1)} = b_i - b_1\frac{a_{i1}}{a_{11}}, \quad i = 2, 3, \ldots, n$$

Step 2: Assume $a_{22}^{(1)} \neq 0$. Multiply the second row of (2) by $\dfrac{a_{i2}^{(1)}}{a_{22}^{(1)}}$ and subtract from the ith row (i = 3, 4, …, n). By this, all elements in the second column except a_{12} and $a_{22}^{(1)}$ are made zero. Thus, (2) reduces to

$$\left[\begin{array}{cccccc|c} a_{11} & a_{12} & a_{13} & \ldots & a_{1n} & b_1 \\ 0 & a_{22}^{(1)} & a_{23}^{(1)} & \ldots & a_{2n}^{(1)} & b_2^{(1)} \\ 0 & 0 & a_{33}^{(2)} & \ldots & a_{3n}^{(2)} & b_3^{(2)} \\ \ldots & \ldots & \ldots & \ldots & \ldots & \ldots \\ 0 & 0 & a_{n3}^{(2)} & \ldots & a_{nn}^{(2)} & b_n^{(2)} \end{array}\right] \tag{3}$$

where
$$a_{ij}^{(2)} = a_{ij}^{(1)} - a_{2j}^{(1)}\frac{a_{i2}^{(1)}}{a_{22}^{(1)}}, \quad i, j = 3, 4, \ldots, n$$

and
$$b_i^{(2)} = b_i^{(1)} - b_2^{(1)}\frac{a_{i2}^{(1)}}{a_{22}^{(1)}}, \quad i = 3, 4, \ldots, n$$

Continuing this process, all elements below the leading diagonal elements of A are made zero and we get

$$\left[a_{ij} \,|\, b_i\right]_{n\times(n+1)} \sim \left[\begin{array}{ccccc|c} a_{11} & a_{12} & a_{13} & \ldots & a_{1n} & b_1 \\ 0 & a_{22}^{(1)} & a_{23}^{(1)} & \ldots & a_{2n}^{(1)} & b_2^{(1)} \\ 0 & 0 & a_{33}^{(2)} & \ldots & a_{3n}^{(2)} & b_3^{(2)} \\ \ldots & \ldots & \ldots & \ldots & \ldots & \ldots \\ 0 & 0 & 0 & \ldots & a_{nn}^{(n-1)} & b_n^{(n-1)} \end{array}\right] \tag{4}$$

The coefficients of the leading terms in (4), i.e., the elements $a_{11}, a_{22}^{(1)}, a_{33}^{(2)}, \ldots, a_{nn}^{(n-1)}$ which have been made to be **non-zero** are known as the **pivotal elements** or **pivots**. In the elimination process, if any one of the pivotal elements $a_{11}, a_{22}^{(1)}, a_{33}^{(2)}, \ldots, a_{nn}^{(n-1)}$ vanishes, then we must attempt to rearrange the rows so as to obtain a non-vanishing pivot. This strategy is called **pivoting**.

Back Substitution In this step, starting from the nth equation, each of the unknowns is found. From (4), we can write the equivalent system as

$$a_{11}x_1 + a_{12}x_2 + \cdots + a_{1n}x_n = b_1$$
$$a_{22}^{(1)}x_2 + \cdots + a_{2n}^{(1)}x_n = b_2^{(1)}$$
$$\ldots$$
$$a_{nn}^{(n-1)}x_n = b_n^{(n-1)}$$

Now,

(i) calculate x_n from the nth equation. Then,
(ii) calculate x_{n-1} from the $(n - 1)$th equation substituting x_n with its value, obtained from the nth equation, *and so on*.

Thus, with this back substitution procedure, the solution for each unknown x_i $(i = n, n-1, n-2, \ldots 2, 1)$ is obtained.

Operational Counts in Gauss Elimination In this method, the total number of multiplications and divisions needed for computations is $\frac{n^3}{3} + n^2 - \frac{n}{3}$ and hence for a large n, the time required for these computations is proportional to n^3.

This method is best understood by examining the following examples.

EXAMPLE 7.1 Solve the system of linear equations by the Gauss elimination method:

$$5x_1 - x_2 = 9$$
$$-x_1 + 5x_2 - x_3 = 4$$
$$-x_2 + 5x_3 = -6$$

Solution: The given system is equivalent to

$$\begin{bmatrix} 5 & -1 & 0 \\ -1 & 5 & -1 \\ 0 & -1 & 5 \end{bmatrix}\begin{bmatrix} x_1 \\ x_2 \\ x_3 \end{bmatrix} = \begin{bmatrix} 9 \\ 4 \\ -6 \end{bmatrix}$$

The augmented matrix is

$$\left[\begin{array}{ccc|c} 5 & -1 & 0 & 9 \\ -1 & 5 & -1 & 4 \\ 0 & -1 & 5 & -6 \end{array}\right]$$

$$\sim \left[\begin{array}{ccc|c} 5 & -1 & 0 & 9 \\ 0 & 24/5 & -1 & 29/5 \\ 0 & -1 & 5 & -6 \end{array}\right], \text{ by } R_2' = R_2 + \frac{R_1}{5}$$

$$\sim \left[\begin{array}{ccc|c} 5 & -1 & 0 & 9 \\ 0 & 24/5 & -1 & 29/5 \\ 0 & 0 & 115/24 & -115/24 \end{array}\right], \text{ by } R_3' = R_3 + \frac{5R_2}{24}$$

Therefore

$$\frac{115}{24}x_3 = -\frac{115}{24}, \text{ i.e., } x_3 = -1$$

$$\frac{24}{5}x_2 - x_3 = \frac{29}{5}, \text{ i.e., } x_2 = 1$$

$$5x_1 - x_2 = 9, \text{ i.e., } x_1 = 2.$$

Hence, the required solution is $(2, 1, -1)$.

EXAMPLE 7.2 Solve the following system of equations by the Gauss elimination method:

$$5x - y + z = 10$$
$$2x + 4y = 12$$
$$x + y + 5z = -1$$

Solution: The given system is equivalent to

$$\begin{bmatrix} 5 & -1 & 1 \\ 2 & 4 & 0 \\ 1 & 1 & 5 \end{bmatrix} \begin{bmatrix} x \\ y \\ z \end{bmatrix} = \begin{bmatrix} 10 \\ 12 \\ -1 \end{bmatrix}$$

The augmented matrix is
$$\left[\begin{array}{ccc|c} 5 & -1 & 1 & 10 \\ 2 & 4 & 0 & 12 \\ 1 & 1 & 5 & -1 \end{array}\right]$$

$$\sim \left[\begin{array}{ccc|c} 5 & -1 & 1 & 10 \\ 0 & 22/5 & -2/5 & 8 \\ 0 & 6/5 & 24/5 & -3 \end{array}\right], \text{ by } R_2' - R_2 - \frac{2R_1}{5}$$

and
$$R_3' = R_3 - \frac{R_1}{5}$$

$$\sim \left[\begin{array}{ccc|c} 5 & -1 & 1 & 10 \\ 0 & 22/5 & -2/5 & 8 \\ 0 & 0 & 54/11 & -57/11 \end{array}\right], \text{ by } R_3' = R_3 - \frac{3R_2}{11}$$

Therefore
$$\frac{54}{11}z = -\frac{57}{11}, \text{ i.e., } z = -1.056$$

$$\frac{22}{5}y - \frac{2}{5}z = 8, \text{ i.e., } y = 1.722$$

$$5x - y + z = 10, \text{ i.e., } x = 2.556.$$

Hence, the required solution is (2.556, 1.722, −1.056).

EXAMPLE 7.3 Use the Gauss elimination method to solve

$$3x - 2y + 2z = 12$$
$$x + 2y + 3z = 11$$
$$2x - 2y - z = 3$$

Solution: The given system is equivalent to

$$\begin{bmatrix} 3 & -2 & 2 \\ 1 & 2 & 3 \\ 2 & -2 & -1 \end{bmatrix} \begin{bmatrix} x \\ y \\ z \end{bmatrix} = \begin{bmatrix} 12 \\ 11 \\ 3 \end{bmatrix}$$

The augmented matrix is

$$\left[\begin{array}{ccc|c} 3 & -2 & 2 & 12 \\ 1 & 2 & 3 & 11 \\ 2 & -2 & -1 & 3 \end{array}\right]$$

$$\sim \left[\begin{array}{ccc|c} 5 & -2 & 2 & 12 \\ 0 & 8/3 & 7/3 & 7 \\ 0 & -2/3 & -7/3 & -5 \end{array}\right], \text{ by } R_2' = R_2 - \frac{R_1}{3}$$

and

$$R_3' = R_3 - \frac{2R_1}{3}$$

$$\sim \left[\begin{array}{ccc|c} 5 & -2 & 2 & 12 \\ 0 & 8/3 & 7/3 & 7 \\ 0 & 0 & -7/4 & -13/4 \end{array}\right], \text{ by } R_3' = R_3 + \frac{R_2}{4}$$

Therefore

$$-\frac{7}{4}z = -\frac{13}{4}, \text{ i.e., } z = 1.86$$

$$\frac{8}{3}y + \frac{7}{3}z = 7, \text{ i.e., } y = 1.00$$

$$5x - 2y + 2z = 12, \text{ i.e., } x = 3.43.$$

Hence, the required solution is (3.43, 1.00, 1.86).

7.2.2 Gauss–Jordan Method

This method is a modified form of the Gauss elimination method, where the coefficient matrix is reduced to a diagonal matrix by applying forward and backward elimination. Thus, in the reduced equivalent system, each equation contains only one unknown. This is why back substitution is not required in this method.

EXAMPLE 7.4 Solve the following system of equations by the Gauss–Jordan method:

$$2x + 8y + 2z = 14$$

$$6x + 6y - z = 13$$

$$2x - y + 2z = 5$$

Solution: The given system is equivalent to

$$\begin{bmatrix} 2 & 8 & 2 \\ 6 & 6 & -1 \\ 2 & -1 & 2 \end{bmatrix} \begin{bmatrix} x \\ y \\ z \end{bmatrix} = \begin{bmatrix} 14 \\ 13 \\ 5 \end{bmatrix}$$

The augmented matrix is

$$\left[\begin{array}{ccc|c} 2 & 8 & 2 & 14 \\ 6 & 6 & -1 & 13 \\ 2 & -1 & 2 & 5 \end{array}\right]$$

$$\sim \left[\begin{array}{ccc|c} 2 & 8 & 2 & 14 \\ 0 & -18 & -7 & -29 \\ 0 & -9 & 0 & -9 \end{array}\right], \text{ by } R_2' = R_2 - 3R_1$$

and

$$R_3' = R_3 - R_1$$

$$\sim \left[\begin{array}{ccc|c} 2 & 8 & 2 & 14 \\ 0 & -18 & -7 & -29 \\ 0 & 0 & 7/2 & 11/2 \end{array}\right], \text{ by } R_3' = R_3 - \frac{3R_2}{11}$$

$$\sim \left[\begin{array}{ccc|c} 2 & 8 & 0 & 76/7 \\ 0 & -18 & 0 & -18 \\ 0 & 0 & 7/2 & 11/2 \end{array}\right], \text{ by } R_1' = R_1 - \frac{4R_3}{7}$$

and

$$R_2' = R_2 + 2R_3$$

$$\sim \left[\begin{array}{ccc|c} 2 & 0 & 0 & 20/7 \\ 0 & -18 & 0 & -18 \\ 0 & 0 & 7/2 & 11/2 \end{array}\right], \text{ by } R_1' = R_1 + \frac{8R_2}{18}$$

Thus, we get

$$2x = \frac{20}{7}, \quad -18y = -18$$

and

$$\frac{7}{2}z = \frac{11}{2}$$

i.e.

$$x = \frac{10}{7}, \quad y = 1$$

$$z = \frac{11}{7}$$

Hence, the required solution is $\left(\frac{10}{7}, 1, \frac{11}{7}\right)$

EXAMPLE 7.5 Solve the following system of equations by the Gauss–Jordan method:

$$10x + y + z = 12$$
$$2x + 10y + z = 13$$
$$x + y + 5z = 7$$

Solution: The given system is equivalent to

$$\begin{bmatrix} 10 & 1 & 1 \\ 2 & 10 & 1 \\ 1 & 1 & 5 \end{bmatrix}\begin{bmatrix} x \\ y \\ z \end{bmatrix} = \begin{bmatrix} 12 \\ 13 \\ 7 \end{bmatrix}$$

The augmented matrix is

$$\left[\begin{array}{ccc|c} 10 & 1 & 1 & 12 \\ 2 & 10 & 1 & 13 \\ 1 & 1 & 5 & 7 \end{array}\right] \sim \left[\begin{array}{ccc|c} 1 & -8 & -44 & -51 \\ 2 & 10 & 1 & 13 \\ 1 & 1 & 5 & 7 \end{array}\right], \text{ by } R_1' = R_1 - 9R_3$$

$$\sim\left[\begin{array}{ccc|c}1 & -8 & -44 & -51\\0 & 26 & 89 & 115\\0 & 9 & 49 & 58\end{array}\right], \text{ by } R_2' = R_2 - 2R_1$$

and $$R_3' = R_3 - R_1$$

$$\sim\left[\begin{array}{ccc|c}1 & -8 & -44 & -51\\0 & -1 & -58 & -59\\0 & 9 & 49 & 58\end{array}\right], \text{ by } R_2' = R_2 - 3R_3$$

$$\sim\left[\begin{array}{ccc|c}1 & -8 & -44 & -51\\0 & 1 & 58 & 59\\0 & 9 & 49 & 58\end{array}\right], \text{ by } R_2' = (-1)R_2$$

$$\sim\left[\begin{array}{ccc|c}1 & -8 & -44 & -51\\0 & 1 & 58 & 59\\0 & 0 & -473 & -473\end{array}\right], \text{ by } R_3' = R_3 - 9R_2$$

$$\sim\left[\begin{array}{ccc|c}1 & -8 & -44 & -51\\0 & 1 & 58 & 59\\0 & 0 & 1 & 1\end{array}\right], \text{ by } R_3' = R_3/(-473)$$

$$\sim\left[\begin{array}{ccc|c}1 & -8 & 0 & -7\\0 & 1 & 0 & 1\\0 & 0 & 1 & 1\end{array}\right], \text{ by } R_1' = R_1 + 44R_3$$

and $$R_2' = R_2 - 58R_3$$

$$\sim\left[\begin{array}{ccc|c}1 & 0 & 0 & 1\\0 & 1 & 0 & 1\\0 & 0 & 1 & 1\end{array}\right], \text{ by } R_1' = R_1 + 8R_2$$

Thus, we get $x = 1, y = 1$ and $z = 1$.
Hence, the required solution is (1, 1, 1).

7.2.3 LU Factorisation Method

This method is based on the fact that if all the principal minors of a given square matrix $A = [a_{ij}]_{n \times n}$ are non-singular, i.e., if

$$a_{11} \neq 0, \begin{vmatrix}a_{11} & a_{12}\\a_{21} & a_{22}\end{vmatrix} \neq 0,$$

$$\begin{vmatrix}a_{11} & a_{12} & a_{13}\\a_{21} & a_{22} & a_{23}\\a_{31} & a_{32} & a_{33}\end{vmatrix} \neq 0, \cdots \begin{vmatrix}a_{11} & a_{12} & a_{13} & \dots & a_{1n}\\a_{21} & a_{22} & a_{23} & \dots & a_{2n}\\a_{31} & a_{32} & a_{33} & \dots & a_{3n}\\\dots & \dots & \dots & \dots & \dots\\a_{n1} & a_{n1} & a_{n1} & \dots & a_{n1}\end{vmatrix} \neq 0$$

then the matrix A can be factorised into the form LU; i.e., $A = LU$ where L is a lower triangular square matrix and U is an upper triangular square matrix.

The method is best understood if we consider $n = 3$. Let us consider a system of three unknowns and three equations as

$$AX = B$$

where
$$A = \begin{bmatrix} a_{11} & a_{12} & a_{13} \\ a_{21} & a_{22} & a_{23} \\ a_{31} & a_{32} & a_{33} \end{bmatrix}$$

$$X = \begin{bmatrix} x_1 \\ x_2 \\ x_3 \end{bmatrix}$$

and
$$B = \begin{bmatrix} b_1 \\ b_2 \\ b_3 \end{bmatrix}$$

such that
$$a_{11} \neq 0, \quad \begin{vmatrix} a_{11} & a_{12} \\ a_{21} & a_{22} \end{vmatrix} \neq 0$$

$$\begin{vmatrix} a_{11} & a_{12} & a_{13} \\ a_{21} & a_{22} & a_{23} \\ a_{31} & a_{32} & a_{33} \end{vmatrix} \neq 0$$

Then, we can write
$$A = LU$$

where $L = \begin{bmatrix} l_{11} & 0 & 0 \\ l_{21} & l_{22} & 0 \\ l_{31} & l_{32} & l_{33} \end{bmatrix}$ and $U = \begin{bmatrix} u_{11} & u_{12} & u_{13} \\ 0 & u_{11} & u_{23} \\ 0 & 0 & u_{11} \end{bmatrix}$ are to be determined and either $l_{ij} = 1$ or $u_{ij} = 1$.

When $l_{ij} = 1$, the factorisation method is called Doolittle's method and when $U_{ij} = 1$, we call it Crout's method. In this context, we discuss Crout's method only.

Let
$$L = \begin{bmatrix} l_{11} & 0 & 0 \\ l_{21} & l_{22} & 0 \\ l_{31} & l_{32} & l_{33} \end{bmatrix}$$

and
$$U = \begin{bmatrix} 1 & u_{12} & u_{13} \\ 0 & 1 & u_{23} \\ 0 & 0 & 1 \end{bmatrix}$$

Then, the system will be
$$LUX = B$$

Let
$$UX = Y \tag{1}$$

where
$$Y = \begin{bmatrix} y_1 \\ y_2 \\ y_3 \end{bmatrix}$$ can be determined by

$$LY = B \tag{2}$$

So, let us determine L and U first. To do this, we have $A = LU$,

i.e.
$$\begin{bmatrix} a_{11} & a_{12} & a_{13} \\ a_{21} & a_{22} & a_{23} \\ a_{31} & a_{32} & a_{33} \end{bmatrix} = \begin{bmatrix} l_{11} & 0 & 0 \\ l_{21} & l_{22} & 0 \\ l_{31} & l_{32} & l_{33} \end{bmatrix} \begin{bmatrix} 1 & u_{12} & u_{13} \\ 0 & 1 & u_{23} \\ 0 & 0 & 1 \end{bmatrix}$$

$$= \begin{bmatrix} l_{11} & l_{11}u_{12} & l_{11}u_{13} \\ l_{21} & l_{21}u_{12} + l_{22} & l_{21}u_{13} + l_{22}u_{23} \\ l_{31} & l_{31}u_{12} + l_{32} & l_{31}u_{13} + l_{32}u_{23} + l_{33} \end{bmatrix}$$

On comparison, it gives

$$l_{11} = a_{11}, l_{21} = a_{21}, l_{31} = a_{31}$$

$$u_{12} = \frac{a_{12}}{l_{11}} = \frac{a_{12}}{a_{11}}$$

$$u_{13} = \frac{a_{13}}{l_{11}} = \frac{a_{13}}{a_{11}}$$

$$l_{22} = a_{22} - l_{21}u_{12} = a_{22} - \frac{a_{21}a_{12}}{a_{11}}$$

$$l_{32} = a_{32} - l_{31}u_{12} = a_{32} - \frac{a_{21}a_{13}}{a_{11}}$$

$$u_{23} = \frac{a_{23} - l_{21}u_{13}}{l_{22}} = \frac{a_{23} - \dfrac{a_{21}a_{13}}{a_{11}}}{a_{22} - \dfrac{a_{21}a_{12}}{a_{11}}}$$

and
$$l_{33} = a_{33} - l_{31}u_{13} - l_{32}u_{23}$$

$$= a_{33} - \frac{a_{31}a_{13}}{a_{11}} - \left(a_{32} - \frac{a_{21}a_{13}}{a_{11}}\right)\left(\frac{a_{23} - \dfrac{a_{21}a_{13}}{a_{11}}}{a_{22} - \dfrac{a_{21}a_{12}}{a_{11}}}\right)$$

Thus, all the coefficients of L and U are known. Therefore, from (2), we get

$$\begin{bmatrix} l_{11} & 0 & 0 \\ l_{21} & l_{22} & 0 \\ l_{31} & l_{32} & l_{33} \end{bmatrix} \begin{bmatrix} y_1 \\ y_2 \\ y_3 \end{bmatrix} = \begin{bmatrix} b_1 \\ b_2 \\ b_3 \end{bmatrix}$$

which gives
$$y_1 = \frac{b_1}{l_{11}}, \quad y_2 = \frac{b_2 - l_{21}y_1}{l_{22}}$$

and
$$y_3 = \frac{b_3 - l_{31}y_1 - l_{32}y_2}{l_{33}}$$

Now, from (1), we can determine X by back substitution, since U and Y are known.

Operations Count for LU Factorisation As before, the total number of multiplications and divisions needed for computations is $\frac{n^3}{3}+\frac{2n^2}{3}-n$, and hence for a large n, the time required for executing these is proportional to n^3.

The following examples could be helpful in understanding this method.

EXAMPLE 7.6 Solve the following system of equations by the LU factorisation method:

$$3x + 4y + 2z = 15$$
$$5x + 2y + z = 18$$
$$2x + 3y + 2z = 10$$

Solution: The given system is of the form $AX = B$

where
$$A = \begin{bmatrix} 3 & 4 & 2 \\ 5 & 2 & 1 \\ 2 & 3 & 2 \end{bmatrix}$$

$$X = \begin{bmatrix} x \\ y \\ z \end{bmatrix}$$

and
$$B = \begin{bmatrix} 15 \\ 18 \\ 10 \end{bmatrix}$$

Let
$$A = LU$$

where
$$L = \begin{bmatrix} l_{11} & 0 & 0 \\ l_{21} & l_{22} & 0 \\ l_{31} & l_{32} & l_{33} \end{bmatrix}$$

and
$$U = \begin{bmatrix} 1 & u_{12} & u_{13} \\ 0 & 1 & u_{23} \\ 0 & 0 & 1 \end{bmatrix}$$

Therefore
$$\begin{bmatrix} 3 & 4 & 2 \\ 5 & 2 & 1 \\ 2 & 3 & 2 \end{bmatrix} = \begin{bmatrix} l_{11} & 0 & 0 \\ l_{21} & l_{22} & 0 \\ l_{31} & l_{32} & l_{33} \end{bmatrix} \begin{bmatrix} 1 & u_{12} & u_{13} \\ 0 & 1 & u_{23} \\ 0 & 0 & 1 \end{bmatrix}$$

$$= \begin{bmatrix} l_{11} & l_{11}u_{12} & l_{11}u_{13} \\ l_{21} & l_{21}u_{12} + l_{22} & l_{21}u_{13} + l_{22}u_{23} \\ l_{31} & l_{31}u_{12} + l_{32} & l_{31}u_{13} + l_{32}u_{23} + l_{33} \end{bmatrix}$$

Comparing the elements, we get

$$l_{11} = 3, l_{21} = 5, l_{31} = 2$$

$$l_{11}u_{12} = 4, \text{ i.e., } u_{12} = 4/3$$

$$l_{11}u_{13} = 2, \text{ i.e., } u_{13} = 2/3$$

$$l_{21}u_{12} + l_{22} = 2, \text{ i.e., } l_{22} = 2 - \frac{20}{3} = -14/3$$

$$l_{31}u_{12} + l_{32} = 3, \text{ i.e., } l_{32} = 3 - 8/3 = 1/3$$

$$l_{21}u_{13} + l_{22}u_{23} = 1, \text{ i.e., } u_{23} = \left(1 - 5 \times \frac{2}{3}\right) \times \left(-\frac{3}{14}\right) = 1/2$$

$$l_{31}u_{13} + l_{32}u_{23} + l_{33} = 2$$

$$\Rightarrow \qquad l_{33} = 2 - 2 \times \frac{2}{3} - \frac{1}{3} \times \frac{1}{2} = 1/2$$

Therefore, we have

$$L = \begin{bmatrix} 3 & 0 & 0 \\ 5 & -14/3 & 0 \\ 2 & 1/3 & 1/2 \end{bmatrix}$$

$$U = \begin{bmatrix} 1 & 4/3 & 2/3 \\ 0 & 1 & 1/2 \\ 0 & 0 & 1 \end{bmatrix}$$

and

$$\begin{bmatrix} 3 & 0 & 0 \\ 5 & -14/3 & 0 \\ 2 & 1/3 & 1/2 \end{bmatrix} \begin{bmatrix} 1 & 4/3 & 2/3 \\ 0 & 1 & 1/2 \\ 0 & 0 & 1 \end{bmatrix} \begin{bmatrix} x \\ y \\ z \end{bmatrix} = \begin{bmatrix} 15 \\ 18 \\ 10 \end{bmatrix}$$

Let

$$UX = Y$$

where

$$Y = \begin{bmatrix} y_1 \\ y_2 \\ y_3 \end{bmatrix}$$

Therefore

$$\begin{bmatrix} 3 & 0 & 0 \\ 5 & -14/3 & 0 \\ 2 & 1/3 & 1/2 \end{bmatrix} \begin{bmatrix} y_1 \\ y_2 \\ y_3 \end{bmatrix} = \begin{bmatrix} 15 \\ 18 \\ 10 \end{bmatrix}$$

which gives

$$3y_1 = 15, \text{ i.e., } y_1 = 5$$

$$5y_1 - \frac{14}{3}y_2 = 18, \text{ i.e., } y_2 = 3/2$$

$$2y_1 - \frac{1}{3}y_2 + \frac{1}{2}y_3 = 10, \text{ i.e., } y_3 = -1$$

Therefore, we get

$$\begin{bmatrix} 1 & 4/3 & 2/3 \\ 0 & 1 & 1/2 \\ 0 & 0 & 1 \end{bmatrix} \begin{bmatrix} x \\ y \\ z \end{bmatrix} = \begin{bmatrix} 5 \\ 3/2 \\ -1 \end{bmatrix}$$

which gives

$$z = -1$$

$$y+\frac{1}{2}z=\frac{3}{2},\ \text{i.e., } y=2$$

$$x+\frac{4}{3}y+\frac{2}{3}z=5,\ \text{i.e., } x=3$$

Hence, the required solution is given by $x = 3, y = 2, z = -1$

EXAMPLE 7.7 Solve the following system of equations by the LU factorisation method:

$$2x - 6y + 8z = 24$$
$$5x + 4y - 3z = 2$$
$$3x + y + 2z = 16$$

Solution: The given system is of the form $AX = B$

where
$$A=\begin{bmatrix}2 & -6 & 8\\ 5 & 4 & -3\\ 3 & 1 & 2\end{bmatrix}$$

$$X=\begin{bmatrix}x\\ y\\ z\end{bmatrix}$$

and
$$B=\begin{bmatrix}24\\ 2\\ 16\end{bmatrix}$$

Let
$$A = LU$$

where
$$L=\begin{bmatrix}l_{11} & 0 & 0\\ l_{21} & l_{22} & 0\\ l_{31} & l_{32} & l_{33}\end{bmatrix}$$

and
$$U=\begin{bmatrix}1 & u_{12} & u_{13}\\ 0 & 1 & u_{23}\\ 0 & 0 & 1\end{bmatrix}$$

Therefore

$$\begin{bmatrix}2 & -6 & 8\\ 5 & 4 & 3\\ 3 & 1 & 2\end{bmatrix}=\begin{bmatrix}l_{11} & 0 & 0\\ l_{21} & l_{22} & 0\\ l_{31} & l_{32} & l_{33}\end{bmatrix}\begin{bmatrix}1 & u_{12} & u_{13}\\ 0 & 1 & u_{23}\\ 0 & 0 & 1\end{bmatrix}$$

$$=\begin{bmatrix}l_{11} & l_{11}u_{12} & l_{11}u_{13}\\ l_{21} & l_{21}u_{12}+l_{22} & l_{21}u_{13}+l_{22}u_{23}\\ l_{31} & l_{31}u_{12}+l_{32} & l_{31}u_{13}+l_{32}u_{23}+l_{33}\end{bmatrix}$$

Comparing the elements, we get

$$l_{11} = 2, l_{21} = 5, l_{31} = 3$$

$$l_{11}u_{12} = -6, \text{ i.e., } u_{12} = -3$$

$$l_{11}u_{13} = 8, \text{ i.e., } u_{13} = 4$$

$$l_{21}u_{12} + l_{22} = 4, \text{ i.e., } l_{22} = 4 + 15 = 19$$

$$l_{31}u_{12} + l_{32} = 1, \text{ i.e., } l_{32} = 1 + 9 = 10$$

$$l_{21}u_{13} + l_{22}u_{23} = -3, \text{ i.e., } u_{23} = \frac{-3 - 5 \times 4}{19} = -23/19$$

$$l_{31}u_{13} + l_{32}u_{23} + l_{33} = 2$$

$$\Rightarrow \quad l_{33} = 2 - 3 \times 4 - 10 \times \left(-\frac{23}{19}\right) = 40/19$$

Therefore, we have

$$L = \begin{bmatrix} 2 & 0 & 0 \\ 5 & 19 & 0 \\ 3 & 10 & 40/19 \end{bmatrix}$$

$$U = \begin{bmatrix} 1 & -3 & 4 \\ 0 & 1 & -23/19 \\ 0 & 0 & 1 \end{bmatrix}$$

and

$$\begin{bmatrix} 2 & 0 & 0 \\ 5 & 19 & 0 \\ 3 & 10 & 40/19 \end{bmatrix} \begin{bmatrix} 1 & -3 & 4 \\ 0 & 1 & -23/19 \\ 0 & 0 & 1 \end{bmatrix} \begin{bmatrix} x \\ y \\ z \end{bmatrix} = \begin{bmatrix} 24 \\ 2 \\ 16 \end{bmatrix}$$

Let $\quad UX = Y$

where

$$Y = \begin{bmatrix} y_1 \\ y_2 \\ y_3 \end{bmatrix}$$

Therefore

$$\begin{bmatrix} 2 & 0 & 0 \\ 5 & 19 & 0 \\ 3 & 10 & 40/19 \end{bmatrix} \begin{bmatrix} y_1 \\ y_2 \\ y_3 \end{bmatrix} = \begin{bmatrix} 24 \\ 2 \\ 16 \end{bmatrix}$$

which gives

$$2y_1 = 24, \text{ i.e., } y_1 = 12$$

$$5y_1 + 19y_2 = 2, \text{ i.e., } y_2 = -58/19$$

$$3y_1 + 10y_2 + \frac{40}{19}y_3 = 16, \text{ i.e., } y_3 = 5$$

Therefore, we get

$$\begin{bmatrix} 1 & -3 & 4 \\ 0 & 1 & -23/19 \\ 0 & 0 & 1 \end{bmatrix} \begin{bmatrix} x \\ y \\ z \end{bmatrix} = \begin{bmatrix} 12 \\ -58/19 \\ 5 \end{bmatrix}$$

which gives

$$z = 5$$

$$y - \frac{23}{19} z = -\frac{58}{19}, \text{ i.e., } y = 3$$

$$x - 3y + 4z = 12, \text{ i.e., } x = 1$$

Hence, the required solution is given by $x = 1, y = 3, z = 5$

EXAMPLE 7.8 Solve the following system of equations by the LU factorisation method:

$$x_1 + x_2 - x_3 = 2$$

$$2x_1 + 3x_2 + 5x_3 = -3$$

$$3x_1 + 2x_2 - 3x_3 = 6$$

Solution: The given system is of the form $AX = B$

where
$$A = \begin{bmatrix} 1 & 1 & -1 \\ 2 & 3 & 5 \\ 3 & 2 & -3 \end{bmatrix}, \quad X = \begin{bmatrix} x_1 \\ x_2 \\ x_3 \end{bmatrix}$$

and
$$B = \begin{bmatrix} 2 \\ -3 \\ 6 \end{bmatrix}$$

Let
$$A = LU$$

where
$$L = \begin{bmatrix} l_{11} & 0 & 0 \\ l_{21} & l_{22} & 0 \\ l_{31} & l_{32} & l_{33} \end{bmatrix}$$

and
$$U = \begin{bmatrix} 1 & u_{12} & u_{13} \\ 0 & 1 & u_{23} \\ 0 & 0 & 1 \end{bmatrix}$$

Therefore

$$\begin{bmatrix} 1 & 1 & -1 \\ 2 & 3 & 5 \\ 3 & 2 & -3 \end{bmatrix} = \begin{bmatrix} l_{11} & 0 & 0 \\ l_{21} & l_{22} & 0 \\ l_{31} & l_{32} & l_{33} \end{bmatrix} \begin{bmatrix} 1 & u_{12} & u_{13} \\ 0 & 1 & u_{23} \\ 0 & 0 & 1 \end{bmatrix}$$

$$= \begin{bmatrix} l_{11} & l_{11}u_{12} & l_{11}u_{13} \\ l_{21} & l_{21}u_{12} + l_{22} & l_{21}u_{13} + l_{22}u_{23} \\ l_{31} & l_{31}u_{12} + l_{32} & l_{31}u_{13} + l_{32}u_{23} + l_{33} \end{bmatrix}$$

Comparing the elements, we get

$$l_{11} = 1, l_{21} = 2, l_{31} = 3$$

$$l_{11}u_{12} = 1, \text{ i.e., } u_{12} = 1$$

$$l_{11}u_{13} = -1, \text{ i.e., } u_{13} = -1$$

$$l_{21}u_{12} + l_{22} = 3, \text{ i.e., } l_{22} = 3 - 2 \times 1 = 1$$

$$l_{31}u_{12} + l_{32} = 2, \text{ i.e., } l_{32} = 2 - 3 \times 1 = -1$$

$$l_{21}u_{13} + l_{22}u_{23} = 5, \text{ i.e., } u_{23} = 5 + 2 \times 1 = 7$$

$$l_{31}u_{13} + l_{32}u_{23} + l_{33} = -3$$

$$\Rightarrow \quad l_{33} = -3 - 3 \times (-1) - (-1) \times 7 = 7$$

Therefore, we have

$$L = \begin{bmatrix} 1 & 0 & 0 \\ 2 & 1 & 0 \\ 3 & -1 & 7 \end{bmatrix}$$

$$U = \begin{bmatrix} 1 & 1 & -1 \\ 0 & 1 & 7 \\ 0 & 0 & 1 \end{bmatrix}$$

and

$$\begin{bmatrix} 1 & 0 & 0 \\ 2 & 1 & 0 \\ 3 & -1 & 7 \end{bmatrix} \begin{bmatrix} 1 & 1 & -1 \\ 0 & 1 & 7 \\ 0 & 0 & 1 \end{bmatrix} \begin{bmatrix} x_1 \\ x_2 \\ x_3 \end{bmatrix} = \begin{bmatrix} 2 \\ -3 \\ 6 \end{bmatrix}$$

Let

$$UX = Y$$

where

$$Y = \begin{bmatrix} y_1 \\ y_2 \\ y_3 \end{bmatrix}$$

Therefore,

$$\begin{bmatrix} 1 & 0 & 0 \\ 2 & 1 & 0 \\ 3 & -1 & 7 \end{bmatrix} \begin{bmatrix} y_1 \\ y_2 \\ y_3 \end{bmatrix} = \begin{bmatrix} 2 \\ -3 \\ 6 \end{bmatrix}$$

which gives

$$y_1 = 2$$

$$2y_1 + y_2 = -3, \text{ i.e., } y_2 = -7$$

$$3y_1 - y_2 + 7y_3 = 6, \text{ i.e., } y_3 = -1$$

Therefore, we get

$$\begin{bmatrix} 1 & 1 & -1 \\ 0 & 1 & 7 \\ 0 & 0 & 1 \end{bmatrix} \begin{bmatrix} x_1 \\ x_2 \\ x_3 \end{bmatrix} = \begin{bmatrix} 2 \\ -7 \\ -1 \end{bmatrix}$$

which gives

$$x_3 = -1$$

$$x_2 + 7x_3 = -7, \text{ i.e., } x_2 = 0$$

$$x_1 + x_2 - x_3 = 2, \text{ i.e., } x_1 = 1$$

Hence, the required solution is given by $x_1 = 1, x_2 = 0, x_3 = -1$

7.2.4 Matrix Inversion Method

You are aware of how to find the inverse of a non-singular square matrix A using the formula

$$A^{-1} = \frac{adjA}{detA}$$

But here, we discuss some suitable and systematic procedures to find A^{-1} by implementing (i) **Gauss elimination** and (ii) **LU factorisation**.

(i) *Method of Gauss elimination*

The procedure is based on the following logic.

We know $$AA^{-1} = I \quad (*)$$

and $$IA^{-1} = A^{-1} \quad (**)$$

The augmented matrix to find A^{-1} from (*) is $[A\,|\,I]$ while from (**), the matrix is $[I\,|\,A^{-1}]$. This shows that

$$[A\,|\,I] = [I\,|\,A^{-1}]$$

Thus, with this concept, the procedure can be formulated as follows:

1. Take a unit matrix I of the same order as that of the given matrix A.
2. Construct the augmented matrix $[A\,|\,I]$.
3. Apply Gauss forward elimination on $[A\,|\,I]$ after reducing the pivot to unity so that all the leading diagonal elements of the upper triangular system, say $[U\,|\,Y]$, become unity.
4. Next, apply Gauss backward elimination on $[U\,|\,Y]$ starting from the nth row. This process will then yield $[I\,|\,A^{-1}]$.
5. Thus, when finally A reduces to a unit matrix, the appended unit matrix will become its inverse, i.e., A^{-1}.

Some examples are given below to illustrate this method.

EXAMPLE 7.9 Find the inverse of the matrix $A = \begin{bmatrix} 8 & -4 & 0 \\ -4 & 8 & -4 \\ 0 & -4 & 8 \end{bmatrix}$ by the Gauss elimination method.

Solution: Let us consider the augmented matrix

$$[A\,|\,I] = \left[\begin{array}{ccc|ccc} 8 & -4 & 0 & 1 & 0 & 0 \\ -4 & 8 & -4 & 0 & 1 & 0 \\ 0 & -4 & 8 & 0 & 0 & 1 \end{array}\right]$$

$$\sim \left[\begin{array}{ccc|ccc} 1 & -1/2 & 0 & 1/8 & 0 & 0 \\ -4 & 8 & -4 & 0 & 1 & 0 \\ 0 & -4 & 8 & 0 & 0 & 1 \end{array}\right], \text{ by } R_1' = \frac{R_1}{8}$$

$$\sim \left[\begin{array}{ccc|ccc} 1 & -1/2 & 0 & 1/8 & 0 & 0 \\ 0 & 6 & -4 & 1/2 & 1 & 0 \\ 0 & -4 & 8 & 0 & 0 & 1 \end{array}\right], \text{ by } R_2' = R_2 + 4R_1$$

$$\sim \left[\begin{array}{ccc|ccc} 1 & -1/2 & 0 & 1/8 & 0 & 0 \\ 0 & 1 & -2/3 & 1/12 & 1/6 & 0 \\ 0 & -4 & 8 & 0 & 0 & 1 \end{array}\right], \text{ by } R_2' = \frac{R_2}{6}$$

$$\sim \left[\begin{array}{ccc|ccc} 1 & -1/2 & 0 & 1/8 & 0 & 0 \\ 0 & 1 & -2/3 & 1/12 & 1/6 & 0 \\ 0 & 0 & 16/3 & 1/3 & 2/3 & 1 \end{array}\right], \text{ by } R_3' = R_3 + 4R_2$$

$$\sim \left[\begin{array}{ccc|ccc} 1 & -1/2 & 0 & 1/8 & 0 & 0 \\ 0 & 1 & -2/3 & 1/12 & 1/6 & 0 \\ 0 & 0 & 1 & 1/16 & 2/16 & 3/16 \end{array}\right], \text{ by } R_3' = \frac{3R_3}{16}$$

$$\sim \left[\begin{array}{ccc|ccc} 1 & -1/2 & 0 & 1/8 & 0 & 0 \\ 0 & 1 & 0 & 1/8 & 1/4 & 1/8 \\ 0 & 0 & 1 & 1/16 & 2/16 & 3/16 \end{array}\right], \ R_2' = R_2 + \frac{2}{3}R_3$$

$$\sim \left[\begin{array}{ccc|ccc} 1 & 0 & 0 & 3/16 & 1/8 & 1/16 \\ 0 & 1 & 0 & 1/8 & 1/4 & 1/8 \\ 0 & 0 & 1 & 1/16 & 2/16 & 3/16 \end{array}\right], \ R_1' = R_1 + \frac{1}{2}R_2$$

$$= \left[I \mid A^{-1}\right]$$

Hence
$$A^{-1} = \begin{bmatrix} 3/16 & 1/8 & 1/16 \\ 1/8 & 1/4 & 1/8 \\ 1/16 & 2/16 & 3/16 \end{bmatrix} = \frac{1}{16}\begin{bmatrix} 3 & 2 & 1 \\ 2 & 4 & 2 \\ 1 & 2 & 3 \end{bmatrix}$$

EXAMPLE 7.10 Find the inverse of the matrix $A = \begin{bmatrix} 1 & 2 & 6 \\ 2 & 5 & 15 \\ 6 & 15 & 46 \end{bmatrix}$ by the Gauss elimination method.

Solution: Let us consider the augmented matrix

$$[A \mid I] = \left[\begin{array}{ccc|ccc} 1 & 2 & 6 & 1 & 0 & 0 \\ 2 & 5 & 15 & 0 & 1 & 0 \\ 6 & 15 & 46 & 0 & 0 & 1 \end{array}\right]$$

$$\sim \left[\begin{array}{ccc|ccc} 1 & 2 & 6 & 1 & 0 & 0 \\ 0 & 1 & 3 & -2 & 1 & 0 \\ 0 & 3 & 10 & -6 & 0 & 1 \end{array}\right], \text{ by } R_2' = R_2 - 2R_1$$

and $R_3' = R_3 - 6R_1$

$$\sim \left[\begin{array}{ccc|ccc} 1 & 2 & 6 & 1 & 0 & 0 \\ 0 & 1 & 3 & -2 & 1 & 0 \\ 0 & 0 & 1 & 0 & -3 & 1 \end{array}\right], \text{ by } R_3' = R_3 - 3R_2$$

$$\sim \left[\begin{array}{ccc|ccc} 1 & 2 & 0 & 1 & 18 & -6 \\ 0 & 1 & 0 & -2 & 10 & -3 \\ 0 & 0 & 1 & 0 & -3 & 1 \end{array}\right], \text{ by } R_2' = R_2 - 3R_3$$

and $$R_1' = R_1 - 6R_3$$

$$\sim \left[\begin{array}{ccc|ccc} 1 & 0 & 0 & 5 & -2 & 0 \\ 0 & 1 & 0 & -2 & 10 & -3 \\ 0 & 0 & 1 & 0 & -3 & 1 \end{array}\right], \text{ by } R_1' = R_1 - 2R_2$$

$$= \left[I \mid A^{-1}\right]$$

Hence $$A^{-1} = \begin{bmatrix} 5 & -2 & 0 \\ -2 & 10 & -3 \\ 0 & -3 & 1 \end{bmatrix}$$

EXAMPLE 7.11 Find the inverse of the matrix $A = \begin{bmatrix} 3 & -1 & 1 \\ -15 & 6 & -5 \\ 5 & -2 & 2 \end{bmatrix}$ by the Gauss elimination method.

Solution: Let us consider the augmented matrix

$$[A \mid I] = \left[\begin{array}{ccc|ccc} 3 & -1 & 1 & 1 & 0 & 0 \\ -15 & 6 & -5 & 0 & 1 & 0 \\ 5 & -2 & 2 & 0 & 0 & 1 \end{array}\right]$$

$$\sim \left[\begin{array}{ccc|ccc} 1 & -1/3 & 1/3 & 1/3 & 0 & 0 \\ -15 & 6 & -5 & 0 & 1 & 0 \\ 5 & -2 & 2 & 0 & 0 & 1 \end{array}\right], \text{ by } R_1' = \frac{R_1}{3}$$

$$\sim \left[\begin{array}{ccc|ccc} 1 & -1/3 & 1/3 & 1/3 & 0 & 0 \\ 0 & 1 & 0 & 5 & 1 & 0 \\ 0 & -1/3 & 1/3 & -5/3 & 0 & 1 \end{array}\right], \text{ by } R_2' = R_2 + 15R_1$$

and $$R_3' = R_3 - 5R_1$$

$$\sim \left[\begin{array}{ccc|ccc} 1 & -1/3 & 1/3 & 1/3 & 0 & 0 \\ 0 & 1 & 0 & 5 & 1 & 0 \\ 0 & 0 & 1/3 & 0 & 1/3 & 1 \end{array}\right], \text{ by } R_3' = R_3 + \frac{R_2}{3}$$

$$\sim \left[\begin{array}{ccc|ccc} 1 & -1/3 & 1/3 & 1/3 & 0 & 0 \\ 0 & 1 & 0 & 5 & 1 & 0 \\ 0 & 0 & 1 & 0 & 1 & 3 \end{array}\right] \text{ by } R_3' = 3R_3$$

$$\sim \left[\begin{array}{ccc|ccc} 1 & -1/3 & 0 & 1/3 & -1/3 & -1 \\ 0 & 1 & 0 & 5 & 1 & 0 \\ 0 & 0 & 1 & 0 & 1 & 3 \end{array}\right] \text{ by } R_1' = R_1 - \frac{R_3}{3}$$

$$\sim \left[\begin{array}{ccc|ccc} 1 & 0 & 0 & 2 & 0 & -1 \\ 0 & 1 & 0 & 5 & 1 & 0 \\ 0 & 0 & 1 & 0 & 1 & 3 \end{array}\right] \text{ by } R_1' = R_1 + \frac{R_2}{3}$$

$$= \left[I \mid A^{-1} \right]$$

Hence
$$A^{-1} = \begin{bmatrix} 2 & 0 & -1 \\ 5 & 1 & 0 \\ 0 & 1 & 3 \end{bmatrix}$$

(ii) *Method of LU factorisation*

The method is based on the following fact:

$A = LU$ and hence $A^{-1} = U^{-1}L^{-1}$ where L^{-1} and U^{-1} are calculated by the formulae $L^{-1} = \dfrac{adjL}{detL}$ and $U^{-1} = \dfrac{adjU}{detU}$, respectively.

This procedure is illustrated in the following examples.

EXAMPLE 7.12 Find the inverse of $A = \begin{bmatrix} 1 & 2 & 6 \\ 2 & 5 & 15 \\ 6 & 15 & 46 \end{bmatrix}$ by LU decomposition.

Solution: Let
$$A = LU$$

where
$$L = \begin{bmatrix} l_{11} & 0 & 0 \\ l_{21} & l_{22} & 0 \\ l_{31} & l_{32} & l_{33} \end{bmatrix}$$

and
$$U = \begin{bmatrix} 1 & u_{12} & u_{13} \\ 0 & 1 & u_{23} \\ 0 & 0 & 1 \end{bmatrix}$$

Therefore
$$\begin{bmatrix} 1 & 2 & 6 \\ 2 & 5 & 15 \\ 6 & 15 & 46 \end{bmatrix} = \begin{bmatrix} l_{11} & 0 & 0 \\ l_{21} & l_{22} & 0 \\ l_{31} & l_{32} & l_{33} \end{bmatrix} \begin{bmatrix} 1 & u_{12} & u_{13} \\ 0 & 1 & u_{23} \\ 0 & 0 & 1 \end{bmatrix}$$

$$= \begin{bmatrix} l_{11} & l_{11}u_{12} & l_{11}u_{13} \\ l_{21} & l_{21}u_{12} + l_{22} & l_{21}u_{13} + l_{22}u_{23} \\ l_{31} & l_{31}u_{12} + l_{32} & l_{31}u_{13} + l_{32}u_{23} + l_{33} \end{bmatrix}$$

Comparing the elements, we get

$$l_{11} = 1, l_{21} = 2, l_{31} = 6$$

$$l_{11}u_{12} = 2, \text{ i.e., } u_{12} = 2$$

$$l_{11}u_{13} = 6, \text{ i.e., } u_{13} = 6$$

$$l_{21}u_{12} + l_{22} = 5, \text{ i.e., } l_{22} = 5 - 2 \times 2 = 1$$

$$l_{31}u_{12} + l_{32} = 15, \text{ i.e., } l_{32} = 15 - 6 \times 2 = 3$$

$$l_{21}u_{13} + l_{22}u_{23} = 15, \text{ i.e., } u_{23} = 15 - 2 \times 6 = 3$$

$$l_{31}u_{13} + l_{32}u_{23} + l_{33} = 46$$

$\Rightarrow$
$$l_{33} = 46 - 6 \times 6 - 3 \times 3 = 1$$

Therefore, we have
$$L=\begin{bmatrix}1 & 0 & 0\\ 2 & 1 & 0\\ 6 & 3 & 1\end{bmatrix}$$
$$U=\begin{bmatrix}1 & 2 & 6\\ 0 & 1 & 3\\ 0 & 0 & 1\end{bmatrix}$$

Therefore, det $(L) = 1$ and det $(U) = 1$.

Again
$$Adj(L)=\begin{bmatrix}1 & -2 & 0\\ 0 & 1 & -3\\ 0 & 0 & 1\end{bmatrix}^T=\begin{bmatrix}1 & 0 & 0\\ -2 & 1 & 0\\ 0 & -3 & 1\end{bmatrix}$$

and
$$Adj(U)=\begin{bmatrix}1 & 0 & 0\\ -2 & 1 & 0\\ 0 & -3 & 1\end{bmatrix}^T=\begin{bmatrix}1 & -2 & 0\\ 0 & 1 & -3\\ 0 & 0 & 1\end{bmatrix}$$

This gives
$$L^{-1}=\frac{Adj(L)}{\det(L)}=\begin{bmatrix}1 & 0 & 0\\ -2 & 1 & 0\\ 0 & -3 & 1\end{bmatrix}$$

and
$$U^{-1}=\frac{Adj(U)}{\det(U)}=\begin{bmatrix}1 & -2 & 0\\ 0 & 1 & -3\\ 0 & 0 & 1\end{bmatrix}$$

Therefore
$$A^{-1}=U^{-1}L^{-1}=\begin{bmatrix}1 & -2 & 0\\ 0 & 1 & -3\\ 0 & 0 & 1\end{bmatrix}\begin{bmatrix}1 & 0 & 0\\ -2 & 1 & 0\\ 0 & -3 & 1\end{bmatrix}$$
$$=\begin{bmatrix}5 & -2 & 0\\ -2 & 10 & -3\\ 0 & -3 & 1\end{bmatrix}$$

EXAMPLE 7.13 Find the inverse of $A=\begin{bmatrix}1 & 3 & 3\\ 1 & 4 & 3\\ 1 & 3 & 4\end{bmatrix}$ by LU factorisation.

Solution: Let
$$A = LU$$

where
$$L=\begin{bmatrix}l_{11} & 0 & 0\\ l_{21} & l_{22} & 0\\ l_{31} & l_{32} & l_{33}\end{bmatrix}$$

and
$$U=\begin{bmatrix}1 & u_{12} & u_{13}\\ 0 & 1 & u_{23}\\ 0 & 0 & 1\end{bmatrix}$$

Therefore
$$\begin{bmatrix}1&3&3\\1&4&3\\1&3&4\end{bmatrix}=\begin{bmatrix}l_{11}&0&0\\l_{21}&l_{22}&0\\l_{31}&l_{32}&l_{33}\end{bmatrix}\begin{bmatrix}1&u_{12}&u_{13}\\0&1&u_{23}\\0&0&1\end{bmatrix}$$
$$=\begin{bmatrix}l_{11}&l_{11}u_{12}&l_{11}u_{13}\\l_{21}&l_{21}u_{12}+l_{22}&l_{21}u_{13}+l_{22}u_{23}\\l_{31}&l_{31}u_{12}+l_{32}&l_{31}u_{13}+l_{32}u_{23}+l_{33}\end{bmatrix}$$

Comparing the elements, we get
$$l_{11}=1,\ l_{21}=1,\ l_{31}=1$$
$$l_{11}u_{12}=3,\text{ i.e., }u_{12}=3$$
$$l_{11}u_{13}=3,\text{ i.e., }u_{13}=3$$
$$l_{21}u_{12}+l_{22}=4,\text{ i.e., }l_{22}=4-1\times 3=1$$
$$l_{31}u_{12}+l_{32}=3,\text{ i.e., }l_{32}=3-1\times 3=0$$
$$l_{21}u_{13}+l_{22}u_{23}=3,\text{ i.e., }u_{23}=3-1\times 3=0$$
$$l_{31}u_{13}+l_{32}u_{23}+l_{33}=4$$
$\Rightarrow$
$$l_{33}=4-1\times 3-0\times 3=1$$

Therefore, we have
$$L=\begin{bmatrix}1&0&0\\1&1&0\\1&0&1\end{bmatrix}$$
$$U=\begin{bmatrix}1&3&3\\0&1&0\\0&0&1\end{bmatrix}$$

Therefore, det $(L)=1$ and det $(U)=1$

Again
$$Adj(L)=\begin{bmatrix}1&-1&-1\\0&1&0\\0&0&1\end{bmatrix}^T=\begin{bmatrix}1&0&0\\-1&1&0\\-1&0&1\end{bmatrix}$$

and
$$Adj(U)=\begin{bmatrix}1&0&0\\-3&1&0\\-3&0&1\end{bmatrix}^T=\begin{bmatrix}1&-3&-3\\0&1&0\\0&0&1\end{bmatrix}$$

This gives
$$L^{-1}=\frac{Adj(L)}{\det(L)}=\begin{bmatrix}1&0&0\\-1&1&0\\-1&0&1\end{bmatrix}$$

and
$$U^{-1}=\frac{Adj(U)}{\det(U)}=\begin{bmatrix}1&-3&-3\\0&1&0\\0&0&1\end{bmatrix}$$

Therefore
$$A^{-1} = U^{-1}L^{-1} = \begin{bmatrix} 1 & -3 & -3 \\ 0 & 1 & 0 \\ 0 & 0 & 1 \end{bmatrix}\begin{bmatrix} 1 & 0 & 0 \\ -1 & 1 & 0 \\ -1 & 0 & 1 \end{bmatrix}$$

$$= \begin{bmatrix} 7 & -3 & -3 \\ -2 & 1 & 0 \\ -1 & 0 & 1 \end{bmatrix}$$

EXAMPLE 7.14 Find the inverse of $A = \begin{bmatrix} 1 & 2 & -2 \\ -1 & 3 & 0 \\ 0 & -2 & 1 \end{bmatrix}$ by LU factorisation.

Solution: Let
$$A = LU$$

where
$$L = \begin{bmatrix} l_{11} & 0 & 0 \\ l_{21} & l_{22} & 0 \\ l_{31} & l_{32} & l_{33} \end{bmatrix}$$

and
$$U = \begin{bmatrix} 1 & u_{12} & u_{13} \\ 0 & 1 & u_{23} \\ 0 & 0 & 1 \end{bmatrix}$$

Therefore
$$\begin{bmatrix} 1 & 2 & -2 \\ -1 & 3 & 0 \\ 0 & -2 & 1 \end{bmatrix} = \begin{bmatrix} l_{11} & 0 & 0 \\ l_{21} & l_{22} & 0 \\ l_{31} & l_{32} & l_{33} \end{bmatrix}\begin{bmatrix} 1 & u_{12} & u_{13} \\ 0 & 1 & u_{23} \\ 0 & 0 & 1 \end{bmatrix}$$

$$= \begin{bmatrix} l_{11} & l_{11}u_{12} & l_{11}u_{13} \\ l_{21} & l_{21}u_{12} + l_{22} & l_{21}u_{13} + l_{22}u_{23} \\ l_{31} & l_{31}u_{12} + l_{32} & l_{31}u_{13} + l_{32}u_{23} + l_{33} \end{bmatrix}$$

Comparing the elements, we get

$$l_{11} = 1,\ l_{21} = -1,\ l_{31} = 0$$
$$l_{11}u_{12} = 2, \text{ i.e., } u_{12} = 2$$
$$l_{11}u_{13} = -2, \text{ i.e., } u_{13} = -2$$
$$l_{21}u_{12} + l_{22} = 3$$

i.e.
$$l_{22} = 3 - (-1) \times 2 = 5$$
$$l_{31}u_{12} + l_{32} = -2$$

i.e.
$$l_{32} = -2 - 0 \times 2 = -2$$
$$l_{21}u_{13} + l_{22}u_{23} = 0$$

i.e.
$$u_{23} = \frac{[0 - (-1) \times (-2)]}{5} = -2/5$$
$$l_{31}u_{13} + l_{32}u_{23} + l_{33} = 1$$

⇒
$$l_{33} = 1 - 0 \times (-2) - (-2) \times (-2/5) = 1/5$$

Therefore, we have
$$L=\begin{bmatrix}1 & 0 & 0\\ -1 & 5 & 0\\ 0 & -2 & 1/5\end{bmatrix}$$

$$U=\begin{bmatrix}1 & 2 & -2\\ 0 & 1 & -2/5\\ 0 & 0 & 1\end{bmatrix}$$

Therefore, $\det(L) = 1$ and $\det(U) = 1$.

Again
$$Adj(L)=\begin{bmatrix}1 & 1/5 & 2\\ 0 & 1/5 & 2\\ 0 & 0 & 5\end{bmatrix}^T=\begin{bmatrix}1 & 0 & 0\\ 1/5 & 1/5 & 0\\ 2 & 2 & 5\end{bmatrix}$$

and
$$Adj(U)=\begin{bmatrix}1 & 0 & 0\\ -2 & 1 & 0\\ 6/5 & 2/5 & 1\end{bmatrix}^T=\begin{bmatrix}1 & -2 & 6/5\\ 0 & 1 & 2/5\\ 0 & 0 & 1\end{bmatrix}$$

This gives
$$L^{-1}=\frac{Adj(L)}{\det(L)}=\begin{bmatrix}1 & 0 & 0\\ 1/5 & 1/5 & 0\\ 2 & 2 & 5\end{bmatrix}$$

and
$$U^{-1}=\frac{Adj(U)}{\det(U)}=\begin{bmatrix}1 & -2 & 6/5\\ 0 & 1 & 2/5\\ 0 & 0 & 1\end{bmatrix}$$

Therefore

$$A^{-1}=U^{-1}L^{-1}=\begin{bmatrix}1 & -2 & 6/5\\ 0 & 1 & 2/5\\ 0 & 0 & 1\end{bmatrix}\begin{bmatrix}1 & 0 & 0\\ 1/5 & 1/5 & 0\\ 2 & 2 & 5\end{bmatrix}$$

$$=\begin{bmatrix}3 & 2 & 6\\ 1 & 1 & 2\\ 2 & 2 & 5\end{bmatrix}$$

Applications of the Matrix Inversion Method The system of linear equations with n unknowns and n equations can be written in matrix form as

$$AX = B \tag{1}$$

where $A = [a_{ij}]_{n\times n}$ is the coefficient matrix, B is the column matrix of the constants b_is and X is the column matrix of the unknowns x_is. If the system has a unique solution, A must be non-singular and so there exists an inverse matrix A^{-1} of A. Pre-multiplying (1) by A^{-1} on both sides, we get

$$A^{-1}(AX) = A^{-1}B$$

or
$$(A^{-1}A)X = A^{-1}B$$

or
$$IX = A^{-1}B, \text{ i.e., } X = A^{-1}B$$

where I is the identity matrix of order n.

Thus, if A^{-1} is obtained from A, the solution vector X can be found by post-multiplying it by B.

EXAMPLE 7.15 Solve the following system by the matrix inversion method:

$$2x + y + z = 10$$
$$3x + 2y + 3z = 18$$
$$x + 4y + 9z = 16$$

Solution: The given system is equivalent to

$$AX = B \tag{1}$$

where
$$A = \begin{bmatrix} 2 & 1 & 1 \\ 3 & 2 & 3 \\ 1 & 4 & 9 \end{bmatrix}, \quad X = \begin{bmatrix} x \\ y \\ z \end{bmatrix}$$

and
$$B = \begin{bmatrix} 10 \\ 18 \\ 16 \end{bmatrix}$$

From (1), we get $X = A^{-1}B$ where A^{-1} is the inverse of A. So, first we find A^{-1} using the Gauss elimination method.

Let us consider the augmented matrix

$$[A \mid I] = \left[\begin{array}{ccc|ccc} 2 & 1 & 1 & 1 & 0 & 0 \\ 3 & 2 & 3 & 0 & 1 & 0 \\ 1 & 4 & 9 & 0 & 0 & 1 \end{array}\right]$$

$$\sim \left[\begin{array}{ccc|ccc} 1 & 1/2 & 1/2 & 1/2 & 0 & 0 \\ 3 & 2 & 3 & 0 & 1 & 0 \\ 1 & 4 & 9 & 0 & 0 & 1 \end{array}\right], \text{ by } R_1' = \frac{1}{2}R_1$$

$$\sim \left[\begin{array}{ccc|ccc} 1 & 1/2 & 1/2 & 1/2 & 0 & 0 \\ 0 & 1/2 & 3/2 & -3/2 & 1 & 0 \\ 0 & 7/2 & 17/2 & -1/2 & 0 & 1 \end{array}\right], \text{ by } R_2' = R_2 - 3R_1$$

and
$$R_3' = R_3 - R_1$$

$$\sim \left[\begin{array}{ccc|ccc} 1 & 1/2 & 1/2 & 1/2 & 0 & 0 \\ 0 & 1 & 3 & -3 & 2 & 0 \\ 0 & 7/2 & 17/2 & -1/2 & 0 & 1 \end{array}\right], \text{ by } R_2' = 2R_2$$

$$\sim \left[\begin{array}{ccc|ccc} 1 & 1/2 & 1/2 & 1/2 & 0 & 0 \\ 0 & 1 & 3 & -3 & 2 & 0 \\ 0 & 0 & -2 & 10 & -7 & 1 \end{array}\right], \text{ by } R_3' = R_3 - \frac{7}{2}R_2$$

$$\sim\left[\begin{array}{ccc|ccc}1 & 1/2 & 1/2 & 1/2 & 0 & 0\\ 0 & 1 & 3 & -3 & 2 & 0\\ 0 & 0 & 1 & -5 & 7/2 & -1/2\end{array}\right], \text{ by } R_3' = -\frac{1}{2}R_3$$

$$\sim\left[\begin{array}{ccc|ccc}1 & 1/2 & 0 & 3 & -7/4 & 1/4\\ 0 & 1 & 0 & 12 & -17/2 & 3/2\\ 0 & 0 & 1 & -5 & 7/2 & -1/2\end{array}\right], \text{ by } R_2' = R_2 - 3R_3$$

and $R_1' = R_1 - \frac{1}{2}R_3$

$$\sim\left[\begin{array}{ccc|ccc}1 & 0 & 0 & -3 & 5/2 & -1/2\\ 0 & 1 & 0 & 12 & -17/2 & 3/2\\ 0 & 0 & 1 & -5 & 7/2 & -1/2\end{array}\right], \text{ by } R_1' = R_1 - \frac{1}{2}R_2$$

$$= [I \mid A^{-1}]$$

Hence
$$A^{-1} = \begin{bmatrix}-3 & 5/2 & -1/2\\ 12 & -17/2 & 3/2\\ -5 & 7/2 & -1/2\end{bmatrix}$$

Therefore
$$\begin{bmatrix}x\\ y\\ z\end{bmatrix} = \begin{bmatrix}-3 & 5/2 & -1/2\\ 12 & -17/2 & 3/2\\ -5 & 7/2 & -1/2\end{bmatrix}\begin{bmatrix}10\\ 18\\ 16\end{bmatrix} = \begin{bmatrix}7\\ -9\\ 5\end{bmatrix}$$

Hence, $x = 7, y = -9, z = 5$.

EXAMPLE 7.16 Solve the following system by the matrix inversion method:

$$2x + 5y + 3z = 9$$
$$3x + y + 2z = 3$$
$$x + 2y - z = 6$$

Solution: The given system is equivalent to

$$AX = B \qquad (1)$$

where
$$A = \begin{bmatrix}2 & 5 & 3\\ 3 & 1 & 2\\ 1 & 2 & -1\end{bmatrix}, \quad X = \begin{bmatrix}x\\ y\\ z\end{bmatrix}$$

and
$$B = \begin{bmatrix}9\\ 3\\ 6\end{bmatrix}$$

From (1), we get $X = A^{-1}B$ where A^{-1} is the inverse of A. So, first we find A^{-1} using the Gauss elimination method.

Let us consider the augmented matrix

$$[A\,|\,I] = \left[\begin{array}{ccc|ccc} 2 & 5 & 3 & 1 & 0 & 0 \\ 3 & 1 & 2 & 0 & 1 & 0 \\ 1 & 2 & -1 & 0 & 0 & 1 \end{array}\right]$$

$$\sim \left[\begin{array}{ccc|ccc} 1 & 5/2 & 3/2 & 1/2 & 0 & 0 \\ 3 & 1 & 2 & 0 & 1 & 0 \\ 1 & 2 & -1 & 0 & 0 & 1 \end{array}\right], \text{ by } R_1' = \frac{1}{2}R_1$$

$$\sim \left[\begin{array}{ccc|ccc} 1 & 5/2 & 3/2 & 1/2 & 0 & 0 \\ 0 & -13/2 & -5/2 & -3/2 & 1 & 0 \\ 0 & -1/2 & -5/2 & -1/2 & 0 & 1 \end{array}\right], \text{ by } R_2' = R_2 - 3R_1$$

and $R_3' = R_3 - R_1$

$$\sim \left[\begin{array}{ccc|ccc} 1 & 5/2 & 3/2 & 1/2 & 0 & 0 \\ 0 & 1 & 5/13 & 3/13 & -2/13 & 0 \\ 0 & -1/2 & -5/2 & -1/2 & 0 & 1 \end{array}\right], \text{ by } R_2' = -\frac{2}{13}R_2$$

$$\sim \left[\begin{array}{ccc|ccc} 1 & 5/2 & 3/2 & 1/2 & 0 & 0 \\ 0 & 1 & 5/13 & 3/13 & -2/13 & 0 \\ 0 & 0 & -30/13 & -5/13 & -1/13 & 1 \end{array}\right], \text{ by } R_3' = R_3 + \frac{1}{2}R_2$$

$$\sim \left[\begin{array}{ccc|ccc} 1 & 5/2 & 3/2 & 1/2 & 0 & 0 \\ 0 & 1 & 5/13 & 3/13 & -2/13 & 0 \\ 0 & 0 & 1 & 1/6 & 1/30 & -13/30 \end{array}\right], \text{ by } R_3' = -\frac{13}{30}R_3$$

$$\sim \left[\begin{array}{ccc|ccc} 1 & 5/2 & 0 & 1/4 & -1/20 & 13/20 \\ 0 & 1 & 0 & 1/6 & -1/6 & 1/6 \\ 0 & 0 & 1 & 1/6 & 1/30 & -13/30 \end{array}\right], \text{ by } R_2' = R_2 - \frac{5}{13}R_3$$

and $R_1' = R_1 - \frac{3}{2}R_3$

$$\sim \left[\begin{array}{ccc|ccc} 1 & 0 & 0 & -1/6 & 11/30 & 7/30 \\ 0 & 1 & 0 & 1/6 & -1/6 & 1/6 \\ 0 & 0 & 1 & 1/6 & 1/30 & -13/30 \end{array}\right], \text{ by } R_1' = R_1 - \frac{5}{2}R_2$$

$$= [I\,|\,A^{-1}]$$

Hence
$$A^{-1} = \begin{bmatrix} -1/6 & 11/30 & 7/30 \\ 1/6 & -1/6 & 1/6 \\ 1/6 & 1/30 & -13/30 \end{bmatrix}$$

Therefore
$$\begin{bmatrix} x \\ y \\ z \end{bmatrix} = \begin{bmatrix} -1/6 & 11/30 & 7/30 \\ 1/6 & -1/6 & 1/6 \\ 1/6 & 1/30 & -13/30 \end{bmatrix} \begin{bmatrix} 9 \\ 3 \\ 6 \end{bmatrix} = \begin{bmatrix} 1 \\ 2 \\ -1 \end{bmatrix}$$

Hence, $x = 1, y = 2, z = -1$.

EXAMPLE 7.17 Solve the following system by the matrix inversion method:

$$x + y + z = 9$$
$$2x - 3y + 4z = 13$$
$$3x + 4y + 5z = 40$$

Solution: The given system is equivalent to

$$AX = B \tag{1}$$

where $$A = \begin{bmatrix} 1 & 1 & 1 \\ 2 & -3 & 4 \\ 3 & 4 & 5 \end{bmatrix}, \quad X = \begin{bmatrix} x \\ y \\ z \end{bmatrix}$$

and $$B = \begin{bmatrix} 9 \\ 13 \\ 40 \end{bmatrix}$$

From (1), we get $X = A^{-1}B$ where A^{-1} is the inverse of A. So, first we find A^{-1} using the Gauss elimination method.

Let us consider the augmented matrix

$$[A \mid I] = \left[\begin{array}{ccc|ccc} 1 & 1 & 1 & 1 & 0 & 0 \\ 2 & -3 & 4 & 0 & 1 & 0 \\ 3 & 4 & 5 & 0 & 0 & 1 \end{array}\right]$$

$$\sim \left[\begin{array}{ccc|ccc} 1 & 1 & 1 & 1 & 0 & 0 \\ 0 & -5 & 2 & -2 & 1 & 0 \\ 0 & 1 & 2 & -3 & 0 & 1 \end{array}\right], \text{ by } R_2' = R_2 - 2R_1$$

and $$R_3' = R_3 - 3R_1$$

$$\sim \left[\begin{array}{ccc|ccc} 1 & 1 & 1 & 1 & 0 & 0 \\ 0 & 1 & -2/5 & 2/5 & -1/5 & 0 \\ 0 & 1 & 2 & -3 & 0 & 1 \end{array}\right], \text{ by } R_2' = -\frac{1}{5}R_2$$

$$\sim \left[\begin{array}{ccc|ccc} 1 & 1 & 1 & 1 & 0 & 0 \\ 0 & 1 & -2/5 & 2/5 & -1/5 & 0 \\ 0 & 0 & 12/5 & -17/5 & 1/5 & 1 \end{array}\right], \text{ by } R_3' = R_3 - R_2$$

$$\sim \left[\begin{array}{ccc|ccc} 1 & 1 & 1 & 1 & 0 & 0 \\ 0 & 1 & -2/5 & 2/5 & -1/5 & 0 \\ 0 & 0 & 1 & -17/12 & 1/12 & 5/12 \end{array}\right], \text{ by } R_3' = \frac{5}{12}R_3$$

$$\sim \left[\begin{array}{ccc|ccc} 1 & 1 & 0 & 29/12 & -1/12 & -5/12 \\ 0 & 1 & 0 & -1/6 & -1/6 & 1/6 \\ 0 & 0 & 1 & -17/12 & 1/12 & 5/12 \end{array}\right], \text{ by } R_2' = R_2 + \frac{2}{5}R_3$$

and $$R_1' = R_1 - R_3$$

$$\sim\left[\begin{array}{ccc|ccc} 1 & 1 & 0 & 31/12 & 1/12 & -7/12 \\ 0 & 1 & 0 & -1/6 & -1/6 & 1/6 \\ 0 & 0 & 1 & -17/12 & 1/12 & 5/12 \end{array}\right], \text{ by } R_1' = R_1 - R_2$$

$$= \left[I \mid A^{-1} \right]$$

Hence $$A^{-1} = \begin{bmatrix} 31/12 & 1/12 & -7/12 \\ -1/6 & -1/6 & 1/6 \\ -17/12 & 1/12 & 5/12 \end{bmatrix}$$

Therefore $$\begin{bmatrix} x \\ y \\ z \end{bmatrix} = \begin{bmatrix} 31/12 & 1/12 & -7/12 \\ -1/6 & -1/6 & 1/6 \\ -17/12 & 1/12 & 5/12 \end{bmatrix} \begin{bmatrix} 9 \\ 13 \\ 40 \end{bmatrix} = \begin{bmatrix} 1 \\ 3 \\ 5 \end{bmatrix}$$

Hence, $x = 1, y = 3, z = 5$

7.3 ITERATIVE METHODS

In these methods, also known as **indirect methods**, the exact solutions are not expected. Rather, we assume initial guesses for the true solutions and improve them step by step up to the desired level of accuracy. A huge number of iterative methods are available in numerical methods. Here, we discuss a few of them.

The convergence of these iterative methods depends on the sufficient condition that the system is **strictly diagonally dominant**. The system

$$a_{11}x_1 + a_{12}x_2 + \cdots + a_{1n}x_n = b_1$$
$$a_{21}x_1 + a_{22}x_2 + \cdots + a_{2n}x_n = b_2$$
$$\cdots$$
$$a_{n1}x_1 + a_{n2}x_2 + \cdots + a_{nn}x_n = b_n$$

is called strictly diagonally dominant if $|a_{ii}| > \sum_{\substack{i,j=1 \\ i \neq j}}^{n} |a_{ij}|, \quad (i, j = 1, 2, \ldots, n)$

For example, consider the following system of linear equations:

$$x_1 + x_2 - x_3 = 2$$
$$2x_1 + 3x_2 + 5x_3 = -3$$
$$3x_1 + 2x_2 - 3x_3 = 6$$

We see that $|a_{ii}| = |1| + |3| + |-3| = 7$ and $\sum_{\substack{i,j=1 \\ i \neq j}}^{3} |a_{ij}| = |1| + |-1| + |2| + |5| + |3| + |2| = 14.$

Clearly, $|a_{ii}| \not> \sum_{\substack{i,j=1 \\ i \neq j}}^{3} |a_{ij}|$. Therefore, the system is not diagonally dominant.

Again, consider the following system:

$$10x_1 + x_2 - 3x_3 = 14$$
$$2x_1 + 10x_2 + x_3 = 17$$
$$x_1 + 7x_2 - 10x_3 = 2$$

We see that $|a_{ii}| = |10| + |10| + |-10| = 30$ and $\sum_{\substack{i,j=1\\ i\neq j}}^{3} |a_{ij}| = |1| + |-3| + |2| + |1| + |1| + |7| = 15.$

Clearly, $|a_{ii}| > \sum_{\substack{i,j=1\\ i\neq j}}^{3} |a_{ij}|$. Therefore, the system is diagonally dominant.

7.3.1 Gauss–Jacobi Iterative Method

This method is, in general, used for a diagonally dominant system. That is why if the given system is not diagonally dominant, we rearrange the system, if possible, to make it diagonally dominant.

Let us consider the system

$$\begin{aligned} a_{11}x_1 + a_{12}x_2 + \cdots + a_{1n}x_n &= b_1 \\ a_{21}x_1 + a_{22}x_2 + \cdots + a_{2n}x_n &= b_2 \\ &\cdots \\ a_{n1}x_1 + a_{n2}x_2 + \cdots + a_{nn}x_n &= b_n \end{aligned}$$

We assume that the above system is diagonally dominant. Then, we re-write the system as

$$x_1 = \frac{1}{a_{11}}[b_1 - a_{12}x_2 - a_{13}x_3 - \cdots - a_{1n}x_n] \quad (1)$$

$$x_2 = \frac{1}{a_{22}}[b_2 - a_{21}x_1 - a_{23}x_3 - \cdots - a_{2n}x_n] \quad (2)$$

$$\cdots$$

$$\cdots$$

$$x_n = \frac{1}{a_{nn}}[b_n - a_{n1}x_1 - a_{n2}x_2 - \cdots - a_{nn-1}x_{n-1}] \quad (n)$$

Now, in the first iteration,

(i) Put $x_2 = x_2^{(0)}, x_3 = x_3^{(0)}, \ldots, x_n = x_n^{(0)}$ as the initial approximation in (1) and calculate $x_1^{(1)}$ for x_1 from it.

(ii) Put $x_1 = x_1^{(0)}, x_3 = x_3^{(0)}, \ldots, x_n = x_n^{(0)}$ in (2) and calculate $x_2^{(1)}$ for x_2 from it.

Similarly, calculate $x_3^{(1)}$ for x_3 from (3), $x_4^{(1)}$ for x_4 from (4),…, $x_n^{(1)}$ for x_n from (n).

Thus, after the first iteration, we have a better set of approximations, namely, $x_1^{(1)}, x_2^{(1)}, \ldots, x_n^{(1)}$.

For the second iteration,

(i) Put $x_2 = x_2^{(1)}, x_3 = x_3^{(1)}, \ldots, x_n = x_n^{(1)}$ in (1) and calculate $x_1^{(2)}$ for x_1 from it.

(ii) Put $x_1 = x_1^{(2)}, x_3 = x_3^{(1)}, \ldots, x_n = x_n^{(1)}$ in (2) and calculate $x_2^{(2)}$ for x_2 from it.

Similarly, calculate $x_3^{(2)}$ for x_3 from (3), $x_4^{(2)}$ for x_4 from (4),…, $x_n^{(2)}$ for x_n from (n).

Thus, after the kth iteration, the iterative formulae are given by

$$x_1^{(k+1)} = \frac{1}{a_{11}}\left[b_1 - a_{12}x_2^{(k)} - a_{13}x_3^{(k)} - \cdots - a_{1n}x_n^{(k)}\right] \quad (1)$$

$$x_2^{(k+1)} = \frac{1}{a_{22}}\left[b_2 - a_{21}x_1^{(k+1)} - a_{23}x_3^{(k)} - \cdots - a_{2n}x_n^{(k)}\right] \quad (2)$$

$$\cdots$$

$$\cdots$$

$$x_n^{(k+1)} = \frac{1}{a_{nn}}\left[b_n - a_{n1}x_1^{(k+1)} - a_{n2}x_2^{(k+1)} - \cdots - a_{nn-1}x_{n-1}^{(k+1)}\right] \quad (n)$$

$(k = 0, 1, 2 \ldots)$

The process is to be continued until we get the solution to the desired degree of accuracy. It should be noted that the initial approximations $x_1^{(0)}, x_2^{(0)}, \ldots, x_n^{(0)}$, if not mentioned, are all set to zero.

The method is best understood by examining the following examples.

EXAMPLE 7.18 Solve the following system of equations by the Gauss–Jacobi method:

$$20x + 2y + 6z = 28$$
$$x + 20y + 9z = -23$$
$$2x - 7y - 20z = -57$$

Solution: As the given system is diagonally dominant, re-write it as

$$x^{(k+1)} = \frac{1}{20}\left[28 - 2y^{(k)} - 6z^{(k)}\right]$$
$$y^{(k+1)} = \frac{1}{20}\left[-23 - x^{(k)} - 9z^{(k)}\right]$$
$$z^{(k+1)} = \frac{1}{20}\left[57 + 2x^{(k)} - 7y^{(k)}\right]$$

where $k = 0, 1, 2, \ldots$

Assuming the initial approximations $x^{(0)} = y^{(0)} = z^{(0)} = 0$, the iterative process is shown in the following table:

k	$x^{(k)}$	$y^{(k)}$	$z^{(k)}$	**Iterative formulae**
0	0	0	0	
1	1.4	−1.15	2.85	
2	0.66	−2.50	3.39	
3	0.633	−2.71	3.79	
4	0.534	−2.887	3.862	$x^{(k+1)} = \frac{1}{20}\left[28 - 2y^{(k)} - 6z^{(k)}\right]$
5	0.5301	−2.9146	3.9139	
6	0.5173	−2.9378	3.9231	$y^{(k+1)} = \frac{1}{20}\left[-23 - x^{(k)} - 9z^{(k)}\right]$
7	0.5169	−2.9413	3.9299	
8	0.5152	−2.9443	3.9312	$z^{(k+1)} = \frac{1}{20}\left[57 + 2x^{(k)} - 7y^{(k)}\right]$
9	0.5151	−2.9448	3.9320	
10	0.5149	−2.9452	3.9322	
11	0.5149	−2.9452	3.9323	

Hence, the required solutions are $x = 0.5149$, $y = -2.9452$, $z = 3.9323$, correct to 4 decimal places.

EXAMPLE 7.19 Using the Gauss–Jacobi method, find the solution of the following system of linear equations correct to 3 decimal places.

$$10x - 5y - 2z = 3$$
$$4x - 10y + 3z = -3$$
$$x + 6y + 10z = -3$$

Solution: We see that the given system is diagonally dominant. The iterative formulae for the Gauss–Jacobi method are given by

$$x^{(k+1)} = \frac{1}{10}\left[3 + 5y^{(k)} + 2z^{(k)}\right]$$
$$y^{(k+1)} = \frac{1}{10}\left[3 + 4x^{(k)} + 3z^{(k)}\right]$$
$$z^{(k+1)} = \frac{1}{10}\left[-3 - x^{(k)} - 6y^{(k)}\right]$$

where $k = 0, 1, 2, \ldots$

Assuming the initial approximations $x^{(0)} = y^{(0)} = z^{(0)}$, the iterative process is shown in the following table:

k	$x^{(k)}$	$y^{(k)}$	$z^{(k)}$	Iterative formulae
0	0	0	0	
1	0.3	0.3	−0.3	
2	0.39	0.33	−0.51	
3	0.363	0.303	−0.537	$x^{(k+1)} = \frac{1}{10}\left[3 + 5y^{(k)} + 2z^{(k)}\right]$
4	0.3441	0.2841	−0.5181	
5	0.33843	0.2822	−0.50487	$y^{(k+1)} = \frac{1}{10}\left[3 + 4x^{(k)} + 3z^{(k)}\right]$
6	0.340126	0.28391	−0.50316	
7	0.341323	0.28511	−0.50436	$z^{(k+1)} = \frac{1}{10}\left[-3 - x^{(k)} - 6y^{(k)}\right]$
8	0.341679	0.28522	−0.50519	
9	0.341572	0.28511	−0.50530	

Hence, the required solutions are $x = 0.342$, $y = 0.285$, $z = -0.505$, correct to 3 decimal places.

7.3.2 Gauss–Seidel Iterative Method

This method is, in general, used for a diagonally dominant system. That is why if the given system is not diagonally dominant, we rearrange it, if possible, to make it diagonally dominant.

Let us consider the system

$$a_{11}x_1 + a_{12}x_2 + \cdots + a_{1n}x_n = b_1$$
$$a_{21}x_1 + a_{22}x_2 + \cdots + a_{2n}x_n = b_2$$
$$\cdots$$
$$a_{11}x_1 + a_{12}x_2 + \cdots + a_{nn}x_n = b_1$$

We assume that the above system is diagonally dominant. Then, we re-write the system as

$$x_1 = \frac{1}{a_{11}}\left[b_1 - a_{12}x_2 - a_{13}x_3 - \cdots - a_{1n}x_n\right] \quad (1)$$

$$x_2 = \frac{1}{a_{22}}\left[b_2 - a_{21}x_1 - a_{23}x_3 - \cdots - a_{2n}x_n\right] \quad (2)$$

$$\cdots$$

$$\cdots$$

$$x_n = \frac{1}{a_{nn}}\left[b_n - a_{n1}x_1 - a_{n2}x_2 - \cdots - a_{nn-1}x_{n-1}\right] \quad (n)$$

Now, in the first iteration,

(iii) Put $x_2 = x_2^{(0)}, x_3 = x_3^{(0)}, \ldots, x_n = x_n^{(0)}$ as the initial approximations in (1) and calculate $x_1^{(1)}$ for x_1 from it.

(iv) Put $x_1 = x_1^{(1)}, x_3 = x_3^{(0)}, \ldots, x_n = x_n^{(0)}$ in (2) and calculate $x_2^{(1)}$ for x_2 from it.

Similarly, calculate $x_3^{(1)}$ for x_3 from (3), $x_4^{(1)}$ for x_4 from (4),..., $x_n^{(1)}$ for x_n from (n). Thus, after the first iteration, we have a better set of approximations, namely, $x_1^{(1)}, x_2^{(1)}, \ldots, x_n^{(1)}$.

For the second iteration,

(iii) Put $x_2 = x_2^{(1)}, x_3 = x_3^{(1)}, \ldots, x_n = x_n^{(1)}$ in (1) and calculate $x_1^{(2)}$ for x_1 from it.

(iv) Put $x_1 = x_1^{(2)}, x_3 = x_3^{(1)}, \ldots, x_n = x_n^{(1)}$ in (2) and calculate $x_2^{(2)}$ for x_2 from it.

Similarly, calculate $x_3^{(2)}$ for x_3 from (3), $x_4^{(2)}$ for x_4 from (4),..., $x_n^{(2)}$ for x_n from (n).

Thus, after k^{th} iteration, the iterative formulae are given by

$$x_1^{(k+1)} = \frac{1}{a_{11}}\left[b_1 - a_{12}x_2^{(k)} - a_{13}x_3^{(k)} - \cdots - a_{1n}x_n^{(k)}\right] \quad (1)$$

$$x_2^{(k+1)} = \frac{1}{a_{22}}\left[b_2 - a_{21}x_1^{(k+1)} - a_{23}x_3^{(k)} - \cdots - a_{2n}x_n^{(k)}\right] \quad (2)$$

$$\cdots$$

$$\cdots$$

$$x_n^{(k+1)} = \frac{1}{a_{nn}}\left[b_n - a_{n1}x_1^{(k+1)} - a_{n2}x_2^{(k+1)} - \cdots - a_{nn-1}x_{n-1}^{(k+1)}\right] \quad (n)$$

$(k = 0, 1, 2, \ldots)$

The process is to be continued until we get the solution to the desired degree of accuracy. It should be noted that the initial approximations $x_2^{(0)}, x_3^{(0)}, \ldots, x_n^{(0)}$, if not mentioned, are all set to zero.

The method is best understood by examining the following examples.

EXAMPLE 7.20 Solve the following system of equations by the Gauss–Seidel method:

$$3x + 4y + 15z = 54.8$$

$$x + 12y + z = 39.66$$

$$10x + y - 2z = 7.74$$

Solution: We see that the given system is not diagonally dominant. We, therefore, rearrange the system as

$$10x + y - 2z = 7.74$$
$$x + 12y + z = 39.66$$
$$3x + 4y + 15z = 54.8$$

We now re-write the diagonally dominant system as

$$x = \frac{1}{10}[7.74 - y + 2z] \quad (1)$$

$$y = \frac{1}{12}[39.66 - x - z] \quad (2)$$

$$z = \frac{1}{15}[54.8 - 3x - 4y] \quad (3)$$

Step 1: Putting the initial approximations $x^{(0)} = z^{(0)} = 0$ in (1), we get $x^{(1)} = 0.774$.
Next, putting $z^{(0)} = 0$ and $x^{(1)} = 0.774$ in (2), we get $y^{(1)} = 3.2405$.
Finally, putting $x^{(1)} = 0.774$ and $y^{(1)} = 3.2405$ in (3), we get $z^{(1)} = 2.6344$.

Step 2: Putting $y^{(1)} = 3.2405$ and $z^{(1)} = 2.6344$ in (1), we get $x^{(2)} = 0.97683$.
Putting $z^{(1)} = 2.6344$ and $x^{(2)} = 0.97683$ in (2), we get $y^{(2)} = 3.00406$.
Putting $x^{(2)} = 0.97683$ and $y^{(2)} = 3.00406$ in (3), we get $z^{(2)} = 2.65688$.

Step 3: Putting $y^{(2)} = 3.00406$ and $z^{(2)} = 2.65688$ in (1), we get $x^{(3)} = 1.00497$
Putting $z^{(2)} = 2.65688$ and $x^{(3)} = 1.00497$ in (2), we get $y^{(3)} = 2.99985$
Putting $x^{(3)} = 1.00497$ and $y^{(3)} = 2.99985$ in (3), we get $z^{(3)} = 2.65238$

Step 4: Putting $y^{(3)} = 2.99985$ and $z^{(3)} = 2.65238$ in (1), we get $x^{(4)} = 1.00449$
Putting $z^{(3)} = 2.65238$ and $x^{(4)} = 1.00449$ in (2), we get $y^{(4)} = 3.00026$
Putting $x^{(4)} = 1.00449$ and $y^{(4)} = 3.00026$ in (3), we get $z^{(4)} = 2.65237$

Thus, the solutions of the given system are $x = 1.004$, $y = 3.000$, $z = 2.652$, correct to 3 decimal places.
The above iterative procedure may be shown by the following table:

k	$x^{(k)}$	$y^{(k)}$	$z^{(k)}$	Iterative formulae
0	–	0	0	$x^{(k+1)} = \frac{1}{10}\left[7.74 - y^{(k)} + 2z^{(k)}\right]$
1	0.774	3.2405	2.6344	
2	0.97683	3.00406	2.65688	$y^{(k+1)} = \frac{1}{12}\left[39.66 - x^{(k+1)} - z^{(k)}\right]$
3	1.00497	2.99985	2.65238	$z^{(k+1)} = \frac{1}{15}\left[54.8 - 3x^{(k+1)} - 4y^{(k+1)}\right]$
4	1.00449	3.00026	2.65237	

EXAMPLE 7.21 Using the Gauss–Seidel method, find the solution of the following system of linear equations correct to 2 decimal places:

$$9x_1 - 2x_2 + x_3 = 50$$
$$x_1 + 5x_2 - 3x_3 = 18$$
$$-2x_1 + 2x_2 + 7x_3 = 19$$

Solution: We see that the given system is diagonally dominant. The iterative formulae for the Gauss–Seidel method are given by

$$x_1^{(k+1)} = \frac{1}{9}\left[50 + 2x_2^{(k)} - x_3^{(k)}\right] \quad (1)$$

$$x_2^{(k+1)} = \frac{1}{5}\left[18 - x_1^{(k+1)} + 3x_3^{(k)}\right] \quad (2)$$

$$x_3^{(k+1)} = \frac{1}{7}\left[19 + 2x_1^{(k+1)} - 2x_2^{(k+1)}\right] \quad (3)$$

where $k = 0, 1, 2, \ldots$

Now, putting $x_2^{(0)} = 0$ and $x_3^{(0)} = 0$ in (1), the iterative process is shown in the table given below:

k	$x_1^{(k)}$	$x_2^{(k)}$	$x_3^{(k)}$	Iterative formulae
0	–	0	0	
1	5.556	2.488	3.591	
2	5.709	4.613	3.027	
3	6.244	4.1674	3.308	$x_1^{(k+1)} = \frac{1}{9}\left[50 + 2x_2^{(k)} - x_3^{(k)}\right]$
4	6.114	4.362	3.215	
5	6.227	4.284	3.269	$x_2^{(k+1)} = \frac{1}{5}\left[18 - x_1^{(k+1)} + 3x_3^{(k)}\right]$
6	6.144	4.333	3.232	
7	6.159	4.307	3.243	$x_3^{(k+1)} = \frac{1}{7}\left[19 + 2x_1^{(k+1)} - 2x_2^{(k+1)}\right]$
8	6.152	4.315	3.239	
9	6.155	4.312	3.241	

Hence, the required solutions are $x_1 = 6.15$, $x_2 = 4.31$, $x_3 = 3.24$, correct to 2 decimal places.

EXAMPLE 7.22 Using the Gauss–Seidel method, find the solution of the following system of linear equations correct to 2 decimal places:

$$3x + y + 5z = 13$$
$$5x - 2y + z = 4$$
$$x + 6y - 2z = -1$$

Solution: We see that the given system is not diagonally dominant. We therefore rearrange the system as

$$5x - 2y + z = 4$$
$$x + 6y - 2z = -1$$
$$3x + y + 5z = 13$$

Now, the Gauss–Seidel iterative formulae for this system are given by

$$x^{(k+1)} = \frac{1}{5}\left[4 + 2y^{(k)} - z^{(k)}\right] \quad (1)$$

$$y^{(k+1)} = \frac{1}{6}\left[-1 - x^{(k+1)} + 2z^{(k)}\right] \tag{2}$$

$$z^{(k+1)} = \frac{1}{5}\left[13 - 3x^{(k+1)} - y^{(k+1)}\right] \tag{3}$$

where $k = 0, 1, 2, \ldots$

Putting the initial approximations $y^{(0)} = z^{(0)} = 0$ in (1), the iterative process is shown in the following table:

k	$x^{(k)}$	$y^{(k)}$	$z^{(k)}$	**Iterative formulae**
0	–	0	0	
1	0.8	−0.3	2.18	
2	0.244	0.5193	1.958	
3	0.616	0.383	2.1538	$x^{(k+1)} = \frac{1}{5}\left[4 + 2y^{(k)} - z^{(k)}\right]$
4	0.522	0.464	2.194	$y^{(k+1)} = \frac{1}{6}\left[-1 - x^{(k+1)} + 2z^{(k)}\right]$
5	0.547	0.474	2.177	$z^{(k+1)} = \frac{1}{5}\left[13 - 3x^{(k+1)} - y^{(k+1)}\right]$
6	0.554	0.467	2.174	
7	0.552	0.466	2.176	
8	0.551	0.467	2.176	

Hence, the required solutions are $x = 0.55, y = 0.47, z = 2.18$, correct to 2 decimal places.

SOLVED EXAMPLES

EXAMPLE 7.23 Solve the following system of equations using the Gaussian elimination method:

$$x + y + z = 9$$
$$2x - 3y + 4z = 13$$
$$3x + 4y + 5z = 40$$

Solution: The given system is equivalent to

$$\begin{bmatrix} 1 & 1 & 1 \\ 2 & -3 & 4 \\ 3 & 4 & 5 \end{bmatrix}\begin{bmatrix} x \\ y \\ z \end{bmatrix} = \begin{bmatrix} 9 \\ 13 \\ 40 \end{bmatrix}$$

The augmented matrix is

$$\left[\begin{array}{ccc|c} 1 & 1 & 1 & 9 \\ 2 & -3 & 4 & 13 \\ 3 & 4 & 5 & 40 \end{array}\right]$$

$$\sim \left[\begin{array}{ccc|c} 1 & 1 & 1 & 9 \\ 0 & -5 & 2 & -5 \\ 0 & 1 & 2 & 13 \end{array}\right], \text{ by } R_2' = R_2 - 2R_1$$

and $$R_3' = R_3 - 3R_1$$

$$\sim \left[\begin{array}{ccc|c} 1 & 1 & 1 & 9 \\ 0 & -5 & 2 & -5 \\ 0 & 0 & 12/5 & 12 \end{array}\right], \text{ by } R_3' = R_3 + \frac{1}{5}R_2$$

Therefore $$\frac{12}{5}z = 12, \text{ i.e., } z = 5$$

$$-5y + 2z = -5, \text{ i.e., } y = 3$$

and $$x + y + z = 9, \text{ i.e., } x = 1.$$

Hence, the required solution is (1, 3, 5).

EXAMPLE 7.24 Solve the following system of linear equations by the Gaussian elimination method:

$$3x + 4y + 5z = 18$$
$$2x - y + 8z = 13$$
$$5x - 2y + 7z = 20$$

Solution: The given system is equivalent to

$$\begin{bmatrix} 3 & 4 & 5 \\ 2 & -1 & 8 \\ 5 & -2 & 7 \end{bmatrix}\begin{bmatrix} x \\ y \\ z \end{bmatrix} = \begin{bmatrix} 18 \\ 13 \\ 20 \end{bmatrix}$$

The augmented matrix is

$$\left[\begin{array}{ccc|c} 3 & 4 & 5 & 18 \\ 2 & -1 & 8 & 13 \\ 5 & -2 & 7 & 20 \end{array}\right]$$

$$\sim \left[\begin{array}{ccc|c} 3 & 4 & 5 & 18 \\ 0 & -11/3 & 14/3 & 1 \\ 0 & 26/3 & -4/3 & -10 \end{array}\right], \text{ by } R_2' = R_2 - \frac{2}{3}R_1$$

and $$R_3' = R_3 - \frac{5}{3}R_1$$

$$\sim \left[\begin{array}{ccc|c} 3 & 4 & 5 & 18 \\ 0 & -11/3 & 14/3 & 1 \\ 0 & 0 & -136/11 & -136/11 \end{array}\right], \text{ by } R_3' = R_3 - \frac{26}{11}R_2$$

Therefore $$-\frac{136}{11}z = -\frac{136}{11}, \text{ i.e., } z = 1$$

$$-\frac{11}{3}y + \frac{14}{3}z = 1, \text{ i.e., } y = 1$$

and $$3x + 4y + 5z = 18, \text{ i.e., } x = 3$$

Hence, the required solution is (3, 1, 1)

EXAMPLE 7.25 Solve the system of equations by the Gauss elimination method, correct to 3 significant figures:

$$2x + 3y + z = 9$$
$$x + 2y + 2z = 6$$
$$3x + y + 2z = 8$$

Solution: The given system is equivalent to

$$\begin{bmatrix} 2 & 3 & 1 \\ 1 & 2 & 3 \\ 3 & 1 & 2 \end{bmatrix}\begin{bmatrix} x \\ y \\ z \end{bmatrix} = \begin{bmatrix} 9 \\ 6 \\ 8 \end{bmatrix}$$

The augmented matrix is

$$\left[\begin{array}{ccc|c} 2 & 3 & 1 & 9 \\ 1 & 2 & 3 & 6 \\ 3 & 1 & 2 & 8 \end{array}\right]$$

$$\sim \left[\begin{array}{ccc|c} 2 & 3 & 1 & 9 \\ 0 & 1/2 & 5/2 & 3/2 \\ 0 & -7/2 & 1/2 & -11/2 \end{array}\right]$$

by $$R_2' = R_2 - \frac{1}{2}R_1$$

and $$R_3' = R_3 - \frac{3}{2}R_1$$

$$\sim \left[\begin{array}{ccc|c} 2 & 3 & 1 & 9 \\ 0 & 1/2 & 5/2 & 3/2 \\ 0 & 0 & 18 & 5 \end{array}\right]$$

by $$R_3' = R_3 + 7R_2$$

Therefore $$18z = 5, \text{ i.e., } z = 0.278$$

$$\frac{1}{2}y + \frac{5}{2}z = \frac{3}{2}, \text{ i.e., } y = 1.61$$

and $$2x + 3y + z = 9, \text{ i.e., } x = 1.95$$

Hence, the required solution is (1.95, 1.61, 0.278), correct to 3 significant figures.

EXAMPLE 7.26 Solve the following system of equations by the LU factorisation method:

$$8x_1 - 3x_2 + 2x_3 = 20$$
$$4x_1 + 11x_2 - x_3 = 33$$
$$6x_1 + 3x_2 + 12x_3 = 36$$

Solution: The given system is of the form

$$AX = B$$

where
$$A=\begin{bmatrix}8 & -3 & 2\\ 4 & 11 & -1\\ 6 & 3 & 12\end{bmatrix},\ X=\begin{bmatrix}x_1\\ x_2\\ x_3\end{bmatrix}$$

and
$$B=\begin{bmatrix}20\\ 33\\ 36\end{bmatrix}$$

Let
$$A=LU$$

where
$$L=\begin{bmatrix}l_{11} & 0 & 0\\ l_{21} & l_{22} & 0\\ l_{31} & l_{32} & l_{33}\end{bmatrix}$$

and
$$U=\begin{bmatrix}1 & u_{12} & u_{13}\\ 0 & 1 & u_{23}\\ 0 & 0 & 1\end{bmatrix}$$

Therefore
$$\begin{bmatrix}8 & -3 & 2\\ 4 & 11 & -1\\ 6 & 3 & 12\end{bmatrix}=\begin{bmatrix}l_{11} & 0 & 0\\ l_{21} & l_{22} & 0\\ l_{31} & l_{32} & l_{33}\end{bmatrix}\begin{bmatrix}1 & u_{12} & u_{13}\\ 0 & 1 & u_{23}\\ 0 & 0 & 1\end{bmatrix}$$
$$=\begin{bmatrix}l_{11} & l_{11}u_{12} & l_{11}u_{13}\\ l_{21} & l_{21}u_{12}+l_{22} & l_{21}u_{13}+l_{22}u_{23}\\ l_{31} & l_{31}u_{12}+l_{32} & l_{31}u_{13}+l_{32}u_{23}+l_{33}\end{bmatrix}$$

Comparing the elements, we get
$$l_{11}=8,\ l_{21}=4,\ l_{31}=6$$
$$l_{11}u_{12}=-3,\text{ i.e., } u_{12}=-3/8$$
$$l_{11}u_{13}=2,\text{ i.e., } u_{13}=1/4$$
$$l_{21}u_{12}+l_{22}=11,\text{ i.e., } l_{22}=11+\frac{3}{2}=25/2$$
$$l_{31}u_{12}+l_{32}=3,\text{ i.e., } l_{32}=3+9/4=21/4$$
$$l_{21}u_{13}+l_{22}u_{23}=-1,\text{ i.e., } u_{23}=\left(-1-4\times\frac{1}{4}\right)\times\left(\frac{2}{25}\right)=-4/25$$
$$l_{31}u_{13}+l_{32}u_{23}+l_{33}=12$$

$\Rightarrow$
$$l_{33}=12-6\times\frac{1}{4}-\frac{21}{4}\times\left(-\frac{4}{25}\right)=567/50$$

Therefore, we have
$$L=\begin{bmatrix}8 & 0 & 0\\ 4 & 25/2 & 0\\ 6 & 21/4 & 567/50\end{bmatrix}$$

$$U = \begin{bmatrix} 1 & -3/8 & 1/4 \\ 0 & 1 & -4/25 \\ 0 & 0 & 1 \end{bmatrix}$$

and
$$\begin{bmatrix} 8 & 0 & 0 \\ 4 & 25/2 & 0 \\ 6 & 21/4 & 567/50 \end{bmatrix}\begin{bmatrix} 1 & -3/8 & 1/4 \\ 0 & 1 & -4/25 \\ 0 & 0 & 1 \end{bmatrix}\begin{bmatrix} x_1 \\ x_2 \\ x_3 \end{bmatrix} = \begin{bmatrix} 20 \\ 33 \\ 36 \end{bmatrix}$$

Let
$$UX = Y$$

where
$$Y = \begin{bmatrix} y_1 \\ y_2 \\ y_3 \end{bmatrix}$$

Therefore
$$\begin{bmatrix} 8 & 0 & 0 \\ 4 & 25/2 & 0 \\ 6 & 21/4 & 567/50 \end{bmatrix}\begin{bmatrix} y_1 \\ y_2 \\ y_3 \end{bmatrix} = \begin{bmatrix} 20 \\ 33 \\ 36 \end{bmatrix}$$

which gives
$$8y_1 = 20, \text{ i.e., } y_1 = 5/2$$

$$4y_1 + \frac{25}{2}y_2 = 33, \text{ i.e., } y_2 = 46/25$$

$$6y_1 + \frac{21}{4}y_2 + \frac{567}{50}y_3 = 36, \text{ i.e., } y_3 = 73/27$$

Therefore, we get
$$\begin{bmatrix} 1 & -3/8 & 1/4 \\ 0 & 1 & -4/25 \\ 0 & 0 & 1 \end{bmatrix}\begin{bmatrix} x_1 \\ x_2 \\ x_3 \end{bmatrix} = \begin{bmatrix} 5/2 \\ 46/25 \\ 73/27 \end{bmatrix}$$

which gives
$$x_3 = 73/27 = 2.704$$

$$x_2 - \frac{4}{25}x_3 = \frac{46}{25}, \text{ i.e., } x_2 = 2.273$$

$$x_1 - \frac{3}{8}x_2 + \frac{1}{4}x_3 = \frac{5}{2}, \text{ i.e., } x_1 = 2.324$$

Hence, the required solution is given by $x_1 = 2.324$, $x_2 = 2.273$, $x_3 = 2.704$, correct to 3 decimal places.

MULTIPLE CHOICE QUESTIONS

1. The Gaussian elimination method does not fail even if one of the pivotal elements is equal to zero.
   (a) true (b) false
2. One of the iterative methods by which we can find the solution of simultaneous equations is
   (a) Gauss–Seidel method (b) Gauss elimination method (c) Crout's method
3. To solve a system of m equations of m unknowns, the total number of multiplications and divisions involved in the Gauss elimination method is of the order $m^3/3$ approximately.
   (a) true (b) false

4. In the Gauss elimination method, the given system of equations represented by $AX = B$ is converted to another system $UX = Y$, where U is
   (a) diagonal matrix (b) upper triangular matrix
   (c) identity matrix (d) null matrix
5. When the Gauss elimination method is used to solve $AX = B$, A is transformed to
   (a) null matrix (b) upper triangular matrix
   (c) identity matrix (d) diagonally dominant matrix
6. An $n \times n$ matrix A is said to be diagonally dominant if
   (a) $|a_{ii}| \le \sum_{\substack{i,j=1 \\ i \ne j}}^{n} |a_{ij}|$ (b) $|a_{ii}| < \sum_{\substack{i,j=1 \\ i \ne j}}^{n} |a_{ij}|$
   (c) $|a_{ii}| > \sum_{\substack{i,j=1 \\ i \ne j}}^{n} |a_{ij}|$ (d) $|a_{ii}| \ge \sum_{\substack{i,j=1 \\ i \ne j}}^{n} |a_{ij}|$
7. When the Gauss elimination method is used to solve $AX = B$, A is transformed to
   (a) unit matrix (b) lower triangular matrix
   (c) diagonally dominant matrix (d) upper triangular matrix
8. Given $A = \begin{bmatrix} 6 & 2 & 3 & 9 \\ 0 & 1 & 2 & 3 \\ 0 & 0 & 4 & 5 \end{bmatrix}$, A is a __________ matrix.
   (a) diagonal (b) identity (c) lower triangular (d) upper triangular
9. The Gauss–Seidel method for the solution of a system of linear simultaneous equations converges if
   (a) $|a_{ii}| \ge \sum_{\substack{i,j=1 \\ i \ne j}}^{n} |a_{ij}|$ (b) $|a_{ii}| > \sum_{\substack{i,j=1 \\ i \ne j}}^{n} |a_{ij}|$
   (c) $\dfrac{|a_{ii}|}{|a_{nn}|} = 1$ (d) none of these
10. A matrix is said to be lower triangular, if and only if,
    (a) all the elements in the principal diagonal are zero
    (b) all the elements above the principal diagonal are zero
    (c) all the elements below the principal diagonal are zero
    (d) none of these
11. Inverse of a matrix A is given by
    (a) $A^{-1} = AdjA/DetA$ (b) $A^{-1} = DetA/Adj\,A$
    (c) $A^{-1} = DetA$ (d) none of these
12. For a system of equation $Ax = b$, a solution exists if and only if A is
    (a) symmetric (b) singular (c) orthogonal (d) diagonal

13. Inverse of a matrix A is given by

 (a) $A^{-1} = \frac{adjA}{detA}$ (b) $A^{-1} = \frac{detA}{adjA}$ (c) $A^{-1} = (detA)^T$ (d) $A^{-1} = (adjA)^T$

14. Diagonal dominance is a must for

 (a) Gauss–Seidel method (b) Gauss–Jordon's matrix inversion method
 (c) Gauss elimination method (d) none of these

15. The iterative method to solve a system of equations is

 (a) Gauss elimination (b) Gauss–Jordon
 (c) Gauss–Seidel (d) none of these

16. Which is the direct method?

 (a) Gauss elimination method (b) Gauss–Jacobi method
 (c) Gauss–Seidel method (d) none of these

17. Pivoting is essential because

 (a) determinant of the coefficient matrix should be greater than zero
 (b) pivot element should not have a very large value compared to the elements of the matrix
 (c) it reduces the possibility of division by zero
 (d) change of convergence is higher

18. Which of the following is an iterative method?

 (a) Gauss elimination method (b) Gauss–Jordon method
 (c) LU decomposition method (d) Gauss–Seidel method

19. The order of convergence of the Gauss–Seidel method is

 (a) 2 (b) 3 (c) 1 (d) 4

20. The order of operations needed for LU decomposition method is

 (a) (b) (c) (d)

21. The Gauss elimination method fails if any one of the pivotal elements becomes

 (a) zero (b) positive (c) negative (d) none of these

22. A matrix A can be factorised into lower and upper triangular matrices if all the principal minors of A are

 (a) non-singular (b) singular (c) null (d) none of these

23. In LU factorisation, a matrix A is factorised into L and U, where L is

 (a) identity matrix (b) upper triangular matrix
 (c) diagonal matrix (d) lower triangular matrix

24. In LU factorisation, a matrix A is factorised into L and U, where U is

 (a) identity matrix (b) upper triangular matrix
 (c) diagonal matrix (d) lower triangular matrix

25. In the Gauss elimination method, the substitution procedure is

 (a) forward (b) backward (c) both (a) and (b) (d) either (a) or (b)

Answers

1. (b) 2. (a) 3. (a) 4. (b) 5. (b) 6. (c) 7. (d) 8. (d) 9. (b) 10. (b) 11. (a)
12. (b) 13. (a) 14. (a) 15. (c) 16. (a) 17. (c) 18. (d) 19. (c) 20. (a) 21. (a) 22. (d)
23. (b) 24. (b)

SHORT ANSWER QUESTIONS

1. Find the inverse of the matrix $A = \begin{bmatrix} 1 & 3 & 3 \\ 1 & 4 & 3 \\ 1 & 3 & 4 \end{bmatrix}$ by the Gauss elimination method.

2. Find the inverse of the matrix $\begin{pmatrix} 1 & 2 & 6 \\ 2 & 5 & 15 \\ 6 & 15 & 46 \end{pmatrix}$ by the Gauss elimination method.

3. Find the inverse of the matrix A given by $A = \begin{bmatrix} 2 & 3 & -1 \\ 4 & 4 & -3 \\ 2 & -3 & 1 \end{bmatrix}$ by the Gauss elimination method.

4. Find A^{-1}, if $A = \begin{bmatrix} 8 & -4 & 0 \\ -4 & 8 & -4 \\ 0 & -4 & 8 \end{bmatrix}$ by the Gauss elimination method.

5. Solve the following system of equations by the Gauss elimination method:

$$5x - y + z = 10$$
$$2x + 4y = 12$$
$$x + y + 5z = -1$$

6. Explain the Gauss–Seidel method for solving a system of linear equations.

7. Solve the following system of linear equations by the Gauss elimination method:

$$x - 2y + 9z = 8$$
$$3x + y - z = 3$$
$$2x - 8y + z = -5$$

8. Find the inverse of the matrix $\begin{bmatrix} 2 & 1 & 0 \\ 4 & 3 & 1 \\ 1 & 1 & 1 \end{bmatrix}$ by the LU factorisation method.

9. Solve the following system by the matrix inverse method:

$$2x + y + z = 10$$
$$3x + 2y + 3z = 18$$
$$x + 4y + 9z = 16$$

10. Solve by Gauss elimination method the following system of equations:

$$2x + y + 4z = 16$$
$$3x + 2y + z = 10$$
$$x + 3y + 3z = 1$$

11. Solve the following system of linear equations by the Gauss elimination method:

$$5x_1 - x_2 = 9$$
$$-x_1 + 5x_2 - x_3 = 4$$
$$-x_2 + 5x_3 = -6$$

12. Find the inverse of the following matrix: $\begin{bmatrix} 3 & -1 & 1 \\ -15 & 6 & -5 \\ 5 & -2 & 2 \end{bmatrix}$

13. Solve the following system of equations using the Gaussian elimination method:

$$x + y + z = 9$$
$$2x - 3y + 4z = 13$$
$$3x + 4y + 5z = 40$$

14. Find the inverse of the matrix by the Gauss elimination method: $\begin{pmatrix} 2 & -2 & 4 \\ 2 & 3 & 2 \\ -1 & 4 & -1 \end{pmatrix}$

15. Use Gauss elimination to solve

$$3x - 2y + 2z = 12$$
$$x + 2y + 3z = 11$$
$$2x - 2y - z = 3$$

16. Solve the following system of equations by the Gauss elimination method correct to 3 significant figures:

$$2x + 3y + z = 9$$
$$x + 2y + 3z = 6$$
$$3x + y + 2z = 8$$

17. Solve by the Gauss elimination method:

$$x + 2y + 3z = 10$$
$$x + 3y - 2z = 7$$
$$x - y + z = 5$$

18. Find the inverse of the matrix $\begin{bmatrix} 2 & 0 & 1 \\ 2 & 1 & 3 \\ 1 & -1 & 0 \end{bmatrix}$ by Gauss's method.

19. Solve the following system of linear equations by the Gaussian elimination method:

$$3x + 4y + 5z = 18$$
$$2x - y + 8z = 13$$
$$5x - 2y + 7z = 20$$

20. Find the inverse of the matrix $A = \begin{bmatrix} 3 & 2 & 1 \\ 2 & 3 & 2 \\ 1 & 2 & 2 \end{bmatrix}$ using the Gaussian elimination method.

Answers

1. $\begin{bmatrix} 7 & -3 & -3 \\ -1 & 1 & 0 \\ -1 & 0 & 1 \end{bmatrix}$ 2. $\begin{pmatrix} 5 & -2 & 0 \\ -2 & 10 & -3 \\ 0 & -3 & 1 \end{pmatrix}$ 3. $\frac{1}{20}\begin{bmatrix} 5 & 0 & 5 \\ 10 & -4 & -2 \\ 20 & -12 & 4 \end{bmatrix}$ 4. $\frac{1}{16}\begin{bmatrix} 3 & 2 & 1 \\ 2 & 4 & 2 \\ 1 & 2 & 3 \end{bmatrix}$

5. $x = 2.56, y = 1.72, z = -1.06$ 6. See 7.3.1 7. $x = 1, y = 1, z = 1$ 8. $\begin{bmatrix} 2 & -1 & 1 \\ -3 & 2 & -2 \\ 1 & -1 & 2 \end{bmatrix}$

9. $x = 7, y = -9, z = 5$ 10. $x = 1, y = 2, z = 3$ 11. $x_1 = 2, x_2 = 1, x_3 = -1$ 12. $\begin{bmatrix} 2 & 0 & -1 \\ 5 & 1 & 0 \\ 0 & 1 & 3 \end{bmatrix}$

13. $x = 1, y = 3, z = 5$ 14. $\frac{1}{22}\begin{pmatrix} -11 & 14 & -16 \\ 0 & 2 & 4 \\ 11 & -6 & 10 \end{pmatrix}$ 15. $x = 3.43, y = 1.00, z = 1.86$

16. $x = 1.94, y = 1.61, z = 0.278$ 17. $x = 5.29, y = 1.12, z = 0.82$ 18. $\frac{1}{3}\begin{bmatrix} 3 & -1 & -1 \\ 3 & -1 & -1 \\ -3 & 2 & 2 \end{bmatrix}$

19. $x = 3, y = 1, z = 1$ 20. $\frac{1}{3}\begin{bmatrix} 2 & -2 & 1 \\ -2 & 5 & -4 \\ 1 & -4 & 5 \end{bmatrix}$

LONG ANSWER QUESTIONS

1. Use the LU decomposition method to solve the following system of equations:

$$\begin{bmatrix} 1 & 2 & 2 \\ 1 & 1 & 1 \\ 2 & 2 & 1 \end{bmatrix}\begin{bmatrix} x \\ y \\ z \end{bmatrix} = \begin{bmatrix} 1 \\ 1 \\ 1 \end{bmatrix}$$

2. Use the Gauss–Seidel method to solve the following system of linear equations, correct to 3 significant figures:

$$x_1 + x_2 + 4x_3 = 9$$
$$9x_1 - 3x_2 + 2x_3 = 20$$
$$4x_1 + 11x_2 - x_3 = 33$$

3. Solve the following system of equations by the LU factorisation method:

$$3x + 4y + 2z = 15$$
$$5x + 2y + z = 18$$
$$2x + 3y + 2z = 10$$

4. Solve the following system of equations by the LU factorisation method:

$$8x_1 - 3x_2 + 2x_3 = 20$$
$$4x_1 + 11x_2 - x_3 = 33$$
$$6x_1 + 3x_2 + 12x_3 = 36$$

5. Solve the system of equations given below by LU factorisation:

$$2x - 3y + 10z = 3$$
$$-x + 4y + 2z = 3$$
$$5x + 2y + z = -12$$

6. Using the Gauss–Seidel method, find the solution of the following system of linear equations correct to 2 decimal places:

$$3x + y + 5z = 13$$
$$5x - 2y + z = 4$$
$$x + 6y - 2z = -1$$

7. Solve the following system of equations by the LU factorisation method:

$$2x + y + z = 3$$
$$x + 3y + z = -2$$
$$x + y + 4z = -6$$

8. Solve the following set of equations by the Gauss–Seidel method correct to 2 decimal places:

$$3y - 2z = 3$$
$$2x - y + 4z = 27$$
$$4x - y - 3z = 3$$

9. Solve the following system of equations by the LU decomposition method:

$$x_1 + x_2 - x_3 = 2$$
$$2x_1 + 3x_2 + 5x_3 = -3$$
$$3x_1 + 2x_2 - 3x_3 = 6$$

10. Using the Gauss–Seidel method, find the solution of the following system of linear equations correct to 2 decimal places:

$$9x_1 - 2x_2 - x_3 = 50$$
$$x_1 + 5x_2 - 3x_3 = 18$$
$$-2x_1 + 2x_2 + 7x_3 = 19$$

11. Check whether the following system of equations is diagonally dominant. If not, re-arrange them and solve by the Gauss–Seidel method.

$$-2x + 3y + 10z = 22$$
$$x + 10y - z = -22$$
$$10x + 2y + z = 9$$

12. Solve the system of linear equations by the Gauss–Seidel method (correct to 3 decimal places):

$$3x + 4y + 15z = 54.8$$
$$x + 12y + 3z = 39.66$$
$$10x + y - 2z = 7.74$$

13. Solve the following system by the LU factorisation method:

$$2x_1 - 3x_2 + 4x_3 = 8$$
$$x_1 + x_2 + 4x_3 = 15$$
$$3x_1 + 4x_2 - x_3 = 8$$

14. Solve the following system of equations by the LU factorisation method:

$$2x - 6y + 8z = 24$$
$$5x + 4y - 3z = 2$$
$$3x + y + 2z = 16$$

15. Solve the following system of equations, correct to 4 decimal places, by the Gauss–Seidel iteration method:

$$x + y + 54z = 110$$
$$27x + 6y - z = 85$$
$$6x + 15y + 2z = 72$$

16. Using the matrix factorisation method, solve the following system of equations:

$$x + 3y + z = 9$$
$$x + 4y + 2z = 4$$
$$x + 2y - 3z = 6$$

17. Solve the system of equations by the Gauss–Seidel method:

$$10x + 2y + z = 9$$
$$2x + 20y - 2z = -44$$
$$-2x + 3y + 10z = 22$$

18. Solve the following system of equations by the LU factorisation method:

$$x + y + z = 9$$
$$2x - 3y + 4z = 13$$
$$3x + 4y + 5z = 40$$

19. Solve the following system of equations using the LU factorisation method:

$$3x - 2y + 2z = 12$$
$$x + 2y + 3z = 11$$
$$2x - 2y - z = 2$$

20. Solve by LU factorisation method:

$$2x - 3y + z = -1$$
$$x + 2y - 3z = -4$$
$$x - 4y + z = -6$$

21. Solve the following system of linear equations by the Gauss–Seidel iterative method:

$$9x + 2y + 3z = -7$$
$$x - 6y + 2z = -2$$
$$x + y + 3z = 5$$

22. Solve the following system of equations by the LU factorisation method:

$$3x + 2y + 7z = 4$$
$$2x + 3y + z = 5$$
$$3x + 4y + z = 7$$

Answers

1. $x = 1, y = -1, z = 1$
2. $x_1 = 3, x_2 = 2, x_3 = 1$
3. $x = 3, y = 2, z = 1$
4. $x_1 = 3, x_2 = 2, x_3 = 1$
5. $x = -2.45, y = -0.23, z = 0.72$
6. $x = 0.55, y = 0.47, z = 2.18$
7. $x = 3, y = -1, z = -2$
8. $x = 5.61, y = 4.36, z = 5.03$
9. $x_1 = 1, x_2 = 0, x_3 = -1$
10. $x_1 = 6.89, x_2 = 4.30, x_3 = 3.46$
11. The system is not diagonally dominant. The diagonally dominant system is

$$10x + 2y + z = 9$$
$$x + 10y - z = -22$$
$$-2x + 3y + 10z = 22$$

Soln. $x = 1, y = -2, z = 3$

12. $x = 1.075, y = 2.524, z = 2.765$
13. $x_1 = 1, x_2 = 2, x_3 = 3$
14. $x = 1, y = 3, z = 5$
15. $x = 2.43, y = 3.57, z = 1.93$
16. $x = 29.33, y = -7.67, z = 2.67$
17. $x = 1, y = -2, z = 3$
18. $x = 1, y = 3, z = 5$
19. $x = 2.71, y = 0.50, z = 2.43$
20. $x = 2, y = 3, z = 4$
21. $x = -1.59, y = 0.72, z = 1.96$

8

Curve Fitting and Spline Interpolation

8.1 CURVE FITTING

8.1.1 Introduction

In many branches of engineering and mathematical sciences, we come across experimental problems that involve two variables, say x and y. The problem is to find a suitable relation that may exist between these two variables. Our aim then is to estimate the best values of the constants involved in that relation. For this purpose, we take several sets of values of x and y.

The process of finding the equation of curve of best fit, which may be the most suitable one for predicting the unknown values, is known as curve fitting. Thus, curve fitting means an exact relationship between two variables.

The equation of the curve may be linear, exponential, logarithmic or quadratic in nature. In this context, we shall discuss some of these equations by applying the method of least squares.

8.1.2 Method of Least Squares

Let us consider the n-paired observations $(x_1, y_1), (x_2, y_2), \ldots, (x_n, y_n)$ of two variables x and y. Suppose we need to fit a polynomial of degree m of the type

$$y = a_0 + a_1x + a_2x^2 + \cdots + a_mx^m \tag{8.1}$$

of these values. Our aim is to determine the constants $a_0, a_1, \ldots, a_m$ so that they represent the curve of best fit of that degree.

Let $y_1^e, y_2^e, \ldots, y_n^e$ be the expected values of y obtained by putting $x_1, x_2, \ldots, x_n$, respectively, in equation (8.1). These values are different from the observed values $y_1, y_2, \ldots, y_n$. The differences $y_i - y_i^e, i = 1, 2, \ldots, n$ are called residuals.

Let us define the quantity

$$S = \sum\left(y_i - y_i^e\right)^2 = \sum\left(y_i - a_0 + a_1x_i + a_2x_i^2 + \cdots + a_mx_i^m\right)^2 \tag{8.2}$$

The constants $a_0, a_1, \ldots, a_m$ are so chosen that S is minimum. The conditions for which S will be minimum are

$$\frac{\partial S}{\partial a_0} = 0, \frac{\partial S}{\partial a_1} = 0, \ldots, \frac{\partial S}{\partial a_m} = 0$$

These give

$$\sum y = na_0 + a_1\sum x + \cdots + a_m\sum x^m$$

$$\sum xy = a_0 \sum x + a_1 \sum x^2 + \cdots + a_m \sum x^{m+1}$$

$$\sum x^2 y = a_0 \sum x^2 + a_1 \sum x^3 + \cdots + a_m \sum x^{m+2}$$

$$\cdots$$

$$\cdots$$

$$\sum x^m y = a_0 \sum x^m + a_1 \sum x^{m+1} + \cdots + a_m \sum x^{2m}$$

These equations, known as **normal equations**, help us find the values of the constants $a_0, a_1, \ldots, a_m$ when the form of the curve has already been chosen. This is explained in the following topics.

8.1.3 Fitting of a Straight Line

Let (x_i, y_i), $i = 1, 2, \ldots, n$ be n observations of the variables x and y

and
$$y = a + bx \tag{8.3}$$

be the straight line to be fitted on the basis of the observed data.

Now, the residual at $x = x_i$ is $\quad y_i - y_i^e = y_i - a - bx_i$

Therefore, the sum of the residuals is given by $S = \sum (y_i - a - bx_i)^2$. Now, by the principle of least squares, S is minimum if $\frac{\partial S}{\partial a} = 0$ and $\frac{\partial S}{\partial b} = 0$, and from these we get

$$\sum y = na + b \sum x \tag{8.4}$$

and
$$\sum xy = a \sum x + b \sum x^2 \tag{8.5}$$

As x_i, y_i are known, equations (8.4) and (8.5) can provide the best values of a and b and (8.3) can then be derived.

EXAMPLE 8.1 Fit a straight line to the following data, considering x as an independent variable:

x	0	1	2	3	4
y	1.1	1.9	2.7	3.5	4.3

Solution: Let the equation of the straight line to be fitted be $y = a+bx$. Then, the normal equations are

$$\sum y = na + b \sum x$$

and
$$\sum xy = a \sum x + b \sum x^2$$

For this, we construct the following table:

x	y	x^2	xy
0	1.1	0	0
1	1.9	1	1.9

Cont'd

Table cont'd

x	y	x^2	xy
2	2.7	4	5.4
3	3.5	9	10.5
4	4.3	16	17.2
$\sum x = 10$	$\sum y = 13.5$	$\sum x^2 = 30$	$\sum xy = 35$

Therefore, we get

$$13.5 = 5a + 10b$$

and

$$35 = 10a + 30b$$

Solving these we get $a = 1.1$ and $b = 0.8$.

Therefore, the required straight line which gives the best fit of the given data is $y = 1.1 + 0.8x$.

EXAMPLE 8.2 Fit a straight line to the following data, considering x as an independent variable:

x	0	2	4	6	8
y	10	12	14	16	18

Solution: To fit a straight line of the form $y = a + bx$, the normal equations are

$$\sum y = na + b\sum x$$

and

$$\sum xy = a\sum x + b\sum x^2$$

So, we construct the following table:

x	y	x^2	xy
0	10	0	0
2	12	4	24
4	14	16	56
6	16	36	96
8	18	64	144
$\sum x = 20$	$\sum y = 70$	$\sum x^2 = 120$	$\sum xy = 320$

Therefore, we get

$$70 = 5a + 20b$$

and

$$320 = 20a + 120b$$

Solving these we get $a = 10$ and $b = 1$.

Therefore, the required straight line which gives the best fit of the given data is $y = 10 + x$.

EXAMPLE 8.3 Fit a straight line to the following data, considering x as an independent variable:

x	2	4	6	8	10
y	10	14	15	16	15

Solution: Let the straight line be $y = a + bx$.

The normal equations are given by

$$\sum y = na + b\sum x$$

and

$$\sum xy = a\sum x + b\sum x^2$$

The computation is shown in the following table:

x	y	x^2	xy
2	10	4	20
4	14	16	56
6	15	36	90
8	16	64	128
10	15	100	150
$\sum x = 30$	$\sum y = 70$	$\sum x^2 = 220$	$\sum xy = 444$

Therefore, we get

$$70 = 5a + 30b$$

and

$$444 = 30a + 220b$$

Solving these we get $a = 10.4$ and $b = 0.6$.

Therefore, the required straight line which gives the best fit of the given data is $y = 10.4 + 0.6x$.

8.1.4 Fitting of the Curve $y = ax + bx^2$

Let (x_i, y_i), $i = 1, 2, \ldots, n$ be n observations of the variables x and y and

$$y = ax + bx^2 \tag{8.6}$$

be the equation of the curve to be fitted on the basis of the observed data.

Now, the residual at $x = x_i$ is $y_i - y_i^e = y_i - ax_i - bx_i^2$.

Therefore, the sum of the residuals is given by $S = \sum\left(y_i - ax_i - bx_i^2\right)^2$. By the principle of least squares, S is minimum if $\frac{\partial S}{\partial a} = 0$ and $\frac{\partial S}{\partial b} = 0$, and from these we get

$$\sum y = a\sum x + b\sum x^2 \tag{8.7}$$

and

$$\sum xy = a\sum x^2 + b\sum x^3 \tag{8.8}$$

As x_i, y_i are known, equations (8.7) and (8.8) can provide the best values of a and b and (8.6) can then be derived.

EXAMPLE 8.4 Fit a curve of the form $y = ax + bx^2$ to the following data, considering x as an independent variable:

x	1	2	3	4	5
y	3	5	6	7	9

Solution: To fit a curve of the form $y = ax + bx^2$, the normal equations are given by

$$\sum y = a\sum x + b\sum x^2$$

and
$$\sum xy = a\sum x^2 + b\sum x^3$$

So, we construct the following table:

x	y	x^2	x^3	xy
1	3	1	1	3
2	5	4	8	10
3	6	9	27	18
4	7	16	64	28
5	9	25	125	45
$\sum x = 15$	$\sum y = 30$	$\sum x^2 = 55$	$\sum x^3 = 225$	$\sum xy = 104$

Therefore, we get

$$30 = 15a + 55b$$

and
$$104 = 55a + 255b$$

Solving these we get $a = 2.943$ and $b = -0.257$.

Therefore, the required curve which gives the best fit of the given data is $y = 2.943x - 0.257x^2$.

EXAMPLE 8.5 Fit a curve of the form $y = ax + bx^2$ to the following data, considering x as an independent variable:

x	0	1	2	3	4
y	0.12	0.17	0.23	0.29	0.37

Solution: The normal equations for the curve of the form $y = ax + bx^2$ are

$$\sum y = a\sum x + b\sum x^2$$

and
$$\sum xy = a\sum x^2 + b\sum x^3$$

For this, we construct the following table:

x	y	x^2	x^3	xy
0	0.12	0	0	0
1	0.17	1	1	0.17
2	0.23	4	8	0.46
3	0.29	9	27	0.87
4	0.37	16	64	1.48
$\sum x = 10$	$\sum x = 1.18$	$\sum x^2 = 30$	$\sum x^3 = 100$	$\sum xy = 2.98$

Therefore, we get

$$1.18 = 10a + 30b$$

and

$$2.98 = 30a + 100b$$

Solving these, we get $a = 0.286$ and $b = -0.056$.

Therefore, the required curve which gives the best fit of the given data is $y = 0.286x - 0.056x^2$.

EXAMPLE 8.6 Fit a curve of the form $y = ax + bx^2$ to the following data, considering x as an independent variable:

x	0	1	2	3	4
y	8	12.5	18	24.5	32

Solution: In this case, the normal equations for the curve $y = ax + bx^2$ are given by

$$\sum y = a\sum x + b\sum x^2$$

and

$$\sum xy = a\sum x^2 + b\sum x^3$$

So, we construct the following table:

x	y	x^2	x^3	xy
0	8	0	0	0
1	12.5	1	1	12.5
2	18	4	8	36
3	24.5	9	27	73.5
4	32	16	64	128
$\sum x = 10$	$\sum x = 95$	$\sum x^2 = 30$	$\sum x^3 = 100$	$\sum xy = 250$

Therefore, we get
$$95 = 10a + 30b$$
and
$$250 = 30a + 100b$$

Solving these we get $a = 20$ and $b = -3.5$.

Therefore, the required curve which gives the best fit of the given data is $y = 20x - 3.5x^2$.

8.1.5 Fitting of the Curve $y = ax + bx + cx^2$

Let (x_i, y_i), $i = 1, 2, \ldots, n$ be n observations of the variables x and y

and
$$y = ax + bx + cx^2 \tag{8.9}$$

be the equation of the curve to be fitted on the basis of the observed data.

Now, the residual at $x = x_i$ is $\quad y_i - y_i^e = y_i - a - bx_i - cx_i^2$

Therefore, the sum of the residuals is given by $S = \sum\left(y_i - a - bx_i - cx_i^2\right)^2$. By the principle of least squares, S is minimum if $\dfrac{\partial S}{\partial a} = 0$, $\dfrac{\partial S}{\partial b} = 0$ and $\dfrac{\partial S}{\partial c} = 0$, and from these we get

$$\sum y = na + b\sum x + c\sum x^2 \tag{8.10}$$

$$\sum xy = a\sum x + b\sum x^2 + c\sum x^3 \tag{8.11}$$

and
$$\sum x^2 y = a\sum x^2 + b\sum x^3 + c\sum x^4 \tag{8.12}$$

As x_i, y_i are known, equations (8.10), (8.11) and (8.12) can provide the best values of a, b and c and (8.9) can then be derived.

EXAMPLE 8.7 Fit a curve of the form $y = a + bx + cx^2$ to the following data, considering x as an independent variable:

x	0	1	2	3	4
y	5.01	5.03	5.04	5.06	5.07

Solution: To fit a curve of the form $y = a + bx + cx^2$, the normal equations are

$$\sum y = na + b\sum x + c\sum x^2,$$

$$\sum xy = a\sum x + b\sum x^2 + c\sum x^3$$

and
$$\sum x^2 y = a\sum x^2 + b\sum x^3 + c\sum x^4$$

Now, we construct the following table:

x	y	x^2	x^3	x^4	xy	x^2y
0	5.01	0	0	0	0	0
1	5.03	1	1	1	5.03	5.03

Cont'd

Table cont'd

x	y	x^2	x^3	x^4	xy	x^2y
2	5.04	4	8	16	10.08	20.16
3	5.06	9	27	81	15.18	45.54
4	5.07	16	64	256	20.28	81.12
$\sum x=10$	$\sum y=25.21$	$\sum x^2=30$	$\sum x^3=100$	$\sum x^4=354$	$\sum xy=50.57$	$\sum x^2y=151.85$

Therefore, we get

$$25.21 = 5a + 10b + 30c$$
$$50.57 = 10a + 30b + 100c$$
$$151.85 = 30a + 100b + 354c$$

Solving these we get $= 5.0106$, $b = 0.0179$ and $c = -0.0007$.
Therefore, the required curve which gives the best fit of the given data is

$$y = 5.0106 + 0.0179x - 0.0007x^2.$$

EXAMPLE 8.8 Fit a curve of the form $y = a + bx + cx^2$ to the following data, considering x as an independent variable:

x	1	2	3	4	5	6	7
y	2	5	9	8	10	11	10

Solution: The normal equations for the curve of the form $y = a + bx + cx^2$ are given by

$$\sum y = na + b\sum x + c\sum x^2,$$

$$\sum xy = a\sum x + b\sum x^2 + c\sum x^3$$

and

$$\sum x^2 y = a\sum x^2 + b\sum x^3 + c\sum x^4$$

We construct the following table:

x	y	x^2	x^3	x^4	xy	x^2y
1	2	1	1	1	2	2
2	5	4	8	16	10	20
3	9	9	27	81	27	81
4	8	16	64	256	32	128
5	10	25	125	625	50	250
6	11	36	216	1296	66	396

Cont'd

Table cont'd

x	y	x^2	x^3	x^4	xy	x^2y
7	10	49	343	2401	70	490
$\sum x = 28$	$\sum y = 55$	$\sum x^2 = 140$	$\sum x^3 = 784$	$\sum x^4 = 4676$	$\sum xy = 257$	$\sum x^2 y = 1367$

Therefore, we get

$$55 = 7a + 28b + 140c$$

$$257 = 28a + 140b + 784c$$

$$1367 = 140a + 784b + 4676c$$

Solving these we get $= -1.57$, $b = 4.08$ and $c = -0.35$.
Therefore, the required curve which gives the best fit of the given data is

$$y = -1.57 + 4.08x - 0.35x^2.$$

EXAMPLE 8.9 Fit a curve of the form $y = a + bx + cx^2$ to the following data, considering x as an independent variable:

x	1	2	3	4	5	6	7
y	4	12	22	32	44	51	60

Solution: Let the curve be $v = a + bu + cu^2$

where $u = x - 4$

and $v = y - 32$

Now, we construct the following table:

$u = x - 4$	$v = y - 32$	u^2	u^3	u^4	uv	u^2v
−3	−28	9	−27	81	84	−252
−2	−20	4	−8	16	40	−80
−1	−10	1	−1	1	10	−10
0	0	0	0	0	0	0
1	12	1	1	1	12	12
2	19	4	8	16	38	76
3	28	9	27	81	84	252
$\sum u = 0$	$\sum v = 1$	$\sum u^2 = 28$	$\sum u^3 = 0$	$\sum u^4 = 196$	$\sum uv = 268$	$\sum u^2 v = -2$

For this case, the normal equations are

$$\sum v = na + b\sum u + c\sum u^2$$

$$\sum uv = a\sum u + b\sum u^2 + c\sum u^3$$

and

$$\sum u^2v = a\sum u^2 + b\sum u^3 + c\sum u^4$$

Therefore, we get

$$1 = 7a + 28c$$

$$268 = 28b$$

$$-2 = 28a + 196c$$

Solving these we get = 0.42, $b = 9.57$ and $c = -0.07$.

Therefore, the required curve which gives the best fit of the given data is

$$v = 0.42 + 9.57u - 0.07u^2$$

$$\Rightarrow \quad y - 32 = 0.42 + 9.57(x - 4) - 0.07(x - 4)^2$$

$$\Rightarrow \quad y = -6.98 + 10.13x - 0.07x^2$$

8.1.6 Fitting of the Curve $y = ab^x$

Taking logarithm on both sides (w.r.t. the base 10), we get

$$\log_{10} y = \log_{10} a + x\log_{10} b$$

i.e.

$$Y = A + Bx \tag{8.13}$$

where

$$Y = \log_{10} y,\ A = \log_{10} a$$

and

$$B = \log_{10} b. \tag{8.14}$$

The normal equations will then be

$$\sum Y = nA + B\sum x \tag{8.15}$$

and

$$\sum xY = A\sum x + B\sum x^2 \tag{8.16}$$

Solving (8.15) and (8.16), we get the values of A and B. Then, from (8.14) we get

$$a = \text{antilog}(A)$$

and

$$b = \text{antilog}(B)$$

Thus, we get the required equation $y = ab^x$.

EXAMPLE 8.10 Fit a curve of the form $y = ab^x$ to the following data, considering x as an independent variable:

x	2	3	4	5	6
y	8	15	33	65	126

Solution: The curve to be fitted is $y = ab^x$

or

$$Y = A + Bx$$

where

$$Y = \log_{10} y,\ A = \log_{10} a$$

and

$$B = \log_{10} b$$

The normal equations are

$$\sum Y = nA + B\sum x$$

and

$$\sum xY = A\sum x + B\sum x^2$$

For this, we construct the following table:

x	y	$Y = \log_{10} y$	x^2	xY
2	8	0.9031	4	1.8062
3	15	1.1761	9	3.5283
4	33	1.5185	16	6.0740
5	65	1.8129	25	9.0645
6	126	2.1004	36	12.6024
$\sum x = 20$	–	$\sum Y = 7.511$	$\sum x^2 = 90$	$\sum xY = 33.0754$

Therefore, from the normal equations and the table, we get

$$7.511 = 5A + 20B$$

and

$$33.0754 = 20A + 90B$$

Solving, we get $A = 0.2896$ and $B = 0.3031$.

$$\therefore \quad a = \text{antilog}(A) = 1.95$$

and

$$b = \text{antilog}(B) = 2.01$$

Therefore, the required equation of the curve is $y = 1.95(2.01)^x$.

EXAMPLE 8.11 Fit a curve of the form $y = ab^x$ to the following data, considering x as an independent variable:

x	2	3	4	5	6
y	144	172	207	248	298

Solution: The curve to be fitted is $\quad y = ab^x$

or $\quad Y = A + Bx$

where $\quad Y = \log_{10} y,\ A = \log_{10} a$

and $\quad B = \log_{10} b$

The normal equations are

$$\sum Y = nA + B\sum x$$

and

$$\sum xY = A\sum x + B\sum x^2$$

For this, we construct the following table:

x	y	$Y = \log_{10} y$	x^2	xY
2	144	2.1584	4	4.3168
3	172	2.2356	9	6.7068
4	207	2.3160	16	9.2640
5	248	2.3945	25	11.9725
6	298	2.4742	36	14.8452
$\sum x = 20$	–	$\sum Y = 11.5787$	$\sum x^2 = 90$	$\sum xY = 47.1053$

Therefore, from the normal equations and the table, we get

$$11.5787 = 5A + 20B$$

and

$$47.1053 = 20A + 90B$$

Solving these we get $A = 1.9995$ and $B = 0.0791$.

$\therefore$

$$a = \text{antilog}(A) = 99.88$$

and

$$b = \text{antilog}(B) = 1.2$$

Therefore, the required equation of the curve is $y = 99.88(1.2)^x$.

EXAMPLE 8.12 Fit a curve of the form $y = ab^x$ to the following data, considering x as an independent variable:

x	2	3	4	5	6
y	1.2	1.8	2.5	3.6	4.7

Solution: The curve to be fitted is

$$y = ab^x$$

or

$$Y = A + Bx$$

where

$$Y = \log_{10} y,\ A = \log_{10} a$$

and

$$B = \log_{10} b$$

The normal equations are

$$\sum Y = nA + B\sum x$$

and

$$\sum xY = A\sum x + B\sum x^2$$

For this, we construct the following table:

x	y	$Y = \log_{10} y$	x^2	xY
2	1.2	0.0792	4	0.1584
3	1.8	0.2553	9	0.7659
4	2.5	0.3979	16	1.5916

Cont'd

Table cont'd

x	y	$Y=\log_{10} y$	x^2	xY
5	3.6	0.5563	25	2.7815
6	4.7	0.6721	36	4.0326
$\sum x = 20$	–	$\sum Y = 1.9608$	$\sum x^2 = 90$	$\sum xY = 9.33$

Therefore, from the normal equations and the table, we get

$$1.9608 = 5A + 20B$$

and

$$9.33 = 20A + 90B$$

Solving these we get $A = -0.2026$ and $B = 0.1487$.

$$\therefore \quad a = \text{antilog}(A) = 0.63$$

and

$$b = \text{antilog}(B) = 1.41$$

Therefore, the required equation of the curve is $y = 0.63(1.41)^x$.

8.1.7 Fitting of the Curve $y = ae^{bx}$

Taking logarithm on both sides (w.r.t. the base e), we get

$$\log_e y = \log_e a + bx$$

i.e.

$$Y = A + bx \tag{8.17}$$

where

$$Y = \log_e y$$

and

$$A = \log_e a \tag{8.18}$$

Then, the normal equations will be

$$\sum Y = nA + b\sum x \tag{8.19}$$

and

$$\sum xY = A\sum x + b\sum x^2 \tag{8.20}$$

Solving (8.19) and (8.20), we get the value of A and b. Then, from (8.18) we get

$$a = \text{antilog}(A)$$

Thus, we get the required equation $y = ae^{bx}$.

EXAMPLE 8.13 Fit a curve of the form $y = ae^{bx}$ to the following data, considering x as an independent variable:

x	0.5	1.0	1.5	2.0	2.5
y	0.45	2.15	9.25	40.35	180.75

Solution: The curve to be fitted is

$$y = ae^{bx}$$

or

$$Y = A + bx$$

where

$$Y = \log_e y$$

and

$$A = \log_e a$$

The normal equations are

$$\sum Y = nA + b\sum x$$

and
$$\sum xY = A\sum x + b\sum x^2$$

For this, we construct the following table:

x	y	$Y = \log_e y$	x^2	xY
0.5	0.45	−0.7985	0.25	−0.3993
1.0	2.15	0.7655	1.00	0.7655
1.5	9.25	2.2246	2.25	3.3369
2.0	40.35	3.6976	4.00	7.3952
2.5	180.75	5.1971	6.25	12.9928
$\sum x = 7.5$	–	$\sum Y = 11.0863$	$\sum x^2 = 13.75$	$\sum xY = 24.0911$

Therefore, from the normal equations and the table, we get

$$11.0863 = 5A + 7.5$$

and
$$24.0911 = 7.5A + 13.75b$$

Solving these we get $A = -2.2597$ and $b = 2.985$.

$$a = \text{antilog}(A) = 0.104$$

Therefore, the required equation of the curve is $y = 0.104e^{2.985x}$.

EXAMPLE 8.14 Fit a curve of the form $y = ae^{bx}$ to the following data, considering x as an independent variable:

x	2	4	6	8	10
y	4	11	30	82	223

Solution: The curve to be fitted is $\quad y = ae^{bx}$

or
$$Y = A + bx$$

where
$$Y = \log_e y$$

and
$$A = \log_e a$$

The normal equations are

$$\sum Y = nA + b\sum x$$

and
$$\sum xY = A\sum x + b\sum x^2$$

For this, we construct the following table:

x	y	$Y = \log_e y$	x^2	xY
2	4	1.3863	4	2.7726
4	11	2.3979	16	9.5916
6	30	3.4012	36	20.4072
8	82	4.4067	64	35.2536
10	223	5.4072	100	54.0720
$\sum x = 30$	–	$\sum Y = 16.9993$	$\sum x^2 = 220$	$\sum xY = 122.097$

Therefore, from the normal equations and the table, we get

$$16.9993 = 5A + 30b$$

and

$$122.097 = 30A + 220b$$

Solving these we get $A = 0.3847$ and $b = 0.5025$.

$$\therefore \quad a - \text{antilog}\,(A) = 1.4692$$

Therefore, the required equation of the curve is $y = 1.4692e^{0.5025x}$.

EXAMPLE 8.15 Fit a curve of the form $y = ae^{bx}$ to the following data, considering x as an independent variable:

x	1	2	3	4	5	6	7	8
y	15.3	20.5	27.4	36.6	49.1	65.6	87.8	117.6

Solution: The curve to be fitted is

$$y = ae^{bx}$$

or

$$Y = A + bx$$

where

$$Y = \log_e y$$

and

$$A = \log_e a$$

The normal equations are

$$\sum Y = nA + b\sum x$$

and

$$\sum xY = A\sum x + b\sum x^2$$

For this, we construct the following table:

x	y	$Y = \log_e y$	x^2	xY
1	15.3	2.7279	1	2.7279
2	20.5	3.0204	4	6.0408
3	27.4	3.3105	9	9.9315

Cont'd

Table cont'd

x	y	$Y = \log_e y$	x^2	xY
4	36.6	3.6000	16	14.4000
5	49.1	3.8939	25	19.4695
6	65.6	4.1836	36	25.1016
7	87.8	4.4751	49	31.3257
8	117.6	4.7673	64	38.1384
$\sum x = 36$	–	$\sum Y = 29.9787$	$\sum x^2 = 204$	$\sum xY = 147.1354$

Therefore, from the normal equations and the table, we get

$$29.9787 = 8A + 36b$$

and

$$147.1354 = 36A + 204b$$

Solving these we get $A = 2.4368$ and $b = 0.2912$.

$$\therefore \quad a = \text{antilog}\,(A) = 11.4364$$

Therefore, the required equation of the curve is $y = 11.4364e^{0.2912x}$.

8.1.8 Fitting of the Curve $y = ax^b$

Taking logarithm on both sides (w.r.t. the base 10), we get

$$\log_{10} y = \log_{10} a + b\log_{10} x$$

i.e.

$$Y = A + bX \tag{8.21}$$

where

$$Y = \log_{10} y,\ A = \log_{10} a$$

and

$$X = \log_{10} x \tag{8.22}$$

The normal equations will then be

$$\sum Y = nA + b\sum X \tag{8.23}$$

and

$$\sum XY = A\sum X + b\sum X^2 \tag{8.24}$$

Solving (8.23) and (8.24), we get the values of A and b. Then, from (8.22) we get

$$a = \text{antilog}\,(A)$$

Thus, we get the required equation $y = ax^b$.

EXAMPLE 8.16 Fit a curve of the form $y = ax^b$ to the following data, considering x as an independent variable:

x	1	2	3	4	5
y	2	3	5	8	12

Solution: The curve to be fitted is

$$y = ax^b$$

or

$$Y = A + bX$$

where $$Y = \log_{10} y,\ A = \log_{10} a$$
and $$X = \log_{10} x$$

The normal equations are

$$\sum Y = nA + b\sum X$$

and $$\sum XY = A\sum X + b\sum X^2$$

For this, we construct the following table:

x	y	$X = \log_{10} x$	$Y = \log_{10} y$	XY	X^2
1	2	0.00	0.30	0.00	0.00
2	3	0.30	0.48	0.14	0.09
3	5	0.48	0.70	0.34	0.23
4	8	0.60	0.90	0.54	0.36
5	12	0.70	1.08	0.76	0.49
–	–	$\sum X = 2.08$	$\sum Y = 3.46$	$\sum XY = 1.78$	$\sum X^2 = 1.17$

Therefore, from the normal equations and the table, we get

$$3.46 = 5A + 2.08b$$

and $$1.78 = 2.08A + 1.17b$$

Solving these we get $A = 0.23$ and $b = 0.1.12$.

$\therefore$ $$a = \text{antilog}\,(A) = 1.70$$

Therefore, the required equation of the curve is $y = 1.7x^{1.12}$.

EXAMPLE 8.17 Fit a curve of the form $y = ax^b$ to the following data, considering x as an independent variable

x	1	2	3	4	5
y	2.98	4.26	5.21	6.12	6.84

Solution: The curve to be fitted is $$y = ax^b$$

or $$Y = A + bX$$

where $$Y = \log_{10} y,\ A = \log_{10} a$$

and $$X = \log_{10} x$$

The normal equations are

$$\sum Y = nA + b\sum X$$

and $$\sum XY = A\sum X + b\sum X^2$$

For this, we construct the following table:

x	y	$X = \log_{10} x$	$Y = \log_{10} y$	XY	X^2
1	2.98	0.00	0.47	0.00	0.00
2	4.26	0.30	0.63	0.19	0.09
3	5.21	0.48	0.72	0.35	0.23
4	6.12	0.60	0.79	0.47	0.36
5	6.84	0.70	0.84	0.59	0.49
–	–	$\sum X = 2.08$	$\sum Y = 3.45$	$\sum XY = 1.6$	$\sum X^2 = 1.17$

Therefore, from the normal equations and the table, we get

$$3.45 = 5A + 2.08b$$

and

$$1.6 = 2.08b + 1.17b$$

Solving these we get $A = 0.46$ and $b = 0.54$.

$$\therefore \quad a = \text{antilog}\,(A) = 2.88$$

Therefore, the required equation of the curve is $y = 2.88x^{0.54}$.

EXAMPLE 8.18 Fit a curve of the form $y = ax^b$ to the following data, considering x as an independent variable:

x	1	2	3	4	5
y	6.1	12.3	18.7	27.4	36.8

Solution: The curve to be fitted is

$$y = ax^b$$

or

$$Y = A + bX$$

where

$$Y = \log_{10} y,\ A = \log_{10}$$

and

$$X = \log_{10} x$$

The normal equations are

$$\sum Y = nA + b\sum X$$

and

$$\sum XY = A\sum X + b\sum X^2$$

For this, we construct the following table:

x	y	$X = \log_{10} x$	$Y = \log_{10} y$	XY	X^2
1	6.1	0.00	0.78	0.00	0.00
2	12.3	0.30	1.09	0.33	0.09
3	18.7	0.48	1.27	0.61	0.23

Cont'd

Table cont'd

x	y	$X = \log_{10} x$	$Y = \log_{10} y$	XY	X^2
4	27.4	0.60	1.44	0.86	0.36
5	36.8	0.70	1.57	1.10	0.49
–	–	$\sum X = 2.08$	$\sum Y = 6.15$	$\sum XY = 2.9$	$\sum X^2 = 1.17$

Therefore, from the normal equations and the table, we get

$$6.15 = 5A + 2.08b$$

and

$$2.9 = 2.08A + 1.17b$$

Solving these we get $A = 0.76$ and $b = 1.12$.

$\therefore$

$$a = \text{antilog}\,(A) = 5.75$$

Therefore, the required equation of the curve is $y = 5.75x^{1.12}$.

8.2 SPLINE INTERPOLATION

8.2.1 Introduction

Before the age of computers, a device was used to draw a smooth curve through a given set of points such that the slope and curvature were continuous along the curve. Such a device was called a **spline** and plotting of the curve was called **spline fitting**.

8.2.2 Spline Function

Suppose a function is defined in the interval $[a, b]$. First, the given interval $[a, b]$ is divided into n subintervals $[x_0, x_1], [x_1, x_2], \ldots, [x_{n-1}, x_n]$ of equal length h where $a = x_0 < x_1 < x_2 < \ldots < x_{n-1} < x_n = b$. The nodes $x_1, x_2, \ldots, x_{n-1}$ are called internal nodes. Then, a spline function of degree n with nodes x_i, $i = 0, 1, 2, \ldots, n$ is a function $F(x)$ satisfying the following properties:

(i) $F(x_i) = f(x_i)$ for $i = 0, 1, 2, \ldots, n$
(ii) On each subinterval $[x_{i-1}, x_i]$, $1 \le i \le n$, $F(x)$ is a polynomial in x of degree at most n
(iii) $F(x)$ and its first $(n - 1)$ derivatives are continuous on $[a, b]$
(iv) $F(x)$ is a polynomial of degree one for $x < a$ and $x > b$

In accordance with the scope of the book, we will discuss only cubic spline interpolation.

8.2.3 Cubic Spline Interpolation

As $F(x)$ is cubic, $F''(x)$ is a linear function and can be written as

$$F''(x) = \frac{x_i - x}{x_i - x_{i-1}} F''(x_{i-1}) + \frac{x - x_{i-1}}{x_i - x_{i-1}} F''(x_i)$$

$$= \frac{x_i - x}{h} F''(x_{i-1}) + \frac{x - x_{i-1}}{h} F''(x_i)$$

$$= \frac{1}{h}\left[(x_i - x) F''(x_{i-1}) + (x - x_{i-1}) F''(x_i)\right] \tag{8.25}$$

Integrating equation (8.25), we get

$$F(x)=\frac{1}{h}\left[\frac{(x_i-x)^3}{6}F''(x_{i-1})+\frac{(x-x_{i-1})^3}{6}F''(x_i)\right]$$
$$+c_1(x_i-x)+c_2(x-x_{i-1}) \tag{8.26}$$

where c_1 and c_2 are arbitrary constants to be determined by the conditions:

$$F(x_i)=f(x_i),\ i=0,1,2,3.$$

Now, putting $x=x_i$ in equation (5.26), we get

$$f_i=\frac{1}{h}\left[\frac{h^3}{6}F''(x_i)\right]+c_2\cdot h \Rightarrow c_2=\frac{f_i}{h}-\frac{h}{6}F''(x_i) \tag{8.27}$$

Again, putting $x=x_{i-1}$ in equation (5.26), we get

$$f_{i-1}=\frac{1}{h}\left[\frac{h^3}{6}F''(x_{i-1})\right]+c_1\cdot h \Rightarrow c_1=\frac{f_{i-1}}{h}-\frac{h}{6}F''(x_{i-1}) \tag{8.28}$$

Therefore, using (8.17) and (8.28) in (8.26), we get

$$F(x)=\frac{1}{h}\left[\theta\frac{(x_i-x)^3}{6}F''(x_{i-1})+\frac{(x-x_{i-1})^3}{6}F''(x_i)+(x_i-x)\right.$$
$$\left.\left\{f_{i-1}-\frac{h^2}{6}F''(x_{i-1})\right\}+(x-x_{i-1})\left\{f_i-\frac{h^2}{6}F''(x_i)\right\}\right] \tag{8.29}$$

Now, denoting $F''(x)=M_i$, we get from equation (8.29)

$$F(x)=\frac{1}{h}\left[\frac{(x_i-x)^3}{6}M_{i-1}+\frac{(x-x_{i-1})^3}{6}M_i+(x_i-x)\right.$$
$$\left.\left\{f_{i-1}-\frac{h^2}{6}M_{i-1}\right\}+(x-x_{i-1})\left\{f_i-\frac{h^2}{6}M_i\right\}\right]$$
$$=\frac{1}{6h}\left[(x_i-x)^3M_{i-1}+(x-x_{i-1})^3M_i+(x_i-x)\left(6f_{i-1}-h^2M_{i-1}\right)\right.$$
$$\left.+(x-x_{i-1})\left(6f_i-h^2M_i\right)\right]$$

Therefore

$$F'(x)=\frac{1}{6h}\left[-3(x_i-x)^2M_{i-1}+3(x-x_{i-1})^2M_i\right.$$
$$\left.-\left(6f_{i-1}-h^2M_{i-1}\right)+\left(6f_i-h^2M_i\right)\right]$$
$$=\frac{1}{6h}\left[3(x-x_{i-1})^2M_i-3(x_i-x)^2M_{i-1}+6(f_i-f_{i-1})+h^2(M_{i-1}-M_i)\right] \tag{8.30}$$

However, according to the assumption, $F'(x)$ is continuous at $x = x_i$,

i.e. $$F'(x_i + 0) = F'(x_i - 0).$$

Now

$$F'(x_i+0)=\frac{1}{6h}\left[-h^2M_{i+1}-2h^2M_i+6(f_{i+1}-f_i)\right] \tag{8.31}$$

and $$F'(x_i-0)=\frac{1}{6h}\left[2h^2M_i+h^2M_{i-1}+6(f_i-f_{i-1})\right] \tag{8.32}$$

Therefore, from (8.31) and (8.32), for the interval $[x_{i-1}, x_i]$, $i = 1, 2, 3$, we get

$$-h^2M_{i+1} - 2h^2M_i + 6(f_{i+1} - f_i) = 2h^2M_i + h^2M_{i-1} + 6(f_i - f_{i-1})$$

$$h^2[M_{i+1} + 4M_i + M_{i-1}] = 6(f_{i+1} + 2f_i + f_{i-1}) \tag{8.33}$$

with the condition $M_0 = M_3 = 0$.

This method will be better understood by examining the following examples.

EXAMPLE 8.19 Derive the cubic spline for the following data:

x	0	1	2	3
y	2	−6	−8	2

Solution: For the interval $[x_{i-1}, x_i]$, $i = 1, 2, 3$, the cubic spline is given by

$$F(x)=\frac{1}{6h}\left[(x_i-x)^3M_{i-1}+(x-x_{i-1})^3M_i+(x_i-x)\left\{6f_{i-1}-h^2M_{i-1}\right\}+(x-x_{i-1})\left\{6f_i-h^2M_i\right\}\right]$$

with the condition $M_0 = M_3 = 0$.

We also have

$$h^2[M_{i+1} + 4M_i + M_{i-1}] = 6(f_{i+1} + 2f_i + f_{i-1})$$

Here $$h = 1$$

Therefore

$$M_2 + 4M_1 + M_0 = 6(f_2 - 2f_1 + f_0) = 6[-8-2(-6) + 2] = 36 \text{ for } 0 \le x \le 1$$

and $$M_3 + 4M_2 + M_1 = 6(f_3 - 2f_2 + f_1) = 6[2-2(-8) + (-6)] = 72 \text{ for } 1 \le x \le 2$$

Putting $M_0 = M_3 = 0$, we get

$$M_2 + 4M_1 = 36 \text{ and } 4M_2 + M_1 = 72.$$

Solving these we get $M_1 = 4.8$ and $M_2 = 16.8$.

Hence, for the interval $0 \le x \le 1$,

$$F(x)=\frac{1}{6}\left[(1-x)^3M_0+(x-0)^3M_1+(1-x)(6f_0-M_0)+(x-0)(6f_1-M_1)\right]$$

$$=\frac{1}{6}\left[4.8x^3+(1-x)(6\times2)+x(6\times(-6)-4.8)\right]=0.8x^3-8.8x+2$$

For the interval $1 \le x \le 2$,

$$F(x)=\frac{1}{6}\left[(2-x)^3M_1+(x-1)^3M_2+(2-x)(6f_1-M_1)+(x-1)(6f_2-M_2)\right]$$

$$= \frac{1}{6}\left[(2-x)^3 \times 4.8 + (x-1)^3 \times 16.8 + (2-x)\,(6\times(-6)-4.8) + (x-1)(6\times(-8)-16.8)\right]$$

$$= 2x^3 - 3.6x^2 - 5.2x + 0.8$$

And for the interval $2 \le x \le 3$,

$$F(x) = \frac{1}{6}\left[(3-x)^3 M_2 + (x-2)^3 M_3 + (3-x)\,(6f_2 - M_2) + (x-2)(6f_3 - M_3)\right]$$

$$= \frac{1}{6}\left[(3-x)^3 \times 16.8 + (3-x)(6\times(-8)-16.8) + (x-2)(6\times 2 - 0)\right]$$

$$= -2.8x^3 + 25.2x^2 - 62.8x + 39.2$$

EXAMPLE 2.20 Derive the cubic spline for the following data:

x	0	1	2	3
y	1	2	33	244

Solution: For the interval $[x_{i-1}, x_i]$, $i = 1, 2, 3$, the cubic spline is given by

$$F(x) = \frac{1}{6h}\Big[(x_i - x)^3 M_{i-1} + (x - x_{i-1})^3 M_i + (x_i - x)$$

$$\{6f_{i-1} - h^2 M_{i-1}\} + (x - x_{i-1})\{6f_i - h^2 M_i\}\Big]$$

with the condition $M_0 = M_3 = 0$.

We also have

$$h^2\,[M_{i+1} + 4M_i + M_{i-1}] = 6\,(f_{i+1} + 2f_i + f_{i-1})$$

Here $h = 1$.

Therefore

$$M_2 + 4M_1 + M_0 = 6\,(f_2 - 2f_1 + f_0) = 6[33-2(2) + 1] = 180 \quad \text{for } 0 \le x \le 1$$

and $$M_3 + 4M_2 + M_1 = 6\,(f_3 - 2f_2 + f_1) = 6[244-2(33) + 2] = 1080 \quad \text{for } 1 \le x \le 2$$

Putting $M_0 = M_3 = 0$, we get

$$M_2 = 4M_1 = 180 \text{ and } 4M_2 + M_1 = 1080.$$

Solving these we get $M_1 = -24$ and $M_2 = 276$.

Hence, for the interval $0 \le x \le 1$,

$$F(x) = \frac{1}{6}\left[(1-x)^3 M_0 + (x-0)^3 M_1 + (1-x)\,(6f_0 - M_0) + (x-0)(6f_1 - M_1)\right]$$

$$= \frac{1}{6}\left[-24x^3 + (1-x)(6\times 1) + x(6\times 2 + 24)\right]$$

$$= -4x^3 + 5x + 1$$

For the interval $1 \le x \le 2$,

$$F(x) = \frac{1}{6}\left[(2-x)^3 M_1 + (x-1)^3 M_2 + (2-x)\,(6f_1 - M_1) + (x-1)(6f_2 - M_2)\right]$$

$$=\frac{1}{6}\left[(2-x)^3\times(-24)+(x-1)^3\times 276+(2-x)\ (6\times 2+24)+(x-1)(6\times 33-276)\right]$$

$$=50x^3-162x^2+167x-53.$$

And for the interval $2\le x\le 3$,

$$F(x)=\frac{1}{6}\left[(3-x)^3 M_2+(x-2)^3 M_3+(3-x)\ (6f_2-M_2)+(x-2)(6f_3-M_3)\right]$$

$$=\frac{1}{6}\left[(3-x)^3\times 276+(3-x)(6\times 33-276)+(x-2)(6\times 244-0)\right]$$

$$=-46x^3+414x^2-985x+715.$$

SHORT ANSWER QUESTIONS

1. Fit a parabola of the form $y = a + bx + cx^2$ to the following data:

x	0	1	2
y	1	6	17

2. Find the best-fit values of a and b so that $y = a + bx$ fits the data given below:

x	0	1	2	3	4
y	1	1.8	3.3	4.5	6.3

3. Fit a straight line $y = a + bx$ to the following data by the method of least squares:

x	0	1	3	6	8
y	1	3	2	5	4

4. By the method of least squares, find the straight line that best fits the following data:

x	1	2	3	4	5
y	14	27	40	55	68

5. Fit a straight line to the following data:

x	1	2	3	4	5	6
y	1200	900	600	200	110	50

6. By the method of least squares, find the straight line that best fits the following data:

x	0	5	10	15	20	25
y	12	15	17	22	24	30

7. Find the best-fit values of a and b so that $y = a + bx$ fits the data given below:

x	0	1	2	3	4
y	1.0	2.9	4.8	6.7	8.6

8. Fit a suitable straight line to the following data:

x	1	2	3	4
y	3	7	13	21

Answers

1. $y = 1 + 2x + 3x^2$	2. $a = 0.72, b = 1.33$	3. $y = 1.6 + 0.38x$	4. $y = 13.6x$
5. $y = 1361.97 - 243.42x$	6. $y = 0.7x + 11.3$	7. $a = 1.9, b = 1.9$	8. $y = -4 + 6x$

LONG ANSWER QUESTIONS

1. Fit a curve of the form $y = ab^x$ to the following data:

x	2	3	4	5	6
y	144	172.8	207.4	248.8	298.5

2. By the method of least squares, fit a curve of the form $y = ab^x$ to the following data:

x	1	2	3	4	5	6	7	8
y	1	1.2	1.8	2.5	3.6	4.7	6.6	9.1

3. By the method of least squares, fit a curve of the form $y = ab^x$ to the following data:

x	2	3	4	5	6
y	8.3	15.4	33.1	65.2	126.4

4. By the method of least squares, fit a curve of the form $y = ax^b$ to the following data:

x	1	2	3	4	5	6
y	2.98	4.26	5.21	6.1	6.8	7.5

5. Fit a curve of the form $y = ax^b$ to the following data:

x	1	2	3	4	5
y	0.5	2	4.5	8	12.5

6. Fit a geometric curve of the form $y = ax^b$ to the following data:

x	1	2	3	4	5
y	7.1	27.8	62.1	110	161

7. Find the curve of best fit of the form $y = ae^{bx}$ to the following data:

x	1	5	7	9	12
y	10	15	12	15	21

8. Find the curve of best fit of the type $y = ae^{bx}$ to the following data:

x	1	2	3	4	5	6
y	1.6	4.5	13.8	40.2	125	300

9. Find the best-fit values of a and b so that $y = ae^{bx}$ fits the data given below:

x	2	4	6	8	10
y	4.077	11.084	30.128	81.897	222.62

10. Find the curve of the type $y = ax + bx^2$ that best fits the following data:

x	1	2	3	4	5
y	1.8	5.1	8.9	14.1	19.8

11. Fit a second-degree parabola of the form $y = a + bx + cx^2$ to the data given below:

x	1	2	3	4	5	6	7
y	3	10	22	35	54	78	104

Answers

1. $y = 99.86(1.2)^x$
2. $y = 0.68(1.38)^x$
3. $y = 2.04(1.2)^x$
4. $y = 2.98x^{0.51}$
5. $y = 0.501x^{1.998}$
6. $y = 7.17x^{1.95}$
7. $y = 9.48e^{0.06x}$
8. $y = 0.56e^{1.06x}$
9. $a = 1.5$, $b = 0.5$
10. $y = 1.52x + 0.49x^2$
11. $y = 0.31 + 0.9x + 1.99x^2$

9

Algorithms and Programs in C Language

9.1 INTRODUCTION

C language is an excellent and user-friendly tool for numerical computations executed on a computer. It was originally developed by Dennis Ritchie in 1972. In the previous chapters, we saw that manual computation of numerical methods can be rather tedious and time consuming. This difficulty can be removed by developing programs in C language. In this chapter, a brief overview of C language and some standard numerical techniques used in the C language environment are discussed, along with algorithms.

9.2 OVERVIEW OF C LANGUAGE

In this context we will discuss the basics of C language which are required to develop programs for numerical techniques. Though C programs will vary for different numerical problems, we can provide a general structure:

```
Links         library functions
Definitions   symbolic constants/user defined functions
main()
{
  Declarations
  Statements and control structure
}
```

Links At the beginning of the program, some header files need to be included. Header files contain definitions of standard functions that can be incorporated into the C program by using the pre-processor `#include` statement. The syntax generally used is as follows:

```
#include<filename>
```

For example, to use the function `printf()` in a program, the line

```
#include<stdio.h>
```

should be inserted at the beginning of the source file, because the declaration for `printf()` is found in the file `stdio.h`. All header files have the extension `.h`.

For developing numerical programs in C, we generally need to include the following header files:

- `stdio.h` which has the syntax `#include<stdio.h>`
- `conio.h` which has the syntax `#include<conio.h>`
- `math.h` which has the syntax `#include<math.h>`

Definition A wide variety of numerical problems such as solving algebraic or transcendental equations, ordinary differential equations and numerical integrations involve the use of functions. Thus, defining functions before the `main()` function is a very important part of C programming. Though there are different ways of defining a function, here we give only the prototype definition, which has the following syntax:

```
Function_type function_name (parameter_declarations)
{
return(function statement);
}
```

For example, consider the real-valued function $f(x) = x^3 - 3x - 5$, where both the independent and dependent variables take only floating numbers. The definition of this function will be

```
float f(float x)
{
return(x*x*x-3*x-5);
}
```

Sometimes, we may require a certain value frequently in the program. In this case, we take advantage of **macro definition**. This is also done before the `main()` function. Some macro definitions are given below:

```
#define PI 3.1415
#define N 3000
#define TAXRATE 0.10
```

The `main()` function The starting point of a C program is identified by the word `main()`. This informs the computer where the program actually starts. The parentheses that follow the keyword `main` indicate that there are no arguments supplied to this program. The syntax of the `main()` function is

```
main()
{
}
```

The two braces, { and }, signify the begin and end segments of the program.

Sometimes, we write `void main()`.The word 'void' indicates that the `main()` function 'does not return anything' after execution is completed.

Declaration All the variables required for a particular problem are declared immediately after the opening braces of the `main()` function. The syntax for this is given below:

data-type variable1,variable2, …;

A variable may be of type integer, float, character, double, etc. The syntax for declaring variables is given below with examples.

- `int i,j,n;` if `i,j` and `n` are integer variables
- `float s,p,r;` when `s,p` and `r` are float variables
- `char a,temp,age;` where `a,temp` and `age` are character variables
- `double t,b,c;` if `t,b` and `c` are double type variables

Declaring an Array An array is a group of elements of similar data type having a common name. In general, we deal with **one-dimensional** and **two-dimensional** arrays.

- **One-dimensional array:** The syntax for declaring a one-dimensional array is as follows:

```
<data_type> < array name>[size]
```

For example,

```
int x[5];
```

The type of this array is integer, the name is x and the size is 5. The computer sets 5 storage locations for integer data, as follows:

x[0]	x[1]	x[2]	x[3]	x[4]

- **Two dimensional array:** The syntax to declare a two-dimensional array is as follows:

```
<data_type> < array name>[row size][column size]
```

For example,

```
int x[3][2];
```

The type of this array is integer, the name is x and it is a matrix with three rows and two columns. The representation of this array in memory is given below:

	column0	column1
row 0	x[0][0]	x[0][1]
row 1	x[1][0]	x[1][1]
row 2	x[2][0]	x[2][1]

Statements and Control Structure This is the most important section for a programmer. In this section, the values of different variables are read or accepted by the program, arithmetic operators, logical operators, relational operators, increment and decrement operators, decision-making statements and loops are used and the result are displayed.

- **Reading the input values:** The function `scanf()` is used to read the value of a variable. To use this function, we need to display a statement on the screen, requesting the user to type the required value. For this, we use the function `printf()`. For example, consider the integer variable n which will take a value, say 5. Then we should write:

```
printf("Enter the value of n:");
scanf("%d",&n);
```

Here, %d is the format for the integer type variable n and '&' is the address of it in memory. Thus input of type integer is read by the program and stored in memory.

A few formats for different types of variables are given below:

Format	Type
%d	Integer
%f	Float
%c	Character

%lf	Double
%ld	Long integer

- **Arithmetic operators:** Six arithmetic operators are used in C language.

Operator	Operation
^	Exponentiation
+	Addition
–	Subtraction
*	Multiplication
/	Division
%	Modulo division

When the / operator is used for integer division, the resulting integer is obtained by discarding the fractional part. For example, 5/3→1, 1/2→0. Modulo division % only works with integers and the resultant is the remainder after the division. For example, 7%2→1, 9/3→0.

- **Logical operators**

<table>
<tr><th>Operator</th><th>Operation</th></tr>
<tr><td>&&</td><td>Logical AND</td></tr>
<tr><td>||</td><td>Logical OR</td></tr>
<tr><td>!</td><td>Logical NOT</td></tr>
</table>

- **Relational operators**

Operator	Operation
<	Less than
>	Greater than
<=	Less than or equal to
>	Greater than or equal to
==	Equal to
!=	Not equal to

- **Increment and decrement operators**

Operator	Operation
++	Increment
--	Decrement

These operators have both prefix and postfix forms. However, we should be very careful while using these operators. There is a significant difference between the prefix and postfix forms when the operators are used as part of a larger expression.

Prefix form: `++i` is equivalent to `i=i+1` and `--i` is equivalent to `i=i-1`.
Postfix form: `i++` is equivalent to `i=i+1` and `i--` is equivalent to `i=i-1`.

If `++k` is used in an expression, `k` is incremented before the expression is evaluated. If `k++` is used in an expression, `k` is incremented after the expression is evaluated. For example, assume that the integer variable `m` has been initialised to 5. Then, the statement

```
a=++m;
```

yields `m`→`6` and then `a`→`6`.
Whereas in this form of the statement,

```
a=m++;
```

we get `a`→`5` and then `m`→`6`.
In the above example, 1 is the increment or decrement value. We can also change this value. For example, the following statement

```
i=i+2;
```
can be written as
```
i+=2;
```

- **Decision-making statements:** Decision-making statements are essential when we need to check certain conditions within the C program. Some important decision-making statements are discussed below:
  - The *if* statement: The syntax of the '*if*' statement is as follows:
    ```
    if (test condition)
      statement;
    ```
    The 'statement' will be executed only if the test condition is TRUE. For example, examine the following program segment:
    ```
    main()
    {
    int i;
    printf("Enter the value of i:");
    scanf("%d",&i);
    if(i%2==0)
    printf("The number %d is an even number",i);
    }
    ```
    The output of this program segment is given below:
    ```
    Enter the value of i:46
    The number 46 is an even number
    ```
  - The *if-else* statement: The syntax for this is given below:
    ```
    if (test condition)
        statement1;
    else
        statement2;
    ```
    If the test condition is TRUE, statement1 is executed; statement2 is skipped. On the other hand, if the test condition is FALSE, statement2 is executed; statement1 is skipped.

    The following program segment is given as an example of the *if–else* statement.
    ```
    main()
    {
    int i;
    printf("Enter the value of i:");
    scanf("%d",&i);
    if(i%2==0)
    printf("The number %d is an even number",i);
    else
    printf("The number %d is an odd number",i);
    }
    ```
    Two different outputs of this program segment are given below:
    ```
    Enter the value of i:38
    The number 38 is an even number
    ```
    and

```
Enter the value of i:37
The number 37 is an odd number
```

- Nested *if–else* statement: This statement has the following syntax:

```
if (test condition 1)
    statement1;
else if (test condition 2)
    statement2;
else
    statement3;
```

For example, examine the following program segment:

```
main()
{
int i;
printf("Enter any year:");
scanf("%d",&i);
if(i%400==0)
printf("The year %d is a leap year",i);
else if(i%4==0)
printf("The year %d is a leap year",i);
else
printf("The year %d is not a leap year",i);
}
```

The output of this program segment is given as:

```
Enter any year:1996
The year 1996 is a leap year
```

- **Loops:** Sometimes, we may need to repetitively perform an action in a program. One simple example is reading the values of an array. This can be done by using a loop. In this section, some important loops are discussed.

  - The *while* loop: The syntax is as follows:

```
while (test condition)
  {
    statement;
    increment counter;
  }
```

The *while* statement works as follows:

1. Test condition is evaluated (entry condition).
2. If it is FALSE, skip over the loop.
3. If it is TRUE, the statement is executed.
4. Go back to step 1.

An example of the *while* loop is given below.

```
int i=1,sum=0,n=10;
while (i<=n)
{
sum= sum+i;
```

```
i=i+1;
}
```

This loop works as follows:

Value of `i`	**Statement condition**	**Value of** `sum`
i=1	1<10 (TRUE)	sum=1
i=2	2<10 (TRUE)	sum=3
i=3	3<10 (TRUE)	sum=6
i=4	4<10 (TRUE)	sum=10
i=5	5<10 (TRUE)	sum=15
i=6	6<10 (TRUE)	sum=21
i=7	7<10 (TRUE)	sum=28
i=8	8<10 (TRUE)	sum=36
i=9	9<10 (TRUE)	sum=45
i=10	10<10 (FALSE)	Skip the loop

- The *do–while* loop: This loop has the following syntax

```
do
   {
     statement;
     increment counter;
   } while (test condition);
```

For example, we can consider the following program segment:

```
float f(float x, float y)
{
return(x*x-y);
}
main()
{
float x0=0,xn=2,y0=0,h=0.5;
do
{
y0=y0+h*f(x,y);
x0=x0+h;
}while(x0<xn);
}
```

- The *for* loop: The general form of the *for* loop is as follows:

```
for (initialise counter; test counter; increment counter)
   {
     statement;
   }
```

Look at the following program segment:

```
sum=10;
for(i=0;i<20;i++)
{
sum=sum+i;
}
```

Displaying Results To display results, we use the `printf()` function. The syntax is as follows:

```
printf(control string, argument list);
```

For example, suppose `sum` is a float type variable and 14.666666 is the resultant value of it. Then, if we write

```
printf("The result is=%f", sum);
```

the computer will display

```
The result is=14.666666
```

Again, if we write

```
printf("The result is=%.3f", sum);
```

the computer will display

```
The result is=14.667
```

Use of Arrays (One- and Two-Dimensional) In programming, it is important that we have the ability to store multiple values of the same data type for a specific problem. We can then use these values for further calculation. For example, suppose a C program is to be constructed for the purpose of implementing Newton's forward interpolation with the help of tabular values. These values need to be stored in arrays. Using these values, a forward difference table is constructed, which is thereafter printed. All these activities are executed using arrays. This is explained by the following example.

Construct the forward difference table using the data given below:

x	1.0	2.0	3.0	4.0	5.0
y	1.1	8.3	27.5	64.7	125.9

Step 1: Reading the number of arguments (n)
Variable declaration and reading of its value is executed by the following program segment:

```
int n;
printf("Enter the number of arguments:");
scanf("%d",&n);
```

Step 2: Reading the values of x
There are 5 (n=5) float values of x and they are required only for interpolation, not for the construction of the difference table. So we have to construct a program segment to read these values using arrays as follows:

```
float x[10];
printf("Enter the values of x:);
for(i=0;i<n;i++)
{
scanf("%f",&x[i]);
}
```

When we enter the values, the computer will store them as follows:

1.000000	2.000000	3.000000	4.000000	5.000000
x[0]	x[1]	x[2]	x[3]	x[4]

Step 3: Reading the values of y

The y values are required for the construction of the difference table, which is two-dimensional. So we have to construct a program segment to read these values using arrays so that these values are stored in a single column. This can be done using the following program segment:

```
float y[10][10];
printf("Enter the values of y:);
for(i=0;i<n;i++)
{
scanf("%f",&y[i][0]);
}
```

If we enter the values of y, the data will be stored as follows:

y[0][0]	1.100000
y[1][0]	8.300000
y[2][0]	27.500000
y[3][0]	64.700000
y[4][0]	125.900000

Step 4: Construction of forward difference table

The horizontal forward difference table for the given data is shown below:

y	Δy	$\Delta^2 y$	$\Delta^3 y$	$\Delta^4 y$
1.1	7.2	12.0	6.0	0.0
8.3	19.2	18.0	6.0	
27.5	37.2	24.0		
64.7	61.2			
125.9				

From the above table, we can derive the working formula for the calculation as

```
y[i] [j] = y[i + 1][j − 1] − y[i][j − 1]
```

However, if we apply this formula, for i = 0, 1, …, n − 1 and j = 0, 1, …., n − 1in C, the empty cells in the difference table will calculate garbage values. This can be easily handled by applying the following program segment:

```
for(j=1;j<n;j++)
{
for(i=0;i<n-j;i++)
{
y[i][j]=y[i+1][j-1]-y[i][j-1];
}
```

```
}
```

To print the difference table, we can write the following program segment:

```
for(i=0;i<n;i++)
{
printf("\n");
for(j=0;j<n-i;j++)
{
printf("%f",y[i][j]);
}
}
```

The `getch()` function `getch()` is a library function that accepts a single byte or character from the keyboard buffer and stores it in memory without echoing the same to the console. This function is defined in the header file `conio.h`.

With this brief discussion on C, we now construct programs for some widely used numerical techniques.

9.3 NEWTON'S FORWARD INTERPOLATION

Method Let a function $y = f(x)$ be known only for a set of $(n + 1)$ distinct equispaced points $x_0, x_1, \ldots, x_n$ such that $x_i = x_0 + ih, (i = 1, 2, \ldots, n)$, where h is the length of each subinterval $[x_i, x_{i+1}]$ and the corresponding values of the function are $y_0 = f(x_0), y_1 = f(x_1), \ldots, y_n = f(x_n)$. Then

$$f(x) = y_0 + \frac{u}{1!}\Delta y_0 + \frac{u(u-1)}{2!}\Delta^2 y_0 + \cdots$$

$$+ \frac{u(u-1)(u-2)\cdots(u-\overline{n-1})}{n!}\Delta^n y_0$$

where $$u = \frac{x - x_0}{h}$$

Algorithm

Step 1. Start the program

Step 2: Read n (no. of arguments)

Step 3: For $i = 0(1)n - 1$
  Read $x[i]\&y[i][0]$
  End i

Step 4: Assign $h = x[1] - x[0]$ (step length)

Step 5: Read a (point of interpolation)

Step 6: Construct the forward difference table
  For $j = 1(1)n - 1$
    For $i = 0(1)\ n - 1 - j$
      $y[i][j] = y[i + 1][j - 1] - y[i][j - 1]$
    End i
  End j

Step 7: Print the forward difference table
For $i = 0(1)n - 1$
For $j = 0(1)n - 1 - i$
Print $y[i][j]$
End j
End i
Step 8: Assign $u = (a - x[0])/h$
Step 9: Assign $sum = y[0][0]$ & $p = 1.0$
Step 10: For $j = 1(1)n - 1$
$p = p * (u - j + 1)/j$
$\text{sum} = \text{sum} + p * y[0][j]$
End j
Step 11: Display a and sum
Step 12: Stop

Program Code

```
#include<stdio.h>
#include<conio.h>
void main()
{
int n,i,j;
float x[10],y[10][10],a,p=1,u,h,sum;
clrscr();
printf("Enter the number of terms:");
scanf("%d",&n);
printf("\n Enter the x values:");
for(i=0;i<n;i++)
{
scanf("%f",&x[i]);
}
printf("\n Enter the y values:");
for(i=0;i<n;i++)
{
scanf("%f",&x[i]);
}
printf("\n Enter the y values:");
scanf("%f",&y[i][0]);
}
for(j=1;j<n;j++)
{
for(i=0;i<n-j;i++)
{
y[i][j]=y[i+1][j-1]-y[i][j-1];
}
}
printf("\n The forward difference table is:\n");
for(i=0;i<n;i++)
```

Cont'd

Program cont'd

```
{
printf("\n");
for(j=0;j<n-i;j++)
{
printf("\t%.3f",y[i][j]);
}
}
printf("\n Enter the value of the interpolating point:");
scanf("%f",&a);
h=x[1]-x[0];
u=(a-x[0])/h;
sum=y[0][0];
for(j=1;j<n;j++)
{
p=p*(u-j+1)/j;
sum=sum+(p*y[0][j]);
}
printf("\n\n The required value of y(%.4f) is =%.4f",a,sum);
getch();
}
```

Input/Output For input/output, let us consider the following example:

Calculate y(0.5) from the following table using Newton's forward interpolation:

x	0	1	2	3
y	1	0	1	10

The input/output for the above program with the given problem will be:

```
Enter the number of terms:4
Enter the x values:0
1
2
3
Enter the y values:1
0
1
10
The forward difference table is:
1.000        -1.000        2.000        6.000
0.000         1.000        8.000
1.000         9.000
10.000
Enter the value of the interpolating point:0.5
The required value of y(0.5000) is =0.6250
```

ASSIGNMENT 1

Write a program in C to implement Newton's forward interpolation for the following:

(a) Calculate $\sqrt{2}$, correct to 4 decimal places using

x	1.9	2.1	2.3	2.5	2.7
y	1.3784	1.4491	1.5166	1.5811	1.6432

(b) Compute sin 32°, correct to 4 decimal places when

x	30°	35°	40°	45°	50°	55°
sin x	0.5000	0.5735	0.6428	0.7071	0.7660	0.8192

(c) Find the area (integral value) of the circle of diameter 82 cm, given

d(cm)	80	85	90	95	100
A(cm²)	5026	5674	6362	7088	7854

(d) Find the number of students (integral value) scoring below 35 marks, given

Marks	30–40	40–50	50–60	60–70
No. of Students	52	36	21	14

9.4 NEWTON'S BACKWARD INTERPOLATION

Method Let a function $y = f(x)$ be known only for a set of $(n + 1)$ distinct equispaced points $x_0, x_1, \ldots, x_n$ such that $x_i = x_0 + ih$, $(i = 1,2, \ldots n)$, where h is the length of each subinterval $[x_i, x_{i+1}]$ and the corresponding values of the function are $y_0 = f(x_0), y_1 = f(x_1), \ldots, y_n = f(x_n)$. Then

$$f(x) = y_n + \frac{u}{1!}\nabla y_n + \frac{u(u+1)}{2!}\nabla^2 y_n + \cdots$$

$$+ \frac{u(u+1)(u+2)\cdots(u+\overline{n-1})}{n!}\nabla^n y_n$$

where $$u = \frac{x - x_n}{h}$$

Algorithm

Step 1: Start the program

Step 2: Read n (no. of arguments)

Step 3: For $i = 0(1)n - 1$

  Read $x[i]$&$y[i][0]$

 End i

Step 4: Assign $h = x[1] - x[0]$ (step length)

Step 5: Read a (point of interpolation)

Step 6: Construct the backward difference table

For $j = 1(1)n - 1$
For $i = j(1)n - 1$
$y[i][j] = y[i][j-1] - y[i-1][j-1]$
End i
End j

Step 7: Print the backward difference table
For $i = 0(1)n - 1$
For $j = 0(1)i$
Print $y[i][j]$
End j
End i

Step 8: Assign $u = (a - x[n-1])/h$

Step 9: Assign $sum = y[n-1][0]$ & $p = 1.0$

Step 10: For $j = 1(1)n - 1$
$p = p * (u + j - 1)/j$
$sum = sum + p * y[n-1][j]$
End j

Step 11: Display a and sum

Step 12: Stop

Program Code

```
#include<stdio.h>
#include<conio.h>
void main()
{
int n,i,j;
float x[10],y[10][10],a,p=1,u,h,sum;
clrscr();
printf("Enter the number of terms:");
scanf("%d",&n);
printf("\n Enter the x values:");
for(i=0;i<n;i++)
{
scanf("%f",&x[i]);
}
printf("\n Enter the y values:");
for(i=0;i<n;i++)
{
scanf("%f",&y[i][0]);
}
for(j=1;j<n;j++)
{
for(i=j;i<n;i++)
{
y[i][j]=y[i][j-1]-y[i-1][j-1];
}}
printf("\n The backward difference table is:");
```

Cont'd

Program cont'd

```
for(i=0;i<n;i++)
{
printf("\n");
for(j=0;j<=i;j++)
{
printf("\t%.3f",y[i][j]);
}}
printf("\n Enter the value of the interpolating point:");
scanf("%f",&a);
h=x[1]-x[0];
u=(a- x[n-1])/h;
sum=y[n-1][0];for(j=1;j<n;j++)
{
p=p*(u+j-1)/j;
sum=sum+(p*y[n-1][j]);
}
printf("\n\n The required value of y(%.4f) is =%.4f",a,sum);
getch();
}
```

Input/Output For input/output, let us consider the following example:

Calculate y(2.5) from the following table by Newton's backward interpolation:

x	0	1	2	3
y	1	0	1	10

The input/output for the above program with the given problem will be:

```
Enter the number of terms:4
Enter the x values:0
1
2
3
Enter the y values:1
0
1
10
The backward difference table is:
1.000
0.000   -1.000
1.000    1.000    2.000
10.000   9.000    8.000    6.000
Enter the value of the interpolating point:2.5
The required value of y(2.5000) is =4.1250
```

ASSIGNMENT 2

Write a program in C to implement Newton's backward interpolation for the following:

(a) Calculate $f(4.3)$, correct up to 4 decimal places using

x	1	2	3	4	5
$f(x)$	7	12	29	64	123

(b) Compute sin52°, correct to 4 decimal places when

x	30°	35°	40°	45°	50°	55°
sin$x$	0.5000	0.5735	0.6428	0.7071	0.7660	0.8192

(c) Find the area (integral value) of the circle of diameter 98 cm, given

***d*(cm)**	80	85	90	95	100
***A*(cm²)**	5026	5674	6362	7088	7854

(d) Find the number of students (integral value) scoring above 65 marks, given

Marks	30–40	40–50	50–60	60–70
No. of Students	52	36	21	14

9.5 LAGRANGE'S INTERPOLATION

Method Let a function $y = f(x)$ be known only for a set of $(n + 1)$ distinct points $x_0, x_1, \ldots, x_n$ and the corresponding values of the function are $y_0 = f(x_0), y_1 = f(x_1), \ldots\ y_n = f(x_n)$.

Then
$$f(x)=\frac{(x-x_1)(x-x_2)\cdots(x-x_n)}{(x_0-x_1)(x_0-x_2)\cdots(x_0-x_n)}y_0$$
$$+\frac{(x-x_0)(x-x_2)\cdots(x-x_n)}{(x_1-x_0)(x_1-x_2)\cdots(x_0-x_n)}y_1+\cdots$$
$$+\frac{(x-x_0)(x-x_1)\cdots(x-x_{n-1})}{(x_n-x_1)(x_n-x_1)\cdots(x_n-x_{n-1})}y_n$$

Algorithm

Step 1: Start the program

Step 2: Read n (no. of arguments)

Step 3: For $i = 0(1)n - 1$

Read $x[i]$ and $y[i]$

End i

Step 4: Read a (point of interpolation)

Step 5: Assign $sum = 0.0$
Step 6: For $i = 0(1)n - 1$
$p = 1.0$
For $j = 0(1)n-1$
If $i \neq j$
$p = p * (a - x[j]/x[i] - x[j])$
End j
$sum = sum + p * y[i]$
End i
Step 7: Display a and sum
Step 8: Stop

Program Code

```
#include<stdio.h>
#include<conio.h>
void main()
{
int n,i,j;
float x[10],y[10],sum=0,p=1,a;
clrscr();
printf("Enter the number of terms:");
scanf("%d",&n);
printf("\n Enter the x values:");
for(i=0;i<n;i++)scanf("%f",&x[i]);
}
printf("\n Enter the y values:");
for(i=0;i<n;i++)
{
scanf("%f",&y[i]);
}
printf("\n Enter the value of the interpolating point:");
scanf("%f",&a);
for(i=0;i<n;i++)
{
p=1;
for(j=0;j<n;j++)
{
if(i!=j)
p=p*(a-x[j])/(x[i]-x[j]);
}
```

Cont'd

Program cont'd

```
sum=sum+p*y[i];
}
printf("\n\n The required value of y(%.4f) is =%.4f",a,sum);
getch();
}
```

Input/Output For input/output, let us consider the following example:

Calculate y(10) from the following table by Lagrange's interpolation:

x	5	6	9	11
y	12	13	14	16

The input/output for the above program with the given problem will be:

```
Enter the number of terms:4
Enter the x values:5
6
9
11
Enter the y values:12
13
14
16
Enter the value of the interpolating point:10
The required value of y(10.0000) is =14.6667
```

ASSIGNMENT 3

Write a program in C to implement Lagrange's interpolation for the following:

(a) Calculate $y(6)$, correct to 4 decimal places using

x	1	2	3	4	7
y	2	4	8	6	128

(b) Compute $f(4)$, correct to 4 decimal places when

x	0	1	2	3	5
$f(x)$	2	3	12	65	147

(c) Find the approximate value of $f(5)$, correct to 2 decimal places, given

x	0	2	3	4	7	8
$f(x)$	4	26	58	112	466	668

(d) Compute $f(0.4)$, correct to 3 decimal places, given

x	0.1	0.3	0.5	0.6	0.7
$f(x)$	0.53	0.61	0.69	0.72	0.75

9.6 TRAPEZOIDAL RULE

Method To evaluate $\int_a^b f(x)dx$, let the interval $[a, b]$ be divided into n equal subintervals of length h, by the points $a = x_0, x_1, \ldots, x_n = b$, where $x_i = x_0 + ih$, $(i = 1, 2, \ldots, n)$, The corresponding values of the function are $y_0 = f(x_0), y_1 = f(x_1), \ldots, y_n = f(x_n)$. Then

$$\int_a^b f(x)dx = \int_{x_0}^{x_0+nh} f(x)dx = \frac{h}{2}[(y_0 + y_n) + 2(y_1 + y_2 + \cdots + y_{n-1})]$$

Algorithm

Step 1: Start the program
Step 2: Define the function $f(x)$
Step 3: Read a (lower limit), b (upper limit) and n (number of subintervals)
Step 4: Assign $h = (b - a)/n$
Step 5: Assign $sum = f(a) + f(b)$
Step 6: For $i = 1(1)n - 1$
 $sum = sum + 2f(a+ih)$
 End i

Step 7: Assign $I = \frac{h}{2} sum$

Step 8: Display I
Step 9: Stop

Program Code The program code requires a function (integrand) declaration. For different functions, the program code will vary in the function declaration section. Therefore, we have to consider a specific function to construct the program. For example, let us consider the following problem:

Evaluate the integral $\int_0^1 dx/(1+x^2)$ *by the trapezoidal rule, dividing the interval into 24 subintervals.*

The program code will be:

```
#include<stdio.h>
#include<conio.h>
float f (float x)
{
return (1/(1+x*x));
}
```

Cont'd

Program cont'd

```
void main()
{
int i,n;
float a,b,h,sum,I;
clrscr();
printf("Enter the number of sub-intervals:");
scanf("%d",&n);
printf("\n Enter the value of lower limit a:");
scanf("%f",&a);
printf("\n Enter the value of upper limit b:");
scanf("%f",&b);
h=(b-a)/n;
sum=f(a)+f(b);
for(i=1;i<n;i++)
{
sum=sum+2* f(a+i*h);
}
I=(sum*h)/2.0;
printf("\n\n The value of the integration is = %.4f",I);
getch();
}
```

Input/Output

```
Enter the number of subintervals:24
Enter the value of lower limit a:0
Enter the value of upper limit b:1
The value of the integration is -0.7853
```

ASSIGNMENT 4

Construct C programs to implement the trapezoidal rule for the following:

(a) Evaluate $\int_0^1 \frac{xdx}{\sin x}$, correct to 3 decimal places taking 15 subintervals, given that $\lim_{x\to 0}\frac{\sin x}{x}=1.$

(b) Evaluate $\int_{-3}^{3} e^{-x^2}dx$, correct to 5 decimal places taking 15 subintervals.

(c) Evaluate $\int_0^1 dx/\left(1+x^2\right)$, correct to 4 decimal places taking 13 subintervals, and hence calculate the approximate value of π from the relation $\frac{\pi}{4}=\int_0^1 dx/\left(1+x^2\right)$, correct to 3 decimal places.

(d) Evaluate $\int_0^{\pi/3} \sqrt{\cos\theta}\,d\theta$, correct to 4 decimal places taking 17 subintervals.

9.7 SIMPSON'S 1/3 RULE

Method To evaluate $\int_a^b f(x)dx$, let the interval $[a, b]$ be divided into n (even) equal subintervals of length h, by the points $a = x_0, x_1, \ldots, x_n = b$, where $x_i = x_0 + ih$, $(i = 1, 2, \ldots, n)$. The corresponding values of the function are $y_0 = f(x_0), y_1 = f(x_1), \ldots, y_n = f(x_n)$. Then

$$\int_a^b f(x)dx = \int_{x_0}^{x_0+nh} f(x)dx = \frac{h}{3}\Big[(y_0 + y_n) + 2(y_2 + y_4 + \cdots + y_{n-2}) + 4(y_1 + y_3 + \cdots + y_{n-1})\Big]$$

Algorithm

Step 1: Start the program
Step 2: Define the function $f(x)$
Step 3: Read a (lower limit), b (upper limit) and n (number of subintervals)
Step 4: Assign $h = (b - a)/n$
Step 5: Assign $sum = f(a) + f(b)$
Step 6: For $i = 1(1)n - 1$
    If((i%2) == 0)
        $sum = sum + 2f(a + ih)$
    Else
        $sum = sum + 4f(a + ih)$
  End i
Step 7: Assign $I = \frac{h}{3} sum$
Step 8: Display I
Step 9: Stop

Program Code The program code requires a function (integrand) declaration. For different functions, the program code will vary in the function declaration section. Therefore, we have to consider a specific function to construct the program. For example, let us consider the following problem:

Evaluate the integral $\int_0^1 dx/(1+x^2)$ *by Simpson's 1/3 rule, dividing the interval into 24 subintervals.*

The program code will be:

```
#include<stdio.h>
#include<conio.h>
float f (float x)
{
return (1/(1+x*x));
}
```

Cont'd

Program cont'd

```
void main()
{
int i,n;
float a,b,h,sum,I;
clrscr();
printf("Enter the number of sub-intervals:");
scanf("%d",&n);
printf("\n Enter the value of lower limit a:");
scanf("%f",&a);
printf("\n Enter the value of upper limit b:");
scanf("%f",&b);
h=(b-a)/n;
sum=f(a)+f(b);
for(i=1;i<n;i++)
{
if(i%2==0)
sum=sum+2*f(a+i*h);
else
sum=sum+4*f(a+i*h);
}
I=(sum*h)/3.0;
printf("\n\ n The value of the integration is = %.4f",I);
getch();
}
```

Input/Output

```
Enter the number of subintervals:24
Enter the value of lower limit a:0
Enter the value of upper limit b:1
The value of the integration is =0.7854
```

ASSIGNMENT 5

Construct C programs to implement Simpson's 1/3 rule for the following:

(a) Evaluate $\int_0^1 \frac{xdx}{\sin x}$, correct to 3 decimal places taking 16 subintervals, given that $\lim_{x\to 0} \frac{\sin x}{x} = 1$.

(b) Evaluate $\int_{-3}^{3} e^{-x^2}\, dx$, correct to 5 decimal places taking 16 subintervals.

(c) Evaluate $\int_0^1 dx/\left(1+x^2\right)$, correct to 4 decimal places taking 12 subintervals, and hence calculate the approximate value of π from the relation $\frac{\pi}{4}=\int_0^1 dx/\left(1+x^2\right)$, correct to 3 decimal places.

(d) Evaluate $\int_0^{\pi/3} \sqrt{\cos\theta}\, d\theta$, correct to 4 decimal places taking 18 subintervals.

9.8 WEDDLE'S RULE

Method To evaluate $\int_a^b f(x)dx$, let the interval $[a, b]$ be divided into n (even) equal subintervals of length h, by the points $a = x_0, x_1, \ldots, x_n = b$, where $x_i = x_0 + ih$, $(i = 1, 2, \ldots, n)$. The corresponding values of the function are $y_0 = f(x_0), y_1 = f(x_1), \ldots, y_n = f(x_n)$. Then

$$\int_a^b f(x)dx = \int_{x_0}^{x_0+nh} f(x)dx = \frac{3h}{10}\left[(y_0+y_n)+5(y_1+y_5+y_7+\cdots)+(y_2+y_4+y_8+y_{14}+\cdots)\right.$$
$$\left.+6(y_3+y_9+y_{15}+\cdots)+2(y_6+y_{12}+y_{18}+\cdots)\right]$$

Algorithm

Step 1: Start the program
Step 2: Define the function $f(x)$
Step 3: Read a (lower limit), b (upper limit) and n (number of subintervals)
Step 4: Assign $h = (b - a)/n$
Step 5: Assign $sum = f(a) + f(b)$
Step 6: For $i = 1(1)n - 1$
        If((i%2) == 0&&(i%3) == 0)
            sum = sum + 2f(a+ih)
              Else if((i%2) == 0&&(i%3)! = 0)
                sum = sum + f(a + ih)
                  Else if((i%3) == 0&&(i%2)! = 0)
                    sum = sum + 6f(a+ih)
                      Else
                        sum = sum + 5f(a + ih)
        End i
Step 7: Assign $I = \frac{3h}{10} sum$
Step 8: Display I
Step 9: Stop

Program Code The program code requires a function (integrand) declaration. For different functions, the program code will vary in the function declaration section. Therefore, we have to consider a specific function to construct the program. For example, let us consider the following problem:

Evaluate the integral $\int_0^1 dx/\left(1+x^2\right)$ *by Simpson's 1/3 rule, dividing the interval into 24 subintervals.*

The program code will be:

```
#include<stdio.h>
#include<conio.h>
float f(float x)
  {
  return (1/(1+x*x));
  }
void main()
{
int i,n;
float a,b,h,sum;
clrscr();
printf("\n Enter the number of sub-intervals: ");
scanf("%d",&n);
printf("\n Enter the value of lower limit a:");
scanf("%f",&a);
printf("\n Enter the value of upper limit b:");
scanf("%f",&b);
h=(b-a)/n;
sum=f(a)+f(b);
for(i=1;i<n;i++)
  {
    if(i%2==0 && i%3!=0)
        sum=sum+f(a+i*h);
    else if(i%6==0)
        sum=sum+2*f(a+i*h);
    else if(i%3==0 && i%2!=0)
        sum=sum+6*f(a+i*h);
    else
        sum=sum+5*f(a+i*h);
  }
I=(sum*3*h)/10.0;
printf("\n\n The value of the integration is =%.4f",I);
getch();
}
```

Input/Output

```
Enter the number of subintervals:24
Enter the value of lower limit a:0
Enter the value of upper limit b:1
The value of the integration is =0.7854
```

ASSIGNMENT 6

Construct C programs to implement Weddle's rule for the following:

(a) Evaluate $\int_0^1 \frac{xdx}{\sin x}$, correct to 3 decimal places taking 18 subintervals, given that $\lim_{x\to 0} \frac{\sin x}{x} = 1$.

(b) Evaluate $\int_{-3}^{3} e^{-x^2} dx$, correct to 5 decimal places taking 18 subintervals.

(c) Evaluate $\int_0^1 dx/\left(1+x^2\right)$, correct to 4 decimal places taking 12 subintervals, and hence calculate the approximate value of π from the relation $\frac{\pi}{4} = \int_0^1 dx/\left(1+x^2\right)$, correct to 3 decimal places.

(d) Evaluate $\int_0^{\pi/3} \sqrt{\cos\theta}\, d\theta$, correct to 4 decimal places taking 18 subintervals.

9.9 BISECTION METHOD

Method To find a real root of $f(x) = 0$ in the interval $[a, b]$ with $f(a)\,f(b) < 0$ and tolerance ε, the iterative formula for a better root is given by

$$x = \frac{(a+b)}{2}$$

If $(|f(x)| < \varepsilon)$, root found and stop iteration
else
if $(f(x)\,f(b) < 0)$ set $a = x$, $b = b$
else set $a = a$, $b = x$
and continue until $|f(x)| < \varepsilon$

Algorithm 1 (when a and b are not given)

Step 1: Start the program
Step 2: Define the function $f(x)$
Step 3: Read n (possible upper limit of the interval containing the root)
Step 4: For $(i = -n, i \le n, i{+}{+})$
    Set $a = i$ and $b = i + 1$
        If $f(a)^* f(b) < 0$
    Break
  End i
Step 5: Print the interval $[a, b]$

Step 6: Read ε (desired accuracy)
Step 7: Do{

$x = (a + b)/2$

If $f(b) f(x) < 0$,

$a = x$

Else

$b = x$

}While ($|f(x)| > \varepsilon$)

Step 8: Print x
Step 9: Stop

Program Code The program code requires the declaration of function $f(x)$. For different functions, the program code will vary in the function declaration section. Therefore, we have to consider a specific function to construct the program. For example, let us consider the following problem:

Find a real root of the equation $x^3 - 3x - 5 = 0$, correct to 3 decimal places by the bisection method. The code for the program is given by:

```
#include<stdio.h>
#include<conio.h>
#include<math.h>
float f(float x)
{
return(pow(x,3)-3*x-5);
}
void main()
{
int i,n;
float a,b,x,e;
clrscr();
printf("\n Enter the possible range:");
scanf("%d",&n);
for(i=--n;i<=n;i++)
{
a=i;
b=i+1;
if(f(a)*f(b)<0)
break;
 }
```

Cont'd

Program cont'd

```
printf("\n The root lies between %.1f and %.1",a,b);
printf("\n\n Enter the permissible error:");
scanf("%f",&e);
do{
x= (a+b)/ 2.0;
if(f(x)*f(a)<0)
b=x;
else
a=x;
} while(fabs(f(x))>e);
printf("\n\n The required root is x= %.3f",x);
getch();
}
```

Input/Output

```
Enter the possible range:50
The root lies between 2.0 and 3.0
Enter the permissible error:0.0001
The required root is x= 2.279
```

Algorithm 2 (when a and b are given)

Step 1: Start the program
Step 2: Define the function $f(x)$
Step 3: Read a and b (lower and upper limits of the interval containing the real root)
Step 4: Read ε (desired accuracy)
Step 5: Do{

$x = (a + b)/2$

If $f(b)\, f(x) < 0$,

$a = x$

Else

$b = x$

}While $(|f(x)| > \varepsilon)$

Step 6: Print x
Step 7: Stop

Program Code The program code requires the declaration of function $f(x)$. For different functions, the program code will vary in the function declaration section. Therefore, we have to consider a specific function to construct the program. For example, let us consider the following problem:

Find a real root of the equation $x^3 - 3x - 5 = 0$ between 2 and 3, correct to 3 decimal places by the bisection method.

The code for the program is given by:

```
#include<stdio.h>
#include<conio.h>
#include<math.h>
float f(float x)
{
return(pow(x,3)-3*x-5);
}
void main()
{
float a,b,x,e;
clrscr();
printf("\n Enter the value of a:");
scanf("%f",&a);
printf("\n Enter the value of b:");
scanf("%f",&b);
printf("\n\n Enter the permissible error:");
scanf("%f",&e);
do{
x= (a+b)/ 2.0;
if(f(x)*f(a)<0)
b=x;
else
a=x;
} while(fabs(f(x))>e);
printf("\n\n The required root is x= %.3f",x);
getch();
}
```

Input/Output

```
Enter the value of a:2
Enter the value of b:3
Enter the permissible error:0.0001
The required root is x= 2.279
```

ASSIGNMENT 7

Write C programs to implement the bisection method for finding the real root of the following equations:

(a) $2x^3 - 5x + 1.06 = 0$, correct to 3 decimal places.

(b) $x^3 - 7x - 9 = 0$, correct to 3 decimal places.
(c) $x^4 + x - 35.4 = 0$, correct to 3 decimal places, between 3 and 4.
(d) $x^3 - 5x - 7 = 0$, correct to 3 decimal places, between 2 and 3.

9.10 REGULA FALSI METHOD

Method To find the real root of $f(x) = 0$ in the interval $[a, b]$ with $f(a)\,f(b) < 0$ and tolerance ε, the iterative formula for a better root is given by

$$x = \frac{af(b) - bf(a)}{f(b) - f(a)}$$

If $(|f(x)| < \varepsilon)$, root found and stop iteration

else
if $(f(x)\,f(b) < 0)$ set $a = x$, $b = b$
else set $a = a$, $b = x$
and continue until $|f(x)| < \varepsilon$

Algorithm 1 (when a and b are not given)

Step 1: Start the program
Step 2: Define the function $f(x)$
Step 3: Read n (possible upper limit of the interval containing the root)
Step 4: For $(i = -n, i \leq n, i{+}{+})$
    Set $a = i$ and $b = i + 1$
        If $f(a) * f(b) < 0$
    Break
  End i
Step 5: Print the interval $[a, b]$
Step 6: Read ε (desired accuracy)
Step 7: Do{
    $x = (a * f(b) - b * f(a))/(f(b) - f(a))$
    If $f(b)\,f(x) < 0$,
        $a = x$
    Else
        $b = x$
    }While($|f(x)| > \varepsilon$))
Step 8: Print x
Step 9: Stop

Program Code The program code requires the declaration of function $f(x)$. For different functions, the program code will vary in the function declaration section. Therefore, we have to consider a specific function to construct the program. For example, let us consider the following problem:

Find a real root of the equation $x^3 - 3x - 5 = 0$, *correct to 3 decimal places by the regula falsi method.*

The code for the program is given by:

```
#include<stdio.h>
#include<conio.h>
#include<math.h>
float f(float x)
{
return(pow(x,3)-3*x-5);
}
void main()
{
int i,n;
float a,b,x,e;
clrscr();
printf("\n Enter the possible range:");
scanf("%d",&n);
for(i=--n;i<=n;i++)
{
a=i;
b=i+1;
if(f(a)*f(b)<0)
break;
 }
printf("\n The root lies between %.1f and %.1f ",a,b);
printf("\n\n Enter the permissible error:");
scanf("%f",&e);
do{
x= (a*f(b)-b*f(a))/ (f(b)-f(a));
if(f(x)*f(a)<0)
b=x;
else
a=x;
} while(fabs(f(x))>e);
printf("\n\n The required root is x= %.3f",x);
getch();
}
```

Input/Output

```
Enter the possible range:50
The root lies between 2.0 and 3.0
Enter the permissible error:0.0001
The required root is x= 2.279
```

Algorithm 2 (when a and b are given)

Step 1: Start the program
Step 2: Define the function $f(x)$
Step 3: Read a and b (lower and upper limits of the interval containing the real root)
Step 4: Read ε (desired accuracy)
Step 5: Do{
$x = (a * f(b) - b * f(a))/(f(b) - f(a))$
If $f(b) f(x) < 0$,
$a = x$
Else
$b = x$
}While($|f(x)| > \varepsilon$)
Step 6: Print x
Step 7: Stop

Program Code The program code requires the declaration of function $f(x)$. For different functions, the program code will vary in the function declaration section. Therefore, we have to consider a specific function to construct the program. For example, let us consider the following problem:

Find a real root of the equation $x^3 - 3x - 5 = 0$, *between 2 and 3, correct to 3 decimal places by the regula falsi method.*

The code for the program is given by:

```
#include<stdio.h>
#include<conio.h>
#include<math.h>
float f(float x)
{
return(pow(x,3)-3*x-5);
}
void main()
{
float a,b,x,e;
clrscr();
```

Cont'd

Program cont'd

```
printf("\n Enter the value of a:");
scanf("%f",&a);
printf("\n Enter the value of b:");
scanf("%f",&b);
printf("\n\n Enter the permissible error:");
do{
x= (a*f(b)-b*f(a))/ (f(b)-f(a));
scanf("%f",&e);
if(f(x)*f(a)<0)
b=x;
else
a=x;
} while(fabs(f(x))>e);
printf("\n\n The required root is x= %.3f",x);
getch();
}
```

Input/Output

```
Enter the value of a:2
Enter the value of b:3
Enter the permissible error:0.0001
The required root is x= 2.279
```

ASSIGNMENT 8

Write C programs to implement the regula falsi method for finding a real root of the following equations:

(a) $x^3 - 3x + 1.06 = 0$, correct to 3 decimal places.
(b) $x^3 - 2.1x - 9 - 0$, correct to 3 decimal places.
(c) $x^4 + x - 35.4 = 0$, correct to 3 decimal places, between 3 and 4.
(d) $x^3 - 5x - 7= 0$, correct to 3 decimal places, between 2 and 3.

9.11 THE SECANT METHOD

Method To find a real root of $f(x) = 0$ in the interval $[a, b]$ with $f(a)\,f(b) < 0$ and tolerance ε, the iterative formula for a better root is given by

$$x = \frac{af(b) - bf(a)}{f(b) - f(a)}$$

If ($|f(x)| < \varepsilon$), root found and stop iteration
else

set $a = x$, $b = b$
and continue until $|f(x)| < \varepsilon$

Algorithm 1 (when a and b are not given)

Step 1: Start the program
Step 2: Define the function $f(x)$
Step 3: Read n (possible upper limit of the interval containing the root)
Step 4: For ($i = -n, i \leq n, i{+}{+}$)
  Set $a = i$ and $b = i + 1$
   If $f(a) * f(b) < 0$
  Break
 End i
Step 5: Print the interval $[a, b]$
Step 6: Read ε (desired accuracy)
Step 7: Do{
  $x = (a * f(b) - b * f(a))/(f(b) - f(a))$
   $a = x$
  }While($|f(x)| > \varepsilon$)
Step 8: Print x
Step 9: Stop

Program Code The program code requires the declaration of function $f(x)$. For different functions, the program code will vary in the function declaration section. Therefore, we have to consider a specific function to construct the program. For example, let us consider the following problem:

Find a real root of the equation $x^3 - 3x - 5 = 0$, correct to 3 decimal places by the secant method.

The code for the program is given by:

```
#include<stdio.h>
#include<conio.h>
#include<math.h>
float f(float x)
{
return(pow(x,3)-3*x-5);
}
void main()
{
int i,n;
float a,b,x,e;
clrscr();
printf("\n Enter the possible range:");
scanf("%d",&n);
for(i=--n;i<=n;i++)
```

Cont'd

Program cont'd

```
{
a=i;
b=i+1;
if(f(a)*f(b)<0)
break;
 }
printf("\n The root lies between %.1f and %.1f ",a,b);
printf("\n\n Enter the permissible error:");
scanf("%f",&e);
do{
x= (a*f(b)-b*f(a))/ (f(b)-f(a));
a=x;
} while(fabs(f(x))>e);
printf("\n\n The required root is x= %.3f",x);
getch();
}
```

Input/Output

```
Enter the possible range:50
The root lies between 2.0 and 3.0
Enter the permissible error:0.0001
The required root is x= 2.279
```

Algorithm 1 (when a and b are given)

Step 1: Start the program
Step 2: Define the function $f(x)$
Step 3: Read a and b (lower and upper limits of the interval containing the real root)
Step 4: Read ε (desired accuracy)
Step 5: Do{

$$x = (a * f(b) - b * f(a))/(f(b) - f(a))$$

$$a = x$$

}While($|f(x)| > \varepsilon$))
Step 6. Print x
Step 7: Stop

Program Code The program code requires the declaration of function $f(x)$. For different functions, the program code will vary in the function declaration section. Therefore, we have to consider a specific function to construct the program. For example, let us consider the following problem:

Find a real root of the equation $x^3 - 3x - 5 = 0$, between 2 and 3, correct to 3 decimal places by the secant method.

The code for the program is given by:

```
#include<stdio.h>
#include<conio.h>
#include<math.h>
float f(float x)
{
return(pow(x,3)-3*x-5);
}
void main()
{
float a,b,x,e;
clrscr();
printf("\n Enter the value of a:");
scanf("%f",&a);
printf("\n Enter the value of b:");
scanf("%f",&b);
printf("\n\n Enter the permissible error:");
scanf("%f",&e);
do{
x= (a*f(b)-b*f(a))/ (f(b)-f(a));
a=x;
} while(fabs(f(x))>e);
printf("\n\n The required root is x= %.3f",x);
getch();
}
```

Input/Output

```
Enter the value of a:2
Enter the value of b:3
Enter the permissible error:0.0001
The required root is x= 2.279
```

ASSIGNMENT 9

Write C programs to implement the secant method for finding the real root of the following equations:

(a) $x^3 - 7x + 1.06 = 0$, correct to 3 decimal places.

(b) $x^3 - 2.1x - 5 = 0$, correct to 3 decimal places.

(c) $x^4 - x - 35.4 = 0$, correct to 3 decimal places, between 3 and 4.
(d) $x^3 - 5x - 10 = 0$, correct to 3 decimal places, between 2 and 3.

9.12 NEWTON–RAPHSON METHOD

Method To find a real root of $f(x) = 0$ with the initial approximation a and tolerance ε, the iterative formula for a better root is given by

$$x = a - \frac{f(a)}{f'(a)}$$

If ($|f(x)| < \varepsilon$), root found and stop iteration
else
set $a = x$
and continue until $|f(x)| < \varepsilon$

Algorithm 1 (initial guess is not given)

Step 1: Start the program
Step 2: Define the function $f(x)$ and its derivative $f'(x)$
Step 3: Read n (possible upper limit of the interval containing the root)
Step 4: For ($i = -n, i \leq n, i{+}{+}$)
Set $a = i$ and $b = i + 1$
If $f(a) * f(b) < 0$
Break
End i
Step 5: Set $x0 = (a + b)/2$ as the initial guess
Step 6: Read ε (desired accuracy)
Step 7: Do{

$$x = x0 - \frac{f(x0)}{f'(x0)}$$

$x0 = x$
} While($|f(x)| > \varepsilon$)
Step 8: Print x
Step 9: Stop

Program Code The program code requires the declaration of function $f(x)$. For different functions, the program code will vary in the function declaration section. Therefore, we have to consider a specific function to construct the program. For example, let us consider the following problem:

Find a real root of the equation: $x^3 - 3x - 5 = 0$, *correct to 3 decimal places by the Newton–Raphson method.*

The code for the program is given by:

```
#include<stdio.h>
#include<conio.h>
#include<math.h>
float f(float x)
{
return(pow(x,3)-3*x-5);
}
void main()
{
int i,n;
float a,b,x,x0,e;
clrscr();
printf("\n Enter the possible range:");
scanf("%d",&n);
for(i=--n;i<=n;i++)
{
a=i;
b=i+1;
if(f(a)*f(b)<0)
break;
}
x0=(a+b)/2.0;
printf("\n\n Enter the permissible error:");
scanf("%f",&e);
do{
x=x0-(f(x0)/f1(x0));
     x0=x;
} while(fabs(f(x))>e);
printf("\n\n The required root is x= %.4f",x);
getch();
}
```

Input/Output

```
Enter the possible range:50
Enter the permissible error:0.0001
The required root is x= 2.279
```

Algorithm 2 (initial guess is given)

Step 1: Start the program

Step 2: Define the function $f(x)$ and its derivative $f'(x)$

Step 3: Read a (initial guessed root) and ε (desired accuracy)

Step 4: Do{

$$x = a - \frac{f(a)}{f'(a)}$$

$$a = x$$

} While ($|f(x)| > \varepsilon$)

Step 5: Print x

Step 6: Stop

Program Code The program code requires the declaration of function $f(x)$. For different functions, the program code will vary in the function declaration section. Therefore, we have to consider a specific function to construct the program. For example, let us consider the following problem:

Find a real root of the equation: $x^3 - 3x - 5 = 0$, *correct to 3 decimal places by the Newton–Raphson method, with* $x_0 = 2$ *as the initial approximation.*

The code for the program is given by:

```
#include<stdio.h>
#include<conio.h>
#include<math.h>
float f(float x)
{
  return(pow(x,3)-3*x-5);
}
float f1(float x)
{
  return(3*pow(x,2)-3);
}
void main()    {
  float a,e,x;
  clrscr();
    printf("\n\n Enter a(initial guess):");
    scanf("%f",&a);
    printf("\n\n Enter the permissible error:");
    scanf("%f",&e);
    do
    {
      x=a-(f(a)/f1(a));
      a=x;
```

Cont'd

Program cont'd

```
    }while(fabs(f(x))>e);
    printf("\n\n The required root is x= %.4f ",x);
    getch();
}
```

Input/Output

```
Enter a(initial guess):2
Enter the permissible error:0.0001
The required root is x= 2.279
```

ASSIGNMENT 10

Write C programs to implement the Newton–Raphson method for finding a real root of the following equations:

(a) $x^3 - 3x + 1.06 = 0$, correct to 3 decimal places.
(b) $x^3 - 2.1x - 9 = 0$, correct to 3 decimal places.
(c) $x^4 + x - 35.4 = 0$, correct to 3 decimal places, between 3 and 4.
(d) $x^3 - 5x - 7 = 0$, correct to 3 decimal places, taking 2 as the initial approximation.

9.13 EULER'S METHOD

Method Let us consider a first order and first degree ODE as $\frac{dy}{dx} = f(x, y)$ with initial condition $y(x_0) = y_0$. Let us suppose that we have to find the approximate numerical value of y, say y_n, at the point $x = x_n$. We divide the range $[x_0, x_n]$ into n equal subintervals of length h by the points $x_0, x_1, \ldots, x_n$, where $x_i = x_0 + ih$, $(i = 1, 2, \ldots, n)$. The corresponding values of the function are $y_0 = f(x_0)$, $y_1 = f(x_1)$, …, $y_n = f(x_n)$.

Then, assuming $f(x, y) \approx f(x_{n-1}, y_{n-1})$ in $[x_{n-1}, x_n]$, the general formula is given by

$y_n = y_{n-1} + hf(x_{n-1}, y_{n-1})$, $n = 1, 2, \ldots, n$

Algorithm

Step 1: Start the program
Step 2: Define the function $f(x, y)$
Step 3: Read x_0, y_0 (initial values of x, y), h (step length), xn (final value of x)
Step 4: Do{

$y0 = y0 + h * f(x0, y0)$

$x0 = x0 + h$

} While $(x0 < xn)$

Step 5: Display $y0$
Step 6: Stop

Program Code The program code requires the declaration of function $f(x, y)$. For different functions, the program code will vary in the function declaration section. Therefore, we have to consider a specific function to construct the program. For example, let us consider the following problem:

Find y when x $= 0.1$, *given y* $= 1$, *when x* $= 0$, *and* $\frac{dy}{dx} = 3x + y^2$, $h = 0.0001$, *by Euler's method.*

The code for the program is given by:

```
#include<stdio.h>
#include<conio.h>
#include<math.h>
float f(float x,float y)
  {
  return(3*x + (pow(y,2)));
  }

void main()
{
float x0,y0,xn,h;
clrscr();
printf("\n Enter the value of x0 :");
scanf("%f",&x0);
printf("\n Enter the value of y0 :");
scanf("%f",&y0);
printf("\n Enter the value of xn :");
scanf("%f",&xn);
printf("\n Enter the value of h :");
scanf("%f",&h);
do
  {
  y0=y0 + h*f(x0,y0);
  x0=x0+h;
} while(x0<xn);
printf("\n The value of y(%.2f) is= %.4f",xn,y0);
getch();
}
```

Input/Output

```
Enter the value of x0 :0
Enter the value of y0 :1
Enter the value of xn :0.1
Enter the value of h :0.0001
The value of y(0.10) is= 1.1272
```

ASSIGNMENT 11

Construct C programs to implement Euler's method for the following initial value problems:

(a) Calculate $y(0.4)$, correct to 4 decimal places, where $\frac{dy}{dx} = \log_{10}(x+y), y(0)=1$, taking $h = 0.001$.

(b) Find $y(0.1)$, correct to 4 decimal places when $\frac{dy}{dx} = \frac{y-x}{y+x}, y(0)=1$, taking $h = 0.0001$.

(c) Evaluate $y(2)$, correct to 4 decimal places from $\frac{dy}{dx} = \frac{1}{2}(x+y)$ with $y(0) = 2$, taking $h = 0.001$.

(d) Obtain y at $x = 0.01$, correct to 4 decimal places from $\frac{dy}{dx} = y + e^x$ with $y(0) = 0$ and $h = 0.001$.

9.14 RUNGE–KUTTA METHOD OF ORDER FOUR

Method Let us consider a first order and first degree ODE as $\frac{dy}{dx} = f(x,y)$ with initial condition $y(x_0) = y_0$. Let us suppose that we have to find the approximate numerical value of y, say y_n, at the point $x = x_n$. We divide the range $[x_0, x_n]$ into n equal subintervals of length h by the points $x_0, x_1, \ldots, x_n$, where $x_i = x_0 + ih$, $(i = 1, 2, \ldots, n)$. The corresponding values of the function are $y_0 = f(x_0)$, $y_1 = f(x_1)$, ..., $y_n = f(x_n)$.

The general formula is given by

$$y_n = y_{n-1} + (k_1 + 2k_2 + 2k_3 + k_4)/6, \qquad n = 1, 2, \ldots, n$$

where
$$k_1 = hf(x_{n-1}, y_{n-1})$$
$$k_2 = hf(x_{n-1} + h/2, y_{n-1} + k_1/2)$$
$$k_3 = hf(x_{n-1} + h/2, y_{n-1} + k_2/2)$$
$$k_4 = hf(x_{n-1} + h, y_{n-1} + k_3)$$

Algorithm

Step 1: Start the program
Step 2: Define the function $f(x, y)$
Step 3: Read $x0$, $y0$ (initial values of x, y), h (step length), xn (final value of x)
Step 4: Do{

$k1 = h * f(x0, y0)$
$k2 = h * f(x0 + h/2, y0 ++ k1/2)$
$k3 = h * f(x0 + h/2, y0 + k2/2)$
$k1 = h * f(x0 + h, y0 + k3)$
$k = (k1 + 2 * k2 + 2 * k3 + k4)/6$
$y0 = y0 + k$
$x0 = x0 + h$
} While $(x0 < xn)$

Step 5: Display $y0$
Step 6: Stop

Program Code The program code requires the declaration of function $f(x, y)$. For different functions, the program code will vary in the function declaration section. Therefore, we have to consider a specific function to construct the program. For example, let us consider the following problem:

Find y when $x = 0.1$, *given* $y = 0$ *when* $x = 0$ *and* $\frac{dy}{dx} = 3x + y^2$, $h = 0.0001$ *by fourth order Runge–Kutta method.*

The code for the program is given by:

```
#include<stdio.h>
#include<conio.h>
float f(float x,float y)
{
  return(3*x+(y*y));
}
void main()
{
    float x0,y0,xn,k,k1,k2,k3,k4;
    clrscr();
    printf("\n Enter the value of x0 :");
    scanf("%f",&x0);
    printf("\n Enter the value of y0 :");
    scanf("%f",&y0);
    printf("\n Enter the value of xn:");
    scanf("%f",&xn);
    printf("\n Enter the value of h :");
    scanf("%f",&h);
    do
    {
    k1=h*f(x0,y0);
    2=h*f(x0+h/2,y0+k1/2);
    k3=h*f(x0+h/2,y0+k2/2);
    k4=h*f(x0+h,y0+k3);
    k=((k1+2*k2+2*k3+k4)/6),
    y0-y0+k;
    x0=x0+h;
    }while(x0<xn);
    printf("\n The value of y(%.2f) is= %.4f",xn,yn);
    getch();
    }
```

Input/Output

```
Enter the value of x0 :0
Enter the value of y0 :1
Enter the value of xn :0.1
Enter the value of h :0.0001
The value of y(0.10) is= 1.1272
```

ASSIGNMENT 12

Construct C programs to implement the fourth order Runge–Kutta method for the following initial value problems:

(a) Calculate $y(0.4)$, correct to 4 decimal places, where $\frac{dy}{dx} = \log_{10}(x+y), y(0)=1$, taking $h = 0.001$.

(b) Find $y(0.1)$, correct to 4 decimal places when $\frac{dy}{dx} = \frac{y-x}{y+x}, y(0)=1$, taking $h = 0.0001$.

(c) Evaluate $y(2)$, correct to 4 decimal places from $\frac{dy}{dx} = \frac{1}{2}(x+y)$ with $y(0) = 2$, taking $h = 0.001$.

(d) Obtain y at $x = 0.01$, correct to 4 decimal places from $\frac{dy}{dx} = y + e^x$ with $y(0) = 0$ and $h = 0.001$.

9.15 GAUSS ELIMINATION METHOD

Method This method is a systematic process of eliminating the unknowns so that the system is reduced to an equivalent upper triangular system which could be solved by back substitution.

Let us consider the following system of n linear equations with n unknowns

$$a_{11}x_1 + a_{12}x_2 + \cdots + a_{1n}x_n = b_1$$
$$a_{21}x_1 + a_{22}x_2 + \cdots + a_{2n}x_n = b_2$$
$$\cdots$$
$$a_{n1}x_1 + a_{n2}x_2 + \cdots + a_{nn}x_n = b_n$$

where the coefficients a_{ij} $(i, j = 1, 2, 3, \ldots, n)$ and the constants b_i $(i = 1, 2, 3, \ldots, n)$ are known. The system can also be written in matrix form as

$$AX = B$$

where $= [a_{ij}]_{n\times n} = \begin{bmatrix} a_{11} & a_{12} & a_{13} & \cdots & a_{1n} \\ a_{21} & a_{22} & a_{23} & \cdots & a_{2n} \\ a_{31} & a_{32} & a_{33} & \cdots & a_{3n} \\ \cdots & \cdots & \cdots & \cdots & \cdots \\ a_{n1} & a_{n1} & a_{n1} & \cdots & a_{n1} \end{bmatrix}$, known as the coefficient matrix and $B = \begin{bmatrix} b_1 \\ b_2 \\ b_3 \\ \vdots \\ b_n \end{bmatrix}$, known as the right-hand side vector are given, while

$X = \begin{bmatrix} x_1 \\ x_2 \\ x_3 \\ \vdots \\ x_n \end{bmatrix}$, known as the solution vector is to be determined.

The Gauss elimination method consists of two steps.
Forward elimination of unknowns: In this step

(i) x_1 is eliminated from each equation except the first equation. After that,
(ii) x_2 is eliminated from each equation except the first and second equations. Next,
(iii) x_3 is eliminated from each equation except the first, second and third equations, and so on.

The process is executed in a systematic way. First, a matrix is constructed with the coefficient matrix A, appended with the constant matrix B. This matrix is known as the **augmented matrix** and is given by

$$[a_{ij} \mid b_i]_{n\times(n+1)} = \left[\begin{array}{ccccc|c} a_{11} & a_{12} & a_{13} & \cdots & a_{1n} & b_1 \\ a_{21} & a_{22} & a_{23} & \cdots & a_{2n} & b_2 \\ a_{31} & a_{32} & a_{33} & \cdots & a_{3n} & b_3 \\ \cdots & \cdots & \cdots & \cdots & \cdots & \cdots \\ a_{n1} & a_{n1} & a_{n1} & \cdots & a_{n1} & b_n \end{array}\right] \tag{1}$$

The aim is to reduce this augmented matrix $[a_{ij}:b_i]_{n\times(n+1)}$ to an upper triangular matrix.

Step 1: Assume $a_{11} \neq 0$. Now, multiply the 1st row of (1) by $\frac{a_{i1}}{a_{11}}$ and subtract from the ith row ($i = 2, 3, \ldots, n$). By this, all elements in the 1st column except a_{11} are made zero. Thus, (1) takes the form

$$\left[\begin{array}{ccccc|c} a_{11} & a_{12} & a_{13} & \cdots & a_{1n} & b_1 \\ 0 & a_{22}^{(1)} & a_{23}^{(1)} & \cdots & a_{2n}^{(1)} & b_2^{(1)} \\ 0 & a_{32}^{(1)} & a_{33}^{(1)} & \cdots & a_{3n}^{(1)} & b_3^{(1)} \\ \cdots & \cdots & \cdots & \cdots & \cdots & \cdots \\ 0 & a_{n2}^{(1)} & a_{n3}^{(1)} & \cdots & a_{nn}^{(1)} & b_n^{(1)} \end{array}\right] \tag{2}$$

where $$a_{ij}^{(1)} = a_{ij} - a_{1j}\frac{a_{i1}}{a_{11}}, i, j = 2, 3, \cdots, n$$

and $$b_i^{(1)} = b_i - b_1\frac{a_{i1}}{a_{11}}, i = 2, 3, \cdots, n$$

Step 2: Assume $a_{22}^{(1)} \neq 0$. Multiply the 2nd row of (2) by $\dfrac{a_{i2}^{(1)}}{a_{22}^{(1)}}$ and subtract from the ith row (i = 3, 4, …, n). By this, all elements in the 2nd column except a_{12} and $a_{22}^{(1)}$ are made zero. Thus, (2) reduces to

$$\left[\begin{array}{ccccc|c} a_{11} & a_{12} & a_{13} & \cdots & a_{1n} & b_1 \\ 0 & a_{22}^{(1)} & a_{23}^{(1)} & \cdots & a_{2n}^{(1)} & b_2^{(1)} \\ 0 & 0 & a_{33}^{(2)} & \cdots & a_{3n}^{(2)} & b_3^{(2)} \\ \cdots & \cdots & \cdots & \cdots & \cdots & \cdots \\ 0 & 0 & a_{n3}^{(2)} & \cdots & a_{nn}^{(2)} & b_n^{(2)} \end{array}\right] \quad (3)$$

where
$$a_{ij}^{(2)} = a_{ij}^{(1)} - a_{2j}^{(1)} \frac{a_{i2}^{(1)}}{a_{22}^{(1)}}, i, j = 3, 4, \cdots, n$$

and
$$b_i^{(2)} = b_i^{(1)} - b_2^{(1)} \frac{a_{i2}^{(1)}}{a_{22}^{(1)}}, i = 3, 4, \cdots, n$$

Continuing this process, all elements below the leading diagonal elements of A are made zero and we get

$$\left[a_{ij} \mid b_i\right]_{n\times(n+1)} \sim \left[\begin{array}{cccccc|c} a_{11} & a_{12} & a_{13} & \cdots & a_{1n} & & b_1 \\ 0 & a_{22}^{(1)} & a_{23}^{(1)} & \cdots & a_{2n}^{(1)} & & b_2^{(1)} \\ 0 & 0 & a_{33}^{(2)} & \cdots & a_{3n}^{(2)} & & b_3^{(2)} \\ \cdots & \cdots & \cdots & \cdots & \cdots & & \cdots \\ 0 & 0 & 0 & \cdots & a_{nn}^{(n-1)} & & b_n^{(n-1)} \end{array}\right] \quad (4)$$

Back substitution: In this step, starting from the nth equation, each of the unknowns is found. From (4), we can write the equivalent system as

$$a_{11}x_1 + a_{12}x_2 + \cdots + a_{1n}x_n = b_1$$

$$a_{22}^{(1)}x_2 + \cdots + a_{2n}^{(1)}x_n = b_2^{(1)}$$

$$\cdots$$

$$a_{nn}^{(n-1)}x_n = b_n^{(n-1)}$$

Now

(i) Calculate x_n from the nth equation.
(ii) Then calculate x_{n-1} from the (n – 1)th equation substituting x_n by its value, obtained from the nth equation, and so on.

Thus, with this back substitution procedure, the solution for each unknown x_i($i = n, n - 1, n - 2, \ldots$ 2, 1) is obtained.

Algorithm

Step 1: Start the program

Step 2: Read n (order of the coefficient matrix)

Step 3: For $i = 0(1)n - 1$

For $j = 0(1)n - 1$

Read $a[i][j]$

End j

End i

Step 4: For $i = 0(1)n - 1$

Read $b[i]$

End i

Step 5: For $k = 0(1)n - 1$

For $i = k + 1(1)n - 1$

$m = a[i][k]/a[k][k]$

$b[i] = b[i] - b[k] * m$

For $j - 0(1)n - 1$

$a[i]\ [j] = a[i]\ [j] - m * a[k]\ [j]$

End j

End i

End k

Step 6: For $k = n - 1(-1)0$

For $i = n - 1(-1)k$

If $i = k$

$x[k] = b[k]/a[k]\ [k]$

else

$b[k] = b[k] - a[k]\ [i] * x[i]$

End i

End k

Step 7: For $i = 0(1)n - 1$

Print $x[i]$

End i

Step 8: Stop

Program Code

```
#include<stdio.h>
#include<conio.h>
void main()
{
int n,i,j,k;
  float a[10][10],b[10],m,x[10],sum;
```

Cont'd

Program cont'd

```
clrscr();
printf("Entre the order of the coefficient matrix A :");
scanf("%d",&n);
printf("\n Enter the elements of the coefficient matrix A :");
for(i=0;i<n;i++)
{
for(j=0;j<n;j++)
{
scanf("%f",&a[i][j]);
}}
printf("\n Enter the elements of the constant matrix B :");
for(i=0;i<n;i++)
{
scanf("%f",&b[i]);
}
for(k=0;k<n;k++)
[
for(i=k+1;i<n;i++)
{
   m=a[i][k]/a[k][k];
b[i]=b[i]-m*b[k];
for(j=0;j<n;j++)
{
a[i][j]=a[i][j]-(m*a[k][j]);
}
}
}
for(k=n-1;k>=0;k--)
{
sum=0;
for(i=n-1;i>=k;i--)
{
   if(i==k)
x[k]=(b[k]-sum)/a[k][k];else
sum=sum+a[k][i]*x[i];
}
}
```

Cont'd

Program cont'd

```
printf("\n The required solutions are: ");
for(i=0;i<n;i++)
{
printf("\nx[%d]=%.3f",x[i]);
}
getch();
}
```

Input/Output For this, let us consider the following example:

Solve the following system of linear equations by the Gauss elimination method, correct to 3 decimal places:

$$10x + y - 2z = 7.74$$
$$x + 12y + 3z = 39.66$$
$$3x + 4y + 15z = 54.8$$

The input/output will be as follows:

```
Entre the order of the coefficient matrix A :3
Enter the elements of the coefficient matrix A :10
1
-2
1
12
3
3
4
15
Enter the elements of the constant matrix B :
7.74
39.66
54.8
The required solutions are:
x[1]=1.075
x[1]=2.523
x[1]=3.765
```

ASSIGNMENT 13

Write C programs to implement the Gauss elimination method to solve the following system of linear equations, correct to 3 decimal places:

(a) $5x - y + z = 10$, $2x + 4y = 12$, $x + y + 5z = -1$
(b) $x + y + z = 9$, $2x - 3y + 4z = 13$, $3x + 4y + 5z = 40$

(c) $3x + 4y + 5z = 18$, $2x - y + 8z = 13$, $5x - 2y + 7z = 20$
(d) $5x_1 - x_2 = 9$, $-x_1 + 5x_2 - x_3 = 4$, $-x_2 + 5x_3 = -6$

9.16 GAUSS–SEIDEL METHOD

Method Let us consider the following diagonally dominant system of n linear equations with n unknowns:

$$a_{11}x_1 + a_{12}x_2 + \cdots + a_{1n}x_n = b_1$$
$$a_{21}x_1 + a_{22}x_2 + \cdots + a_{2n}x_n = b_2$$
$$\cdots$$
$$a_{n1}x_1 + a_{n2}x_2 + \cdots + a_{nn}x_n = b_n$$

We can re-write the system as

$$x_1 = \frac{1}{a_{11}}\left[b_1 - a_{12}x_2 - a_{13}x_3 - \cdots - a_{1n}x_n\right] \tag{1}$$

$$x_2 = \frac{1}{a_{22}}\left[b_2 - a_{21}x_1 - a_{23}x_3 - \cdots - a_{2n}x_n\right] \tag{2}$$

$$\cdots$$
$$\cdots$$

$$x_n = \frac{1}{a_{nn}}\left[b_n - a_{n1}x_1 - a_{n2}x_2 - \cdots - a_{nn-1}x_{n-1}\right] \tag{n}$$

Now, in the first iteration,

(i) Put $x_2 = x_2^{(0)}, x_3 = x_3^{(0)}, \cdots, x_n = x_n^{(0)}$ as the initial approximation in (1) and calculate $x_1^{(1)}$ for x_1 from it.

(ii) Put $x_1 = x_1^{(1)}, x_3 = x_3^{(0)}, \cdots, x_n = x_n^{(0)}$ in (2) and calculate $x_2^{(1)}$ for x_2 from it.

Similarly, calculate $x_3^{(1)}$ for x_3 from (3), $x_4^{(1)}$ for x_4 from (4), …, $x_n^{(1)}$ for x_n from (n).

Thus, after the first iteration, we have a better set of approximations, namely, $x_1^{(1)}, x_2^{(1)}, \cdots, x_n^{(1)}$
For the second iteration,

(i) Put $x_2 = x_2^{(1)}, x_3 = x_3^{(1)}, \cdots, x_n = x_n^{(1)}$ in (1) and calculate $x_1^{(2)}$ for x_1 from it.

(ii) Put $x_1 = x_1^{(2)}, x_3 = x_3^{(1)}, \cdots, x_n = x_n^{(1)}$ in (2) and calculate $x_2^{(2)}$ for x_2 from it.

Similarly, calculate $x_3^{(2)}$ for x_3 from (3), $x_4^{(2)}$ for x_4 from (4), …, $x_n^{(2)}$ for x_n from (n).

Thus, after the kth iteration, the iterative formulae are given by

$$x_1^{(k+1)} = \frac{1}{a_{11}}\left[b_1 - a_{12}x_2^{(k)} - a_{13}x_3^{(k)} - \cdots - a_{1n}x_n^{(k)}\right] \tag{1}$$

$$x_2^{(k+1)} = \frac{1}{a_{22}}\left[b_2 - a_{21}x_1^{(k+1)} - a_{23}x_3^{(k)} - \cdots - a_{2n}x_n^{(k)}\right] \tag{2}$$

...

...

$$x_n^{(k+1)} = \frac{1}{a_{nn}}\left[b_n - a_{n1}x_1^{(k+1)} - a_{n2}x_2^{(k+1)} - \cdots - a_{nn-1}x_{n-1}^{(k+1)}\right] \quad (n)$$

(K = 0, 1, 2,...)

The process should to be continued till we get the solution to the desired degree of accuracy. It should be noted that the initial approximations $x_2^{(0)}, x_3^{(0)}, \ldots, x_n^{(0)}$, if not mentioned, are all set to be zero.

Algorithm

Step 1: Start the program

Step 2: Read *n* (order of the coefficient matrix)

Step 3: For $i = 0(1)n - 1$

    For $j = 0(1)n - 1$

        Read $a[i][j]$

    End *j*

    End *i*

Step 4: For $i = 0(1)n - 1$

    Read $b[i]$

    End *i*

Step 5: while ($t == 0$)

    For $i = 0(1)n - 1$

        $x[i] = x0[i]$

        $sum == b[i]$

        For $j = 0(1)n - 1$

            If $j \neq i$

                $sum = sum - a[i]\,[j] * x0[j]$

                $x0[i] = sum/a[i]\,[i]$

            For $i = 0(1)n - 1$

                If($fabs(x0[i] - x[i]) > \varepsilon$)

                $t = 0$

                    else

                    $t = 1$

Step 6: For $i = 0(1)n - 1$

    Print $x[i]$

Step 7: Stop

Program Code

```
#include<stdio.h>
#include<conio.h>
#include<math.h>
```

Cont'd

Program cont'd

```
void main()
{
int i,j,n,t=0;
float a[20][20],b[20],x[20],x0[20],e,sum=0.0;
clrscr();
printf("Enter the order of the coefficient matrix A:");
scanf("%d",&n);
printf("\n Enter the elements of the coefficient matrix A:");
for(i=0;i<n;i++)
{
for(j=0;j<n;j++)
{
scanf("%f",&a[i][j]);
}
}
printf("\n Enter the elements of the constant matrix B:");
for(i=0;i<n;i++)
{
scanf("%f",&b[i]);
}
printf("\n Enter the desired accuracy:");
scanf("%f",&e);
while(t==0)
{
for(i=0;i<n;i++)
{
x[i]=x0[i];
sum=b[i];
for(j=0;j<n;j++)
{
if(j!=i)
{
sum=sum-a[i][j]*x0[j];
}
x0[i]=sum/a[i][i];
}
}
```

Cont'd

Program cont'd

```
for(i=0;i<n;i++)
{
if(fabs(x0[i]-x[i])>e)
t=0;
else
t=1;
}
}
printf("\n The required solutions are: ");
for(i=0;i<n;i++)
{
printf("\nx[%d]=%.3f",x[i]);
}
getch();
}
```

Input/Output For this, let us consider the following example:

Solve the following system of linear equations by the Gauss–Seidel iterative method, correct to 3 decimal places:

$$10x + y - 2z = 7.74$$
$$x + 12y + 3z = 39.66$$
$$3x + 4y + 15z = 54.8$$

The input/output will be as follows:

```
Enter the order of the coefficient matrix A :3
Enter the elements of the coefficient matrix A :10
1
-2
1
12
3
3
4
15
Enter the elements of the constant matrix B :
7.74
39.66
54.8
```

Cont'd

Program cont'd

```
Enter the desired accuracy:0.0001
The required solutions are:
x[1]=1.075
x[1]=2.523
x[1]=2.765
```

ASSIGNMENT 14

Write C programs to implement the Gauss–Seidel iterative method to solve the following system of linear equations, correct to 3 decimal places:

(a) $5x - y + z = 10, 2x + 4y = 12, x + y + 5z = -1$
(b) $9x_1 - 3x_2 + 2x_3 = 20, 4x_1 + 11x_2 - x_3 = 33, x_1 + x_2 + 4x_3 = 9$
(c) $9x_1 - 2x_2 - x_3 = 50, x_1 + 5x_2 - 3x_3 = 18, -2x_1 + 2x_2 + 7x_3 = 19$
(d) $5x_1 - x_2 = 9, -x_1 + 5x_2 - x_3 = 4, -x_2 + 5x_3 = -6$

9.17 FITTING THE STRAIGHT LINE $y = a + bx$

Method Let (x_i, y_i), $i = 1, 2, \ldots, n$ be n observations of the variables x and y and $\boldsymbol{y = a + bx}$ be the straight line to be fitted on the basis of the observed data. By the principle of least squares, we get the normal equations as

$$\sum y = na + b\sum x \tag{1}$$

and

$$\sum xy = a\sum x + b\sum x^2 \tag{2}$$

Solving (1) and (2), we obtain

$$a = \left(\sum x^2 \sum y - \sum x \sum xy\right) / \left(n\sum x^2 - \left(\sum x\right)^2\right)$$

and

$$b = \left(n\sum xy - \sum x \sum y\right) / \left(n\sum x^2 - \left(\sum x\right)^2\right)$$

Therefore, the required equation is $y = a + bx$.

Algorithm

Step 1: Start the program
Step 2: Read n (no. of points)
Step 3: For $i = 0(1)n - 1$
  Read $x[i]$ & $y[i]$
 End i
Step 4: Assign $sumx = 0.0$, $sumy = 0.0$, $sumxy = 0.0$, $sumx2 = 0.0$
Step 5: For $i = 0(1)n - 1$
  $sumx = sumx + x[i]$
  $sumy = sumy + y[i]$
  $sumxy = sumxy + x[i] * y[i]$

sumx2 = *sum x*2 + *x*[*i*] * *x*[*i*]

End *i*

Step 6: *a* = (*sumx*2 * *sumy* – *sumx***sumxy*) / (*n* * *sumx2* – *sumx***sumx*)

b = (*n* * *sumxy* – *sumx***sumy*) / (*n* * *sumx2* – *sumx***sumx*)

Step 7: Print the straight line *y* = *a* + *b* * *x*

Step 8: Stop

Program Code

```
#include<stdio.h>
#include<conio.h>
void main()
{
int I,n;
float x[20],y[20],a,b;
clrscr();
printf("Enter the number of points:");
scanf("%d",&n);
printf("\n Enter the values of X:");
for(i=0;i<n;i++)
{
scanf("%f",&x[i]);
}
printf("\n Enter the values of Y:");
for(i=0;i<n;i++)
{
scanf("%f",&y[i]);
}
for(i-0;i<n;i++)
{
p=1;
for(j=0;j<n;j++)
{
sumx=sumx+x[i];
sumy=sumy+y[i];
sumxy=sumxy+x[i]*y[i];
sumx2=sumx2+x[i]*x[i];
}
a=(sumx2*sumy-sumx*sumxy)/(n*sumx2-sumx*sumx);
b=(n*sumxy-sumx*sumy) )/(n*sumx2-sumx*sumx);
```

Cont'd

Program Cont'd

```
printf("\n The line of best fit is:");
printf("\n\n Y= %.1f + %.1fX",a,b);
getch();
}
```

Input/Output For input/output, let us consider the following example:

Fit a straight line to the following data, considering x as an independent variable:

x	0	5	10	15	20	25
y	12	15	17	22	24	30

The input/output for the above program with the given problem will be:

```
Enter the number of points:6
Enter the values of X:0
5
10
15
20
25
Enter the values of Y:12
15
17
22
24
30
The line of best fit is:
Y= 11.3 + 0.7X
```

ASSIGNMENT 15

Write a program in C to fit a straight line for each of the following data:

(a)

x	0	1	2	3	4
y	1.1	1.9	2.7	3.5	4.3

(b)

x	0	2	4	6	8
y	10	12	14	16	18

(c)

x	0	2	3	4	7	8
y	4	26	58	112	466	668

9.18 FITTING THE CURVE $y = ax + bx^2$

Method Let (x_i, y_i), $i = 1, 2, \ldots, n$ be n observations of the variables x and y and $\mathbf{y = ax + bx^2}$ be the curve to be fitted on the basis of the observed data. By the principle of least squares, we get the normal equations as

$$\sum y = a\sum x + b\sum x^2 \tag{1}$$

and

$$\sum xy = a\sum x^2 + b\sum x^3 \tag{2}$$

Solving (1) and (2), we obtain

$$a = \left(\sum x^3 \sum y - \sum xy \sum x^2\right) / \left(\sum x^3 \sum x - \left(\sum x^2\right)^2\right)$$

and

$$b = \left(\sum xy \sum x - \sum x^2 \sum y\right) / \left(\sum x^3 \sum x - \left(\sum x^2\right)^2\right)$$

Therefore, the required equation is $y = a + bx$.

Algorithm

Step 1: Start the program

Step 2: Read n (no. of points)

Step 3: For $i = 0(1)n - 1$

Read $x[i]$ & $y[i]$

End i

Step 4: Assign $sumx = 0.0$, $sumy = 0.0$, $sumxy = 0.0$, $sumx2 = 0.0$, $sumx3 = 0.0$

Step 5: For $i = 0(1)n - 1$

$sumx = sumx + x[i]$

$sumy = sumy + y[i]$

$sumxy = sumxy + x[i] * y[i]$

$sumx2 = sum\ x2 + x[i] * x[i]$

$sumx3 = sum\ x3 + x[i] * x[i] * x[i]$

End i

Step 6: $a = (sumx3 * sumy - sumxy * sumx2)/(sumx3 * sumx - sumx2 * sumx2)$

$b = (sumxy * sumx - sumx2 * sumy)/(sumx3 * sumx - sumx2 * sumx2)$

Step 7: Print the straight line $y = a * x + b * x\perp 2$

Step 8: Stop

Program Code

```
#include<stdio.h>
#include<conio.h>
void main()
{
int I,n;
float x[20],y[20],a,b;
float sumx=0.0,sumy=0.0,sumxy=0.0,sumx2=0.0,sumx3=0.0;
clrscr();
printf("Enter the number of points:");
scanf("%d",&n);
printf("\n Enter the values of X:");
for(i=0;i<n;i++)
{
scanf("%f",&x[i]);
}
printf("\n Enter the values of Y:");
for(i=0;i<n;i++)
{
scanf("%f",&y[i]);
}
for(i=0;i<n;i++)
{
p=1;
for(j=0;j<n;j++)
{
sumx=sumx+x[i];
sumy=sumy+y[i];
sumxy=sumxy+x[i]*y[i];
sumx2=sumx2+x[i]*x[i];
sumx3=sumx3+x[i]*x[i]*x[i];
}
a=(sumx3*sumy-sumxy*sumx2)/(sumx3*sumx-sumx2*sumx2);
b=(sumxy*sumx-sumx2*sumy) )/(sumx3*sumx-sumx2*sumx2);
 printf("\n The curve of best fit is:");
printf("\n\n Y= %.1f X+ %.1fX^2",a,b);
getch();
}
```

Input/Output For input/output, let us consider the following example:

Fit a curve of the form $y = ax + bx^2$ to the following data, considering x as an independent variable:

x	0	1	2	3	4
y	8	12.5	18	24.5	32

The input/output for the above program with the given problem will be:

```
Enter the number of points:5
Enter the values of X:0
1
2
3
4
Enter the values of Y:8
12.5
18
24.5
32
The curve of best fit is:
Y= 20.0X - 3.5X^2
```

ASSIGNMENT 16

Write a program in C to a curve of the form $y = ax + bx2$ for the following:

(a)

x	1	2	3	4	7
y	2	4	8	6	128

(b)

x	0	1	2	3	5
$f(x)$	2	3	12	65	147

(c)

x	0.1	0.3	0.5	0.6	0.7
y	0.53	0.61	0.69	0.72	0.75

MULTIPLE CHOICE QUESTIONS

1. A two-dimensional array may be declared as `name[10,10]`.
   (a) true (b) false
2. Minimum number of functions in any C program is
   (a) 1 (b) 2 (c) 3 (d) 4
3. If `a` is an integer variable then `a=5/2` will return a value
   (a) 2.5 (b) 3 (c) 2 (d) 0
4. If `i=6` and `t=++i` then the value of `t` is
   (a) 7 (b) 6 (c) 5 (d) 8

5. The output of the following program segment is

```
x = 5;
y = x++;
printf("%d%d", x, y);
```

(a) 5, 6 (b) 5, 5 (c) 6, 5 (d) 6, 6

6. The following program segment will give the output

```
main() {
long i = 30000;
printf("%d", i); }
```

(a) 3000 (b) 3 (c) 30000 (d) 0

7. C language includes a header file for standard input and output in the library

(a) `stdlib.h` (b) `stdio.h` (c) `conio.h` (d) `math.h`

8. What is the output of the following program segment?

```
main()
{
long i = 65536;
printf("%d", i);
}
```

(a) 0 (b) 65 (c) 65536 (d) –1

9. What is the output of the following program segment?

```
main()
{
int i = ++2;
printf("%d", i);
}
```

(a) 0 (b) 1 (c) 2 (d) 3

10. The output of the following program segment is

```
int i;
main()
{
printf("%d", i);
}
```

(a) –1 (b) 0 (c) 1 (d) 2

Answers

1. (b) 2. (a) 3. (c) 4. (a) 5. (c) 6. (c) 7. (b) 8. (a) 9. (d) 10. (c)

SHORT ANSWER QUESTIONS

1. Write a C program to read and print a 3*3 matrix.
2. Write a C program to find the average of the first 100 natural numbers.
3. State the basic difference between the *do–while* loop and *for* loops with examples.
4. What is the basic feature of the *if* statement in C? Give suitable examples.

5. Construct a C program to check whether a year is a leap year or not.
6. Write a program in C to check whether a number is prime or not.
7. Consider the following table:

x	1	2	3	4	5
y	1.3784	1.4491	1.5166	1.5811	1.6432

Write a program in C to print the forward difference table using the data given above.
8. Write a C program to calculate the standard deviation of the numbers 4, 8, 9.2, 10, 11.1, 11.8, 12.5 and 13.2.
9. Consider the function $y = x^3 - 2x - 9$. Write a program to print the functional values from $x = 0$ to $x = 10$ with step length $h = 1$.

LONG ANSWER QUESTIONS

1. Write the algorithm for Newton's forward interpolation formula. Construct a C program to implement this method for calculating $y(1.5)$ from the following table:

x	1	2	3	4	5
y	1	8	27	64	125

2. Write the algorithm for Newton's backward interpolation formula. Construct a C program to implement this method for calculating $y(3.5)$ from the following table:

x	0	1	2	3	4
y	41	43	47	53	61

3. Give the algorithm for Lagrange's interpolation. Write a C program for this method to calculate $y(5)$ from the following table:

x	1	2	3	4	7
y	2	4	8	16	128

4. Write a program in C for the trapezoidal rule to evaluate the integral $\int_0^1 (4x - 3x^2)dx$, taking 24 subintervals, and hence calculate the absolute error.
5. Write the algorithm of Simpson's 1/3 rule and construct a C program to evaluate $\int_0^1 xe^x dx$, taking 12 subintervals.
6. Write the algorithm for the regula falsi and Newton–Raphson methods to find the real root of an equation.
7. Write C programs for the Euler and Runge–Kutta methods to calculate $y(0.1)$, given that

$$\frac{dy}{dx} = \frac{y-x}{y+x}, \quad y(0) = 1 \text{ and } h = 0.0001.$$

10

Introduction to Software Packages

10.1 INTRODUCTION TO MATLAB

The word MATLAB is an abbreviation of **MAT**rix **LAB**oratory. It is like an encyclopedia of various computing tools. Being a high-performance software package, it provides an interactive environment to the user to perform numerical computation, data manipulation, data visualisation and algorithm development. It has a huge number of well-developed toolboxes that simplify computational problems. Technical computing problems can be solved and data analysed graphically according to various aspects such as **image processing, neural networks, time series, control system, fuzzy logic and optimisation** using MATLAB.

10.1.1 MATLAB Overview

The MATLAB package has a wealth of built-in functions. The MATLAB environment is user friendly and its general structure is as follows:

(a) MATLAB windows: The MATLAB package, in general, works through three basic windows:

- **(i) Command window:** This is the main window of MATLAB. If we start MATLAB, the command prompt '>>' appears.
- **(ii) Graphics window:** This is a separate grey window. The user can create different types of graphs to analyse technical problems using this window.
- **(iii) Edit window:** In this window, the user can create, edit, save and record technical problems in **M-files**.

(b) Input/Output: MATLAB does not consider data type declarations. The basic unit of data storage in MATLAB is an array (or more precisely, matrices). It can read input files and write output files. MATLAB has three types of file formats: **M-files, MAT-files** and **Mex-files**.

(c) Online help: This is helpful to users of the MATLAB environment. This may be called the **self-learning section**. The user can operate MATLAB without any professional training. The commands `lookfor`, `help` and `helpwin` bring up the online help. In addition, MATLAB has a demonstration program that guides users on different aspects of the software. By typing `demo` at the MATLAB prompt, you can invoke the demonstration program.

10.1.2 Operating the MATLAB Package

The MATLAB package primarily involves the construction of matrices. To construct a matrix, first start MATLAB. The command prompt '>>' will appear. A scalar in MATLAB is a $\mathbf{1 \times 1}$ matrix while a vector is a $\mathbf{1 \times n}$ or $\mathbf{n \times 1}$ matrix. To construct a matrix in MATLAB, we use the matrix constructor operator `[]`. Now, we discuss the procedure used to construct different types of matrices.

(i) One-dimensional row matrix: Create a row using the matrix constructor operator `[]` by entering elements separated by a **comma** or **space**. For example, if we want to create a one-row matrix of four elements, such as A = (1, 2, 3, 4) in MATLAB, we follow the procedure given below:

1. Type `A = [1 2 3 4]` just after the command prompt '>>', i.e., `>> A = [1 2 3 4]`.
2. Press the ENTER button. The screen will display

```
A=
    1  2  3  4
```

(ii) One-dimensional column matrix: This is quite similar to the previous type. The only difference is that in this case, each element is separated by a semi-colon.

For example, the column matrix $C = \begin{pmatrix} 1 \\ 2 \\ 3 \\ 4 \end{pmatrix}$ is stored in MATLAB by typing `C = [1;2;3;4]` just after the command prompt '>>', as shown below:

1. `>> C=[1;2;3;4]`
2. Press the ENTER button. The screen will display

```
C=
    1
    2
    3
    4
```

(iii) Multi-dimensional matrix: This is a combination of procedures (i) and (ii). In this case, each row is separated by a semi-colon. However, it should be noted that all rows must contain the same number of elements.

For example, consider the matrix $A = \begin{pmatrix} 1 & 2 & 3 & 4 \\ 5 & 6 & 7 & 8 \\ 9 & 10 & 11 & 12 \end{pmatrix}$. Clearly the matrix has 3 rows and 4 columns and its representation in MATLAB can be accomplished by typing `A = [1 2 3 4;5 6 7 8;9 10 11 12]` after the command prompt '>>', as shown below:

1. `>> A=[1 2 3 4;5 6 7 8;9 10 11 12]`
2. Press the ENTER button. The screen will display

```
A=
    1   2   3   4
    5   6   7   8
    9  10  11  12
```

Note: If we do not want to display the matrices, we need to insert a **semi-colon** at the end of `[]`. Thus, if we type `A = [1 2 3 4;5 6 7 8;9 10 11 12];` after the command prompt '>>', MATLAB will store the elements of the matrix but will not display it.

10.1.3 Operations on Matrices

Before learning about various operations such as addition, subtraction, multiplication and division on matrices, we should become familiar with some basic operators and functions.

- **Arithmetic operators in MATLAB**

`+`	Addition
`-`	Subtraction
`*`	Multiplication
`/`	Division
`^`	Exponent

- **Array operators in MATLAB**

`.*`	Element by element multiplication
`./`	Element by element division
`.^`	Element by element exponentiation

- **Relation operators in MATLAB**

`<`	Less than
`>`	Greater than
`<=`	Less than or equal
`>=`	Greater than or equal
`==`	Equal
`~=`	Not equal

- **Logical operators in MATLAB**

`&`	Logical AND
`|`	Logical OR
`~`	Logical NOT
`xor`	Exclusive OR

- **Matrix functions in MATLAB**

`det(A)`	Determinant of matrix A
`A'`	Transpose of matrix A
`expm(A)`	Exponential of matrix A
`inv(A)`	Inverse of matrix A
`norm(A)`	Norm of matrix A
`eig(A)`	Eigen values of matrix A
`imag(A)`	Imaginary part of matrix A
`real(A)`	Real part of matrix A

- **Mathematical functions in MATLAB**

`logm(x)`	Logarithm of x to the base m
`sqrt(x)`	Square root of x
`factorial(x)`	Factorial of x
`min(x)`	Minimum element of an array x
`abs(x)`	Absolute value of x
`rem`	Remainder after division
`sin`	Sine function
`cos`	Cosine function
`tan`	Tangent function
`asin`	Inverse sine function
`acos`	Inverse cosine function
`atan`	Inverse tangent function
`sinh`	Hyperbolic sine function

`cosh`	Hyperbolic cosine function
`asinh`	Inverse hyperbolic sine function
`acosh`	Inverse hyperbolic cosine function
`int(f,x)`	Indefinite integral of f w.r.t. x
`int(f,x,a,b)`	Definite integral of f w.r.t. x from a to b
`diff(f, 'x')`	Differentiation of f w.r.t. x

We will apply these operators on matrix operations in the following examples.

EXAMPLE 10.1 Evaluate $C = A + B$, $D = A - B$ and $E = 3A - 2B$, where $A = (5\ -1\ 3\ 8)$ and $B = (1\ 4\ 2\ 1)$.

Solution:
```
>>A=[5 -1 3 8]

A=
   5  -1  3  8
>>B=[1 4 2 1]

B=
   1  4  2  1
>>C=A+B

C=
   6  3  5  9
>>D=A-B

D=
   4  -5  1  7
>>E=3*A-2*B

E=
   13  -11  5  22
```

EXAMPLE 10.2 Using MATLAB, evaluate $C = AB$ and $D = BA$, where $A = \begin{pmatrix} 5 & 2 & 6 \\ 1 & 4 & 2 \end{pmatrix}$ and $B = \begin{pmatrix} 1 & 7 \\ 6 & -2 \\ 5 & 4 \end{pmatrix}$.

Solution:
```
>>A=[5 2 6;1 4 2]

A=
   5  2  6
   1  4  2
>>B=[1 7;6 -2;5 4]

B=
   1   7
   6  -2
   5   4
>>C=A*B
```

```
C=
    47  55
    35   7
>>D=B*A
D=
    12  30  20
    28   4  32
    29  26  38
```

Note: If we use `'.*'` in place of `'*'`, it will yield a different result because `.*` is used for element-by-element multiplication; that is why the two matrices A and B must be of the same order.

For illustration, consider the following example.

EXAMPLE 10.3 If $A = \begin{pmatrix} 1 & 2 \\ 3 & 4 \end{pmatrix}$ and $B = \begin{pmatrix} 1 & 0 \\ 2 & 1 \end{pmatrix}$, then using MATLAB find C = A*B and D = A.*B.

Solution:

```
>>A=[1 2;3 4];
>>B=[1 0;2 1];
>>C=A*B
C=
    5   2
   11   4
>>D=A.*B
D=
   1  0
   6  4
```

Clearly `A*B` and `A.*B` are distinct.

Note: Evaluation of `A.*B` is not possible for Example 10.2 because A and B are not of the same order.

We can also find the inverse and transpose of a given matrix. This is illustrated in the following examples.

EXAMPLE 10.4 Find the transpose and inverse of the matrix $A = \begin{pmatrix} 5 & 2 \\ 7 & 3 \end{pmatrix}$.

Solution:

```
>>A=[5 2;7 3]
A=
    5  2
    7  3
>>B=A'
B=
    5  7
    2  3
>>C=inv(A)
C=
    3   -2
   -7    5
```

EXAMPLE 10.5 Find A^{-1} when $A = \begin{pmatrix} 3 & 4 \\ 8 & 11 \end{pmatrix}$ using MATLAB, and check the relation $AA^{-1} = I$, where I is the identity matrix of order 2.

Solution:

```
>>A=[3 4;8 11]

A=
    3    4
    8   11

>>B=inv(A)

B=
   11   -4
   -8    3

>>C=A*B

C=
    1  0
    0  1
```

10.1.4 Well-known Matrices

In MATLAB, we can generate some well-known and frequently used matrices such as null matrix, identity matrix and matrix of random numbers. The functions are listed below.

`zeros`	Creates a null matrix
`eye`	Creates an identity matrix
`ones`	Creates a matrix of ones
`rand`	Creates a matrix of random numbers

For example, if we require an identity matrix of order 3, we write `A = eye(3, 3)` just after the '>>' prompt and it will generate a 3 × 3 identity matrix, as shown below.

```
>>A=eye(3,3)

A=
    1  0  0
    0  1  0
    0  0  1
```

EXAMPLE 10.6 Using MATLAB, show that $AI = IA = A$, where $A = \begin{pmatrix} 4 & 1 & 3 \\ 1 & 5 & 7 \\ 2 & 6 & 9 \end{pmatrix}$, by generating the identity matrix I of order 3 × 3.

Solution:

```
>>A=[4 1 3;1 5 7;2 6 9]

A=
    4  1  3
    1  5  7
    2  6  9

>>I=eye(3,3)
```

```
I=
    1  0  0
    0  1  0
    0  0  1
>>C=A*I
C=
    4  1  3
    1  5  7
    2  6  9
>>D=I*A
D=
    4  1  3
    1  5  7
    2  6  9
```

Therefore $AI = IA = A.$

EXAMPLE 10.7 Generate a null matrix O of order 3×3 and show that $A + O = A$, where $A = \begin{pmatrix} 2 & 1 & 2 \\ 1 & 5 & 1 \\ 9 & 2 & 9 \end{pmatrix}$.

Solution: `>>A=[2 1 2;1 5 1;9 2 9]`

```
A=
    2  1  2
    1  5  1
    9  2  9
>>O=zeros(3,3)
O=
    0  0  0
    0  0  0
    0  0  0
>>C=A+O
C=
    2  1  2
    1  5  1
    9  2  9
```

Hence $A + O = A.$

10.1.5 Solution of Algebraic Equations

MATLAB has tools of different standard methods that can be used to solve algebraic equations. Here, we discuss how to construct a polynomial and how to the find roots of that polynomial. In MATLAB, first insert the coefficients or constants in a row matrix, say `x`. Then use `roots(x)` to find the roots, as given below:

```
>>roots(x)
```

This is best understood by examining the following examples.

EXAMPLE 10.8 Solve the equations $x^3 - 3x - 5 = 0$ using MATLAB.

Solution: We see that the coefficients of x^3, x^2, x, x^0 are, respectively, 1, 0, −3, −5. Now, we put these values in a row matrix x as

```
>>x=[1 0 -3 -5];
```

To find all the solutions, we apply

```
>>roots(x)
```

This will give

```
ans =
      2.2790
     -1.1395 + 0.9463i
     -1.1395 - 0.9463i
```

Note: If we need to find only the real roots of a polynomial equation, we write it in a different manner, as shown in the following example.

EXAMPLE 10.9 Find the real roots of the equation $x^3 - 3x - 7 = 0$ using MATLAB.

Solution:

```
>>A = [1 0 -3 -7];
>>r = roots(A);
>>r = r(imag(r)==0)
r =
    2.4260
```

10.1.6 Solution of System of Linear Equations

The system of linear equation AX = B can be solved using MATLAB. First, write the coefficient matrix A and the constant matrix B. Then use `linsolve(A,B)`. This will give the solution for the system. For illustration, consider the following examples.

EXAMPLE 10.10 Solve the equations $3x - 2y = 8$, $x + 5y = 14$ using MATLAB.

Solution: The system is equivalent to AX = B where $A = \begin{pmatrix} 3 & -2 \\ 1 & 5 \end{pmatrix}$, $X = \begin{pmatrix} x \\ y \end{pmatrix}$ and $B = \begin{pmatrix} 8 \\ 14 \end{pmatrix}$. To solve this, we write it as follows:

```
>>A = [3 -2;1 5];
>>B = [8;14];
>>X = linsolve(A,B)
```

This will give

```
X =
      4
      2
```

Therefore $x = 4$

and $y = 2$

EXAMPLE 10.11 Solve the following system of linear equations using MATLAB:

$$5x - y + z = 10$$
$$2x + 4y = 12$$
$$x + y + 5z = -1$$

Solution: We first construct the coefficient matrix A and the constant matrix B as

```
>> A = [5 -1 1;2 4 0;1 1 5];
>> B = [10;12;-1];
```

Now, we apply

```
>> X = linsolve(A,B)
```

This gives

```
X =
      2.5556
      1.7222
     -1.0556
```

Therefore $x = 2.5556, y = 1.7222$

and $z = -1.0556.$

10.1.7 Numerical Programs in MATLAB

To solve numerical problems, we can develop programs in the MATLAB environment.

Newton's Forward Interpolation

Method: Let a function $y = f(x)$ be known only for a set of $(n + 1)$ distinct equispaced points x_0, x_1, ..., x_n such that $x_i = x_0 + ih$, $(i = 1, 2, \ldots, n)$, where h is the length of each subinterval $[x_i, x_{i+1}]$. The corresponding values of the function are $y_0 = f(x_0)$, $y_1 = f(x_1)$, ..., $y_n = f(x_n)$. Then

$$f(x) = y_0 + \frac{u}{1!}\Delta y_0 + \frac{u(u-1)}{2!}\Delta^2 y_0 + \cdots + \frac{u(u-1)(u-2)\ldots(u-\overline{n-1})}{n!}\Delta^n y_0$$

where $u = \dfrac{x - x_0}{h}$

Algorithm

Step 1: Start the program

Step 2: Read n (no. of arguments)

Step 3: For $i = 0(1)n - 1$

Read $x[i]$ & $y[i][0]$

End i

Step 4: Assign $h = x[1] - x[0]$ (step length)

Step 5: Read a (point of interpolation)

Step 6: Construct the forward difference table

For $j = 1(1)n - 1$

For $i = 0(1)n - 1 - j$

$y[i][j] = y[i + 1][j - 1] - y[i][j - 1]$

End i

End j

Step 7: Print the forward difference table
  For $i = 0(1)n - 1$
    For $j = 0(1)n - 1 - i$
      Print $y[i][j]$
    End j
  End i

Step 8: Assign $u = (a - x[0]) / h$

Step 9: Assign $sum = y[0][0]$ & $p = 1.0$

Step 10: For $j = 1(1)n - 1$
    $p = p^* (u - j + 1) / j$
    $sum = sum + p^* y[0][j]$
  End j

Step 11: Display a and sum

Step 12: Stop

Program Code

```
clc;
n=input('Enter the number of arguments n:');
y=zeros(n+1,n+1);
for i=1:n
    x(i)=input('Enter the value of x:');
end
for i=1:n
    y(i,1)=input('Enter the value of y:');
end
a=input('Enter the value of the interpolating point a:');
h=x(2)-x(1);
u=(a-x(1))/h;
for j=2:1:n
    for i=1:n-j+1
        y(i,j)=y(i+1,j-1)-y(i,j-1);
    end
end
for i=1:n
    fprintf('\n')
    for j=1:n-i+1
        fprintf('\t%0.1f',y(i,j))
    end
end
s=y(1,1);
p=1.0;
for j=2:1:n
    p=p*(u-j+2)/(j-1);
    s=s+p*y(1,j);
end
fprintf('\n The value of y at x=%0.3f is= %0.3f',a,s)
```

Output: For output, let us consider the following example:

Calculate y(0.5) from the following table using Newton's forward interpolation:

x	0	1	2	3
y	1	0	1	10

The output for the above program with the given problem will be:

```
Enter the number of arguments n:4
Enter the value of x:0
Enter the value of x:1
Enter the value of x:2
Enter the value of x:3
Enter the value of y:1
Enter the value of y:0
Enter the value of y:1
Enter the value of y:10
Enter the value of the interpolating point a:0.5
1.0      -1.0       2.0       6.0
0.0       1.0       0.0
1.0       9.0
10.0
The value of y at x=0.500 is= 0.625
```

ASSIGNMENT 1

Write a program in MATLAB to implement Newton's forward interpolation for the following:

(a) Calculate $\sqrt{2}$, correct to 4 decimal places using

x	1.9	2.1	2.3	2.5	2.7
y	1.3784	1.4491	1.5166	1.5811	1.6432

(b) Compute sin 32°, correct to 4 decimal places when

x	30°	35°	40°	45°	50°	55°
sin x	0.5000	0.5735	0.6428	0.7071	0.7660	0.8192

(c) Find the area (integral value) of the circle of diameter 82 cm, given

d(cm)	80	85	90	95	100
A(cm²)	5026	5674	6362	7088	7854

(d) Find the number of students (integral value) scoring below 35 marks, given

Marks	30–40	40–50	50–60	60–70
No. of Students	52	36	21	14

Newton's Backward Interpolation

Method: Let a function $y = f(x)$ be known only for a set of $(n + 1)$ distinct equispaced points x_0, x_1, ..., x_n such that $x_i = x_0 + ih$, $(i = 1, 2, ..., n)$, where h is the length of each subinterval $[x_i, x_{i+1}]$. The corresponding values of the function are $y_0 = f(x_0)$, $y_1 = f(x_1)$, ..., $y_n = f(x_n)$. Then

$$f(x) = y_n + \frac{u}{1!}\nabla y_n + \frac{u(u+1)}{2!}\nabla^2 y_n + \cdots + \frac{u(u+1)(u+2)\ldots(u+n-1)}{n!}\nabla^n y_n$$

where $$u = \frac{x - x_n}{h}$$

Algorithm

```
Step 1:  Start the program
Step 2:  Read n (no. of arguments)
Step 3:  For i = 0(1)n – 1
             Read x[i] & y[i][0]
         End i
Step 4:  Assign h = x[1] – x[0] (step length)
Step 5:  Read a (point of interpolation)
Step 6:  Construct the backward difference table
         For j = 1(1)n – 1
             For i = j (1)n – 1
                 y[i][j] = y[i][j – 1] – y[i – 1][j – 1]
             End i
         End j
Step 7:  Print the backward difference table
         For i = 0(1)n – 1
             For j = 0(1) i
                 Print y[i][j]
             End j
         End i
Step 8:  Assign u = (a – x[n – 1]) / h
Step 9:  Assign sum = y[n – 1][0] & p = 1.0
Step 10: For j = 1(1)n – 1
             p = p* (u + j – 1) / j
             sum = sum + p * y[n – 1][j]
         End j
Step 11: Display a and sum
Step 12  Stop
```

Program Code

```
clc;
n=input('Enter the number of arguments n:');
y=zeros(n+1,n+1);
for i=1:n
    x(i)=input('Enter the value of x:');
end
for i=1:n
    y(i,1)=input('Enter the value of y:');
end
a=input('Enter the value of the interpolating point a:');
h=x(2)-x(1);
u=(a-x(n))/h;
for j=2:1:n
    for i=j:1:n
        y(i,j)=y(i,j-1)-y(i-1,j-1);
    end
end
for i=1:n
    fprintf('\n')
    for j=1:1:i
        fprintf('\t%0.1f',y(i,j))
    end
end
s=y(n,1);
p=1.0;
for j=2:1:n
    p=p*(u+j-2)/(j-1);
    s=s+p*y(n,j);
end
fprintf('\n The value of y at x=%0.2f is= %0.3f',a,s)
```

Output: For output, let us consider the following example:

Calculate y(2.5) from the following table by Newton's backward interpolation:

x	0	1	2	3
y	1	0	1	10

The output for the above program with the given problem will be:

```
Enter the number of arguments n:4
Enter the value of x:0
Enter the value of x:1
Enter the value of x:2
Enter the value of x:3
```

Cont'd

Output cont'd

```
Enter the value of y:1
Enter the value of y:0
Enter the value of y:1
Enter the value of y:10
Enter the value of the interpolating point a:2.5
1.0
0.0      -1.0
1.0       1.0       2.0
10.0      9.0       8.0       6.0
The value of y at x=2.50 is= 4.125
```

ASSIGNMENT 2

Write a program in MATLAB to implement Newton's backward interpolation for the following:

(a) Calculate *f*(4.3), correct to 4 decimal places using

x	1	2	3	4	5
f(*x*)	7	12	29	64	123

(b) Compute sin 52°, correct to 4 decimal places when

x	30°	35°	40°	45°	50°	55°
Sin *x*	0.5000	0.5735	0.6428	0.7071	0.7660	0.8192

(c) Find the area (integral value) of the circle of diameter 98 cm, given

***d*(*cm*)**	80	85	90	95	100
***A*(*cm*²)**	5026	5674	6362	7088	7854

(d) Find the number of students (integral value) scoring above 65 marks, given

Marks	30–40	40–50	50–60	60–70
No. of Students	52	36	21	14

Lagrange's Interpolation

Method: Let a function $y = f(x)$ be known only for a set of $(n + 1)$ distinct points $x_0, x_1, \ldots, x_n$ and the corresponding values of the function be $y_0 = f(x_0), y_1 = f(x_1), \ldots, y_n = f(x_n)$. Then

$$f(x) = \frac{(x-x_1)(x-x_2)\ldots(x-x_n)}{(x_0-x_1)(x_0-x_2)\ldots(x_0-x_n)}y_0 + \frac{(x-x_0)(x-x_2)\ldots(x-x_n)}{(x_1-x_0)(x_1-x_2)\ldots(x_1-x_n)}y_1 + \cdots$$

$$+ \frac{(x-x_0)(x-x_1)\ldots(x-x_{n-1})}{(x_n-x_1)(x_n-x_1)\ldots(x_n-x_{n-1})}y_n$$

Algorithm

Step 1: Start the program
Step 2: Read n (no. of arguments)
Step 3: For $i = 0(1)n - 1$
    Read $x[i]$ and $y[i]$
  End i
Step 4: Read a (point of interpolation)
Step 5: Assign $sum = 0.0$
Step 6: For $i = 0(1)n - 1$
    $p = 1.0$
      For $j = 0(1)n - 1$
        If $i \neq j$
          $p = p^* (a - x[j])/(x[i] - x[j])$
      End j
          $sum = sum + p^* y[i]$
  End i
Step 7: Display a and sum
Step 8: Stop

Program Code

```
clc;
n=input('Enter the number of arguments n:');
y=zeros(n+1,n+1);
for i=1:n
    x(i)=input('Enter the value of x:');
end
for i=1:n
    y(i)=input('Enter the value of y:');
end
a=input('Enter the value of the interpolating point a:');
s=0;
for i=1:n
    p=1;
    for j=1:n
     if(i~=j)
         p=p*(a-x(j))/(x(i)-x(j));
     end
    end
    s=s+p*y(i);
end
fprintf('The value of y at x=%0.1f is= %0.3f',a,s)
```

Output: For input/output, let us consider the following example:

Calculate y(10) from the following table by Lagrange's interpolation:

x	5	6	9	11
y	12	13	14	16

The input/output for the above program with the given problem will be:

```
Enter the number of arguments n:4
Enter the value of x:5
Enter the value of x:6
Enter the value of x:9
Enter the value of x:11
Enter the value of y:12
Enter the value of y:13
Enter the value of y:14
Enter the value of y:16
Enter the value of the interpolating point a:10
The value of y at x=10.0 is= 14.667
```

ASSIGNMENT 3

Write a program in MATLAB to implement Lagrange's interpolation for the following:

(a) Calculate $y(6)$, correct to 4 decimal places using

x	1	2	3	4	7
y	2	4	8	6	128

(b) Compute $f(4)$, correct to 4 decimal places when

x	0	1	2	3	5
$f(x)$	2	3	12	65	147

(c) Find the approximate value of $f(5)$, correct to 2 decimal places, given

x	0	2	3	4	7	8
$f(x)$	4	26	58	112	466	668

(d) Compute $f(0.4)$, correct to 3 decimal places, given

x	0.1	0.3	0.5	0.6	0.7
$f(x)$	0.53	0.61	0.69	0.72	0.75

Trapezoidal Rule

Method: To evaluate $\int_a^b f(x)dx$, let the interval $[a, b]$ be divided into n equal subintervals of length h, by the points $a = x_0, x_1, \ldots, x_n = b$, where $x_i = x_0 + ih$, $(i = 1, 2, \ldots, n)$. The corresponding values of the function are $y_0 = f(x_0), y_1 = f(x_1), \ldots, y_n = f(x_n)$. Then

$$\int_a^b f(x)dx = \int_{x_0}^{x_0+nh} f(x)dx = \frac{h}{2}\left[(y_0 + y_n) + 2(y_1 + y_2 + \cdots + y_{n-1})\right]$$

Algorithm

Step 1: Start the program
Step 2: Define the function $f(x)$
Step 3: Read a (lower limit), b (upper limit) and n (number of subintervals)
Step 4: Assign $h = (b - a) / n$
Step 5: Assign $sum = f(a) + f(b)$
Step 6: For $i = 1(1)n - 1$
$sum = sum + 2f(a + ih)$
End i
Step 7: Assign $I = \frac{h}{2} sum$
Step 8: Display I
Step 9: Stop

Program Code: The program code requires a function (integrand) declaration. For different functions, the program code will vary in the function declaration section. Therefore, we have to consider a specific function to construct the program. For example, let us consider the following problem:

Evaluate the integral $\int_0^1 dx/(1+x^2)$ *by the trapezoidal rule, dividing the interval into 6 subintervals.*

The program code will be:

```
clc;
a=input('Enter the lower limit a:');
b=input('Enter the upper limit b:');
n=input('Enter the number of subintervals n:');
f=@(x)[1/(1+x^2)];
h=(b-a)/n;
s=f(a)+f(b);
for i=1:n-1
    s=s+2*f(a+i*h);
end
s=(h/2)*s;
fprintf('The value of integral is= %0.4f',s)
```

Output

```
Enter the lower limit a:0
Enter the upper limit b:1
Enter the number of subintervals n:6
The value of integral is= 0.7842
```

ASSIGNMENT 4

Construct MATLAB programs to implement the trapezoidal rule for the following:

(a) Evaluate $\int_0^1 \frac{xdx}{\sin x}$, correct to 3 decimal places taking 15 subintervals, given that $\lim_{x\to 0} \frac{\sin x}{x} = 1$.

(b) Evaluate $\int_{-3}^{3} e^{-x^2}\,dx$, correct to 5 decimal places taking 15 subintervals.

(c) Evaluate $\int_0^1 dx/\left(1+x^2\right)$, correct to 4 decimal places taking 13 subintervals, and hence calculate the approximate value of π from the relation $\frac{\pi}{4}=\int_0^1 dx/\left(1+x^2\right)$, correct to 3 decimal places.

(d) Evaluate $\int_0^{\pi/3} \sqrt{\cos\theta}\,d\theta$, correct to 4 decimal places taking 17 subintervals.

Simpson's 1/3 Rule

Method: To evaluate $\int_a^b f(x)\,dx$, let the interval $[a, b]$ be divided into n (even) equal subintervals of length h, by the points $a = x_0, x_1, \ldots, x_n = b$, where $x_i = x_0 + ih$, $(i = 1, 2, \ldots, n)$. The corresponding values of the function are $y_0 = f(x_0), y_1 = f(x_1), \ldots, y_n = f(x_n)$. Then

$$\int_a^b f(x)\,dx = \int_{x_0}^{x_0+nh} f(x)\,dx = \frac{h}{3}\left[(y_0 + y_n) + 2(y_2 + y_4 + \cdots + y_{n-2}) + 4(y_1 + y_3 + \cdots + y_{n-1})\right]$$

Algorithm

Step 1: Start the program

Step 2: Define the function $f(x)$

Step 3: Read a (lower limit), b (upper limit) and n (number of subintervals)

Step 4: Assign $h = (b - a) / n$

Step 6: Assign $sum = f(a) + f(b)$

Step 6: For $i = 1(2)n-1$

If((*i*%2)! = 0)

$sum = sum + 4f(a + ih)$

Else

$sum = sum + 2f(a + ih)$

End *i*

Step 7: Assign $I - \frac{h}{3} sum$

Step 8: Display I

Step 9: Stop

Program Code: The program code requires a function (integrand) declaration. For different functions, the program code will vary in the function declaration section. Therefore, we have to consider a specific function to construct the program. For example, let us consider the following problem:

Evaluate the integral $\int_0^1 dx/\left(1+x^2\right)$ *by Simpson's 1/3 rule, dividing the interval into 24 subintervals.*

The program code will be:

```
clc;
a=input('Enter the lower limit a:');
b=input('Enter the upper limit b:');
```

Cont'd

Program cont'd

```
n=input('Enter the number of subintervals n:');
f=@(x)[1/(1+x^2)];
h=(b-a)/n;
s=f(a)+f(b);
for i=1:n-1
    if(mod(i,2)~=0)
        s=s+4*f(a+i*h);
    else
        s=s+2*f(a+i*h);
    end
end
s=(h/3)*s;
fprintf('The value of integral is= %0.4f',s)
```

Output

```
Enter the lower limit a:0
Enter the upper limit b:1
Enter the number of subintervals n:6
The value of integral is= 0.7854
```

ASSIGNMENT 5

Construct MATLAB programs to implement Simpson's 1/3 rule for the following:

(a) Evaluate $\int_0^1 \frac{x dx}{\sin x}$, correct to 3 decimal places taking 16 subintervals, given that $\lim_{x\to 0} \frac{\sin x}{x} = 1$.

(b) Evaluate $\int_{-3}^{3} e^{-x^2} dx$, correct to 5 decimal places taking 16 subintervals.

(c) Evaluate $\int_0^1 dx/\left(1+x^2\right)$, correct to 4 decimal places taking 12 subintervals, and hence calculate the approximate value of π from the relation $\frac{\pi}{4} = \int_0^1 dx/\left(1+x^2\right)$, correct to 3 decimal places.

(d) Evaluate $\int_0^{\pi/3} \sqrt{\cos\theta}\, d\theta$, correct to 4 decimal places taking 18 subintervals.

Weddle's Rule

Method: To evaluate $\int_a^b f(x) dx$, let the interval $[a, b]$ be divided into n (even) equal subintervals of length h, by the points $a = x_0, x_1, \ldots, x_n = b$, where $x_i = x_0 + ih$, $(i = 1, 2, \ldots, n)$. The corresponding values of the function are $y_0, f(x_0), y_1 = f(x_1), \ldots, y_n = f(x_n)$. Then

$$\int_a^b f(x) dx = \int_{x_0}^{x_0+nh} f(x) dx = \frac{3h}{10}\left[(y_0 + y_n) + 5(y_1 + y_5 + y_7 + \cdots) + (y_2 + y_4 + y_8 + y_{14} + \cdots) + 6(y_3 + y_9 + y_{15} + \cdots) + 2(y_6 + y_{12} + y_{18} + \cdots)\right]$$

Algorithm

Step 1: Start the program

Step 2: Define the function $f(x)$

Step 3: Read a (lower limit), b (upper limit) and n (number of subintervals)

Step 4: Assign $h = (b - a) / n$

Step 5: Assign $sum= f(a) + f(b)$

Step 6: For $i = 1(1)n - 1$

If((i%2) == 0&& (i%3) == 0)

$sum = sum+2f(a + ih)$

Else if((i%2) == 0&& (i%3)! = 0)

$sum = sum + f(a + ih)$

Else if((i%3) == 0&& (i%2)! = 0)

$sum = sum + 6f(a + ih)$

Else

$sum = sum + 5f(a + ih)$

End i

Step 7: Assign $I = \frac{3h}{10} sum$

Step 8: Display I

Step 9: Stop

Program Code: The program code requires a function (integrand) declaration. For different functions, the program code will vary in the function declaration section. Therefore, we have to consider a specific function to construct the program. For example, let us consider the following problem:

Evaluate the integral $\int_0^1 dx/\left(1+x^2\right)$ *by Simpson's 1/3 rule, dividing the interval into 24 subintervals.*

The program code will be:

```
clc;
a=input('Enter the lower limit a:');
b=input('Enter the upper limit b:');
n=input('Enter the number of subintervals n:');
f=@(x)[1/(1+x^2)];
h=(b-a)/n;
s=f(a)+f(b);
for i=1:n-1
    if(mod(i,2)==0 && mod(i,3)==0)
        s=s+2*f(a+i*h);
    elseif(mod(i,2)==0 && mod(i,3)~=0)
        s=s+f(a+i*h);
    elseif(mod(i,2)~=0 && mod(i,3)==0)
        s=s+6*f(a+i*h);
    else
        s=s+5*f(a+i*h);
    end
end
s=(3*h/10)*s;
fprintf('The value of integral is= %0.4f',s)
```

Output

```
Enter the lower limit a:0
Enter the upper limit b:1
Enter the number of subintervals n:6
The value of integral is= 0.7854
```

ASSIGNMENT 6

Construct MATLAB programs to implement Weddle's rule for the following:

(a) Evaluate $\int_0^1 \frac{xdx}{\sin x}$, correct to 3 decimal places taking 18 subintervals, given that $\lim_{x\to 0} \frac{\sin x}{x} = 1$.

(b) Evaluate $\int_{-3}^{3} e^{-x^2}\, dx$, correct to 5 decimal places taking 18 subintervals.

(c) Evaluate $\int_0^1 dx/\left(1+x^2\right)$, correct to 4 decimal places taking 12 subintervals, and hence calculate the approximate value of π from the relation $\frac{\pi}{4} = \int_0^1 dx/\left(1+x^2\right)$, correct to 3 decimal places.

(d) Evaluate $\int_0^{\pi/3} \sqrt{\cos\theta}\, d\theta$, correct to 4 decimal places taking 18 subintervals.

Bisection Method

Method: To find a real root of $f(x) = 0$ in the interval $[a, b]$ with $f(a) f(b) < 0$ and tolerance ε, the iterative formula for a better root is given by

$$x = \frac{(a+b)}{2}$$

If ($|f(x)| < \varepsilon$), root found and stop iteration
else
if ($f(x) f(b) < 0$) set $a = x$, $b = b$
else set $a = a$, $b = x$ and continue until $|f(x)| < \varepsilon$

Algorithm 1 (a and b are not given)

Step 1: Start the program
Step 2: Define the function $f(x)$
Step 3: Read m (possible upper limit of the interval containing the root)
Step 4: For ($i = -m, i \le m, i{+}{+}$)
    Set $a = i$ and $b = i + 1$
        If $f(a) * f(b) < 0$
    Break
  End i
Step 5: Read n (number of iterations)
Step 6: For ($i = 1, i \le n + 1, i{+}{+}$) $x = (a + b)/2$
    If $f(b) f(x) < 0$,
      $a = x$
    Else
      $b = x$

Step 7: Print x
Step 8: Stop

Program Code: The program code requires the declaration of function $f(x)$. For different functions, the program code will vary in the function declaration section. Therefore, we have to consider a specific function to construct the program. For example, let us consider the following problem:

Find a real root of the equation $x^3 - 3x - 5 = 0$, *correct to* 3 *decimal places by the bisection method.*

The code for the program is given by:

```
clc;
f=@(x)[x^3-3*x-5];
m=input('Enter the range m:');
for i=-m:1:m
    a=i;
    b=i+1;
    if(f(a)*f(b)<0)
        break
    end
end
n=input('Enter the number of iteration n:');
for i=1:n+1
    x=(a+b)/2;
    if(f(a)*f(x)<0)
        b=x;
    else
        a=x;
    end
end
fprintf('The root of the equation is= %.3f',x)
```

Output

```
Enter the range m:50
Enter the number of iteration n:15
The root of the equation is= 2.279
```

Algorithm 2 (a and b are given)

Step 1: Start the program
Step 2: Define the function $f(x)$
Step 3: Read a and b (lower and upper limits of the interval containing the real root)
Step 4: Read n (number of iterations)
Step 5: For ($i = 1, i \leq n + 1, i{+}{+}$)
    $x = (a + \text{b}) / 2$
    If $f(b) f(x) < 0$
        $a = x$
    Else
        $b = x$

Step 6: Print x
Step 7: Stop

Program Code: The program code requires the declaration of function $f(x)$. For different functions, the program code will vary in the function declaration section. Therefore, we have to consider a specific function to construct the program. For example, let us consider the following problem:

Find a real root of the equation $x^3 - 3x - 5 = 0$ *between 2 and 3, correct to 3 decimal places by the bisection method.*

The code for the program is given by:

```
clc;
f=@(x)[x^3-3*x-5];
a=input('Enter the value of a:');
b=input('Enter the value of b:');
n=input('Enter the number of iteration n:');
for i=1:n+1
    x=(a+b)/2;
    if(f(a)*f(x)<0)
        b=x;
    else
        a=x;
    end
end
fprintf('The root of the equation is= %0.3f',x')
```

Output

```
Enter the value of a:2
Enter the value of b:3
Enter the number of iteration n:15
The root of the equation is= 2.279
```

ASSIGNMENT 7

Write MATLAB programs to implement the bisection method to find the real root of the following equations:

(a) $x^3 - 3x + 1.06 = 0$, correct to 3 decimal places.
(b) $x^3 - 2.1x - 9 = 0$, correct to 3 decimal places.
(c) $x^4 + x - 35.4 = 0$, correct to 3 decimal places between 3 and 4.
(d) $x^3 - 5x - 7 = 0$, correct to 3 decimal places between 2 and 3.

Regula Falsi Method

Method: To find a real root of $f(x) = 0$ in the interval [a, b] with $f(a)f(b) < 0$ and tolerance ε, the iterative formula for a better root is given by

$$x = \frac{af(b) - bf(a)}{f(b) - f(a)}$$

If ($|f(x)| < \varepsilon$), root found and stop iteration
else
if ($f(x)f(b) < 0$) set $a = x$, $b = b$
else set $a = a$, $b = x$
and continue until $|f(x)| < \varepsilon$

Algorithm 1 (a and b are not given)

Step 1: Start the program
Step 2: Define the function $f(x)$
Step 3: Read m (possible upper limit of the interval containing the root)
Step 4: For ($i = -m$, $i \leq m$, i++)
Set $a = i$ and $b = i + 1$
If $f(a)*f(b) < 0$
Break
End i
Step 5: Read n (number of iterations)
Step 6: For ($i = 1$, $i \leq n + 1$, i++)
$x = (a*f(b) - b*f(a))/(f(b) - f(a))$
If $f(b)f(x) < 0$,
$a = x$
Else
$b = x$
Step 7: Print x
Step 8: Stop

Program Code: The program code requires the declaration of function $f(x)$. For different functions, the program code will vary in the function declaration section. Therefore, we have to consider a specific function to construct the program. For example, let us consider the following problem:

Find a real root of the equation $x^3 - 3x - 5 = 0$, *correct to* 3 *decimal places by the regula falsi method.*

The code for the program is given by:

```
clc;
f=@(x)[x^3-3*x-5];
m=input('Enter the range m:');
for i=-m:1:m
    a=i;
    b=i+1;
    if(f(a)*f(b)<0)
        break
    end
end
n=input('Enter the number of iteration n:');
for i=1:n+1
    x=(a*f(b)-b*f(a))/(f(b)-f(a));
    if(f(a)*f(x)<0)
        b=x;
    else
```

Cont'd

Program cont'd

```
            a=x;
        end
end
fprintf('The root of the equation is= %0.3f',x)
```

Output

```
Enter the range m:50
Enter the number of iteration n:15
The root of the equation is= 2.279
```

Algorithm 2 (a and b are given)

Step 1: Start the program
Step 2: Define the function $f(x)$
Step 3: Read a and b (lower and upper limits of the interval containing the real root)
Step 4: Read n (number of iterations)
Step 5: For $(i = 1, i \le n + 1, i{+}{+})$
$x = (a*f(b) - b*f(a))/(f(b) - f(a))$
If $f(b)f(x) < 0$,
$a = x$
Else
$b = x$
Step 6: Print x
Step 7: Stop

Program Code: The program code requires the declaration of function $f(x)$. For different functions, the program code will vary in the function declaration section. Therefore, we have to consider a specific function to construct the program. For example, let us consider the following problem:

Find a real root of the equation $x^3 - 3x - 5 = 0$ *between* 2 *and* 3, *correct to* 3 *decimal places by the regula falsi method.*

The code for the program is given by:

```
clc;
f=@(x)[x^3-3*x-5];
a=input('Enter the value of a:');
b=input('Enter the value of b:');
n=input('Enter the number of iteration n:');
for i=1:n+1
    x=(a*f(b)-b*f(a))/(f(b)-f(a));
    if(f(a)*f(x)<0)
        b=x;
    else
        a=x;
    end
end
fprintf('The root of the equation is= %0.3f',x)
```

Output

```
Enter the value of a:2
Enter the value of b:3
Enter the number of iteration n:15
The root of the equation is= 2.279
```

ASSIGNMENT 8

Write MATLAB programs to implement the regula falsi method to find the real root of the following equations:

(a) $x^3 - 3x + 1.06 = 0$, correct to 3 decimal places.
(b) $x^3 - 2.1x - 9 = 0$, correct to 3 decimal places.
(c) $x^4 + x - 35.4 = 0$, correct to 3 decimal places between 3 and 4.
(d) $x^3 - 5x - 7 = 0$, correct to 3 decimal places between 2 and 3.

The Secant Method

Method: To find the real root of $f(x) = 0$ in the interval [a, b] with $f(a)f(b) < 0$ and tolerance ε, the iterative formula for a better root is given by

$$x = \frac{af(b) - bf(a)}{f(b) - f(a)}$$

If ($|f(x)| < \varepsilon$), root found and stop iteration

Else
Set $a = x$, $b = b$
Continue until $|f(x)| < \varepsilon$

Algorithm 1 (a and b are not given)
Step 1: Start the program
Step 2: Define the function $f(x)$
Step 3: Read m (possible upper limit of the interval containing the root)
Step 4: For ($i = -m$, $i \le m$, i++)
Set $a = i$ and $b = i + 1$
If $f(a)*f(b) < 0$
Break
End i
Step 5: Read n (number of iterations)
Step 6: For ($i = 1$, $i \le n + 1$, i++)
$x = (a*f(b) - b*f(a))/(f(b) - f(a))$
$a = x$
Step 7: Print x
Step 8: Stop

Program Code: The program code requires the declaration of function $f(x)$. For different functions, the program code will vary in the function declaration section. Therefore, we have to consider a specific function to construct the program. For example, let us consider the following problem:

Find a real root of the equation $x^3 - 3x - 5 = 0$, *correct to* 3 *decimal places by the regula falsi method.*

The code for the program is given by:

```
clc;
f=@(x)[x^3-3*x-5];
m=input('Enter the range m:');
for i=-m:1:m
    a=i;
    b=i+1;
    if(f(a)*f(b)<0)
        break
    end
end
n=input('Enter the number of iteration n:');
for i=1:n+1
    x=(a*f(b)-b*f(a))/(f(b)-f(a));
    a=x;
end
fprintf('The root of the equation is= %0.3f',x)
```

Output

```
Enter the range m:50
Enter the number of iteration n:15
The root of the equation is= 2.279
```

Algorithm 2 (a and b are given)

Step 1: Start the program
Step 2: Define the function $f(x)$
Step 3: Read a and b (lower and upper limits of the interval containing the real root)
Step 4: Read n (number of iterations)
Step 5: For ($i = 1, i \leq n + 1, i$++)
$$x = (a*f(b) - b*f(a))/(f(b) - f(a))$$
$$a = x$$
Step 6: Print x
Step 7: Stop

Program Code: The program code requires the declaration of function $f(x)$. For different functions, the program code will vary in the function declaration section. Therefore, we have to consider a specific function to construct the program. For example, let us consider the following problem:

Find a real root of the equation $x^3 - 3x - 5 = 0$ *between* 2 *and* 3, *correct to* 3 *decimal places by the secant method.*

The code for the program is given by:

```
clc;
f=@(x)[x^3-3*x-5];
a=input('Enter the value of a:');
b=input('Enter the value of b:');
```

Cont'd

Program cont'd

```
n=input('Enter the number of iteration n:');
for i=1:n+1
    x=(a*f(b)-b*f(a))/(f(b)-f(a));
    a=x;
end
fprintf('The root of the equation is= %0.3f',x)
```

Output

```
Enter the value of a:2
Enter the value of b:3
Enter the number of iteration n:15
The root of the equation is= 2.279
```

ASSIGNMENT 9

Write MATLAB programs to implement the secant method to find the real root of the following equations:

(a) $x^3 - 3x + 1.06 = 0$, correct to 3 decimal places.
(b) $x^3 - 2.1x - 9 = 0$, correct to 3 decimal places.
(c) $x^4 + x - 35.4 = 0$, correct to 3 decimal places between 3 and 4.
(d) $x^3 - 5x - 7 = 0$, correct to 3 decimal places between 2 and 3.

Newton–Raphson Method

Method: To find a real root of $f(x) = 0$ with the initial guess a and tolerance ε, the iterative formula for a better root is given by

$$x = a - \frac{f(a)}{f'(a)}$$

If ($|f(x)| < \varepsilon$), root found and stop iteration
Else
Set $a - x$
Continue until $|f(x)| < \varepsilon$

Algorithm (initial guess is not given)

Step 1: Start the program
Step 2: Define the function $f(x)$ and its derivative $f'(x)$
Step 3: Read m (possible upper limit of the interval containing the root)
Step 4: For ($i = -m, i \leq m, i{+}{+}$)
  Set $a = i$ and $b = i + 1$
   If $f(a)*f(b) < 0$
  Break
 End i
Step 5: Set $x0 = (a + b)/2$ as the initial guess
Step 6: Read n (number of iterations)
Step 7: For ($i = 1, i \leq n + 1, i{+}{+}$)

$$x = x0 - \frac{f(x0)}{f'(x0)}$$

$x0 = x$

Step 8: Print x

Step 9: Stop

Program Code: The program code requires the declaration of function $f(x)$. For different functions, the program code will vary in the function declaration section. Therefore, we have to consider a specific function to construct the program. For example, let us consider the following problem:

Find a real root of the equation $x^3 - 3x - 5 = 0$, correct to 3 decimal places by the Newton–Raphson method.

The code for the program is given by:

```
clc;
f=@(x)[x^3-3*x-5];
g=@(x)[3*(x^2)-3];
m=input('Enter the range m:');
for i=-m:1:m
    a=i;
    b=i+1;
    if(f(a)*f(b)<0)
        break
    end
end
x0=(a+b)/2;
n=input('Enter the number of iteration n:');
for i=1:n+1
    x=x0-f(x)/g(x);
    x0=x;
end
fprintf('The root of the equation is= %0.3f',x)
```

Output

```
Enter the range m:50
Enter the number of iteration n:15
The root of the equation is= 2.279
```

ASSIGNMENT 10

Write MATLAB programs to implement the Newton–Raphson method to find the real root of the following equations:

(a) $x^3 - 3x + 1.06 = 0$, correct to 3 decimal places.

(b) $x^3 - 2.1x - 9 = 0$, correct to 3 decimal places.

(c) $x^4 + x - 35.4 = 0$, correct to 3 decimal places between 3 and 4.

(d) $x^3 - 5x - 7 = 0$, correct to 3 decimal places taking 2 as the initial approximation.

Euler's Method

Method: Let us consider a first order and first degree ODE as $\frac{dy}{dx} = f(x, y)$ with initial condition $y(x_0) = y_0$. Let us suppose that we have to find the approximate numerical value of y, say y_n at the point $x = x_n$. We divide the range $[x_0, x_n]$ into n equal subintervals of length h by the points $x_0, x_1, \ldots, x_n$, where $x_i = x_0 + ih$, $(i = 1, 2, \ldots, n)$. The corresponding values of the function are $y_0 = f(x_0)$, $y_1 = f(x_1)$, …, $y_n = f(x_n)$.

Then, assuming $f(x, y) \approx f(x_{n-1}, y_{n-1})$ in $[x_{n-1}, x_n]$, the general formula is given by

$$y_n = y_{n-1} + hf(x_{n-1}, y_{n-1}),\ n = 1, 2, \ldots, n$$

Algorithm

Step 1: Start the program
Step 2: Define the function $f(x, y)$
Step 3: Read $x0$, $y0$ (initial values of x, y), h (step length), xn (final value of x)
Step 4: If($x0 <= xn$)
$y0 = y0 + h * f(x0, y0)$
$x0 = x0 + h$
Step 5: Display $y0$
Step 6: Stop

Program Code: The program code requires the declaration of function $f(x, y)$. For different functions, the program code will vary in the function declaration section. Therefore, we have to consider a specific function to construct the program. For example, let us consider the following problem:

Find y when $x = 0.1$, *given* $y = 1$ *when* $x = 0$ *and* $\frac{dy}{dx} = 3x + y^2$, $h = 0.0001$ *by Euler's method.*
The code for the program is given by:

```
clc;
f=@(x,y)[(3*x)+(y^2)];
x0=input('Enter the initial value of x:');
y0=input('Enter the initial value of y:');
xn=input('Enter the final value of x:');
n=input('Enter the number of subintervals n:');
h=(xn-x0)/n;
for i=1:n
    if(x0<=xn)
    y1=y0+h*f(x0,y0);
    x0=x0+h;
    y0=y1;
    else
        break
    end
end
fprintf('The value of y at x= %0.3f is= %0.4f',xn,y0)
```

Output

```
Enter the initial value of x:0
Enter the initial value of y:1
Enter the final value of x:0.1
Enter the number of subintervals n:120
The value of y at x= 0.100 is= 1.1270
```

ASSIGNMENT 11

Construct MATLAB programs to implement Euler's method for the following initial value problems:

(a) Calculate $y(0.4)$, correct to 4 decimal places, where $\frac{dy}{dx} = \log_{10}(x+y), y(0)=1$, taking $h = 0.001$.

(b) Find $y(0.1)$, correct to 4 decimal places when $\frac{dy}{dx} = \frac{y-x}{y+x}, y(0)=1$, taking $h = 0.001$.

(c) Evaluate $y(2)$, correct to 4 decimal places from $\frac{dy}{dx} = \frac{1}{2}(x+y)$ with $y(0) = 2$, taking $h = 0.001$.

(d) Obtain y at $x = 0.01$, correct to 4 decimal places from $\frac{dy}{dx} = y + e^x$ with $y(0) = 0$ and $h = 0.001$.

Runge–Kutta Method

Method: Let us consider a first order and first degree ODE as $\frac{dy}{dx} = f(x,y)$ with initial condition $y(x_0) = y_0$.Let us suppose that we have to find the approximate numerical value of y, say y_n, at the point $x = x_n$. We divide the range $[x_0, x_n]$ into n equal subintervals of length h by the points $x_0, x_1, \ldots, x_n$, where $x_i = x_0 + ih$, $(i = 1, 2, \ldots, n)$. The corresponding values of the function are $y_0 = f(x_0)$, $y_1 = f(x_1)$, …, $y_n = f(x_n)$.

The general formula is given by

$$y_n = y_{n-1} + (k_1 + 2k_2 + 2k_3 + k_4)/6, n = 1,2, \ldots, n$$

where
$$k_1 = hf(x_{n-1}, y_{n-1})$$
$$k_2 = hf(x_{n-1} + h/2, y_{n-1} + k_1/2)$$
$$k_3 = hf(x_{n-1} + h/2, y_{n-1} + k_1/2)$$
$$k_4 = hf(x_{n-1} + h, y_{n-1} + k_3)$$

Algorithm

Step 1: Start the program

Step 2: Define the function $f(x, y)$

Step 3: Read $x0$, $y0$ (initial values of x, y), h (step length), xn (final value of x)

Step 4: If($x0$ < = xn)

$k1 = h * f(x0, y0)$

$k2 = h * f(x0 + h/2, y0 + k1/2)$

$k3 = h * f(x0 + h/2, y0 + k2/2)$

$k = h * f(x0 + h, y0 + k3)$

$k = (k1 + 2 * k2 + 2 * k3 + k4)/6$
$y0 = y0 + k$
$x0 = x0 + h$

Step 5: Display $y0$
Step 6: Stop

Program Code: The program code requires the declaration of function $f(x, y)$. For different functions, the program code will vary in the function declaration section. Therefore, we have to consider a specific function to construct the program. For example, let us consider the following problem:

Find y when $x = 0.1$*, given* $y = 1$ *when* $x = 0$ *and* $\frac{dy}{dx} = 3x + y^2$, $h = 0.0001$ *by fourth order Runge–Kutta method.*

The code for the program is given by:

```
clc;
f=@(x,y) [(3*x)+(y^2)];
x0=input('Enter the initial value of x:');
y0=input('Enter the initial value of y:');
xn=input('Enter the final value of x:');
n=input('Enter the number of subintervals n:');
h=(xn-x0)/n;
for i=1:n
    if(x0<=xn)
    k1=h*f(x0,y0);
    k2=h*f(x0+h/2,y0+k1/2);
    k3=h*f(x0+h/2,y0+k2/2);
    k4=h*f(x0+h,y0+k3);
    k=(k1+2*k2+2*k3+k4)/6;
    y1=y0+k;
    x0=x0+h;
    y0=y1;
    else
       break
    end
end
fprintf('The value of y at x= %0.3f is= %0.4f',xn,y0)
```

Output

```
Enter the initial value of x:0
Enter the initial value of y:1
Enter the final value of x:0.1
Enter the number of subintervals n:120
The value of y at x= 0.100 is= 1.1273
```

ASSIGNMENT 12

Construct MATLAB programs to implement fourth order Runge–Kutta method for the following initial value problems:

(a) Calculate $y(0.4)$, correct to 4 decimal places, where $\dfrac{dy}{dx} = \log_{10}(x+y), y(0)=1$, taking $h = 0.001$.

(b) Find $y(0.1)$, correct to 4 decimal places when $\dfrac{dy}{dx} = \dfrac{y-x}{y+x}, y(0)=1$, taking $h = 0.0001$.

(c) Evaluate $y(2)$, correct to 4 decimal places from $\dfrac{dy}{dx} = \dfrac{1}{2}(x+y)$ with $y(0) = 2$, taking $h = 0.001$.

(d) Obtain y at $x = 0.01$, correct to 4 decimal places from $\dfrac{dy}{dx} = y + e^x$ with $y(0) = 0$ and $h = 0.001$.

Fitting the Straight Line $y = a + bx$

Method: Let (x_i, y_i), $i = 1, 2, \ldots, n$ be n observations of the variables x and y and $\boldsymbol{y = a + bx}$ be the straight line to be fitted on the basis of the observed data. By the principle of least squares, we get the normal equations as

$$\sum y = na + b\sum x \tag{1}$$

and

$$\sum xy = a\sum x + b\sum x^2 \tag{2}$$

Solving (1) and (2), we obtain

$$a = \left(\sum x^2 \sum y - \sum x \sum xy\right) \Big/ \left(n\sum x^2 - \left(\sum x\right)^2\right)$$

and

$$b = \left(n\sum xy - \sum x \sum y\right) \Big/ \left(n\sum x^2 - \left(\sum x\right)^2\right)$$

Therefore, the required equation is $y = a + bx$

Algorithm

Step 1: Start the program
Step 2: Read n (no. of points)
Step 3: For $i = 0(1)n - 1$
  Read $x[i]$ & $y[i]$
 End i
Step 4: Assign $sumx = 0.0$, $sumxy = 0.0$, $sumx2 = 0.0$
Step 5: For $i = 0(1)n - 1$
  $sumx = sumx + x[i]$
  $sumy = sumy + y[i]$
  $sumxy = sumxy + x[i] * y[i]$
  $sumx2 = sumx2 + x[i] * x[i]$
 End i
Step 6: $a = (sumx2 * sumy - sumx * sumxy)/(n * sumx2 - sumx * sumx)$
 $b = (n * sumxy - sumx * sumy)/(n * sumx2 - sumx * sumx)$
Step 7: Print the straight line $y = a + b * x$
Step 8: Stop

Program Code

```
clc;
n=input('Enter the number of points n:');
x=zeros(n+1,n+1);
y=zeros(n+1,n+1);
for i=1:n
   x(i)=input('Enter the value of x:');
end
for i=1:n
   y(i)=input('Enter the value of y:');
end
sx=0;sy=0;sxy=0;sx2=0;
for i=1:n
   sx=sx+x(i);
   sy=sy+y(i);
   sxy=sxy+x(i)*y(i);
   sx2=sx2+x(i)*x(i);
end
a=(sx2*sy-sx*sxy)/(n*sx2-sx*sx);
b=(n*sxy-sx*sy)/(n*sx2-sx*sx);
fprintf('The straight line is: Y=%0.1f + %0.1fX',a,b)
```

Output: For output, let us consider the following problem:

Fit a straight line to the following data considering x as an independent variable:

x	0	5	10	15	20	25
y	12	15	17	22	24	30

The output will be:

```
Enter the number of points n:6
Enter the value of x:0
Enter the value of x:5
Enter the value of x:10
Enter the value of x:15
Enter the value of x:20
Enter the value of x:25
Enter the value of y:12
Enter the value of y:15
Enter the value of y:17
Enter the value of y:22
Enter the value of y:24
Enter the value of y:30
The straight line is: Y=11.3 + 0.7X
```

ASSIGNMENT 13

Write a program in MATLAB to fit a straight line for each of the following:

(a)

x	0	1	2	3	4
y	1.1	1.9	2.7	3.5	4.3

(b)

x	0	2	4	6	8
y	10	12	14	16	18

(c)

x	0	2	3	4	7	8
y	4	26	58	112	466	668

Fitting the Curve $y = ax + bx^2$

Method: Let (x_i, y_i), $i = 1, 2, \ldots, n$ be n observations of the variables x and y and $\mathbf{y = ax + bx^2}$ be the curve to be fitted on the basis of the observed data. By the principle of least squares, we get the normal equations as

$$\sum y = a\sum x + b\sum x^2 \qquad (1)$$

and

$$\sum xy = a\sum x^2 + b\sum x^3 \qquad (2)$$

Solving (1) and (2), we obtain

$$a = \left(\sum x^3 \sum y - \sum xy \sum x^2\right) \Big/ \left(\sum x^3 \sum x - \left(\sum x^2\right)^2\right)$$

and

$$b = \left(\sum xy \sum x - \sum x^2 \sum y\right) \Big/ \left(\sum x^3 \sum x - \left(\sum x^2\right)^2\right)$$

Therefore, the required equation is $y = a + bx$.

Algorithm

Step 1: Start the program

Step 2: Read n (no. of points)

Step 3: For i = 0(1)n – 1

    Read $x[i]$ & $y[i]$

  End i

Step 4: Assign *sumx* = 0.0, *sumy* = 0.0, *sumxy* = 0.0, *sumx2* = 0.0, *sumx3* = 0.0

Step 5: For i = 0(1)n – 1

    sumx = *sumx* + *x*[*i*]

    sumy = *sumy* + *y*[*i*]

    sumxy = *sumxy* + *x*[*i*]**y*[*i*]

*sumx*2 = *sumx*2 + *x*[*i*]**x*[*i*]
*sumx*3 = *sumx*3 + *x*[*i*]**x*[*i*]**x*[*i*]
End *i*

Step 6: *a* = (*sumx*3**sumy* – *sumxy***sumx*2)/(*sumx*3**sumx* – *sumx*2**sumx*2)
b = (*sumxy***sumx*–*sumx*2**sumy*)/(*sumx*3**sumx*–*sumx*2**sumx*2)

Step 7: Print the straight line *y* = *a***x* + *b***x*^2

Step 8: Stop

Program Code

```
clc;
n=input('Enter the number of points n:');
x=zeros(n+1,n+1);
y=zeros(n+1,n+1);
for i=1:n
   x(i)=input('Enter the value of x:');
end
for i=1:n
   y(i)=input('Enter the value of y:');
end
sx=0;sy=0;sxy=0;sx2=0;sx3=0;
for i=1:n
   sx=sx+x(i);
   sy=sy+y(i);
   sxy=sxy+x(i)*y(i);
   sx2=sx2+x(i)*x(i);
   sx3=sx3+x(i)*x(i)*x(i);
end
a=(sx3*sy-sxy*sx2)/(sx3*sx-sx2*sx2);
b=(sxy*sx-sx2*sy)/(sx3*sx-sx2*sx2);
fprintf('The straight line is: Y=%0.1fX + %0.1fX^2',a,b)
```

Output: For output, let us consider the following problem:

Fit a curve of the form $y = ax + bx^2$ to the following data considering x as an independent variable:

x	0	1	2	3	4
y	8	12.5	18	24.5	32

The output will be:

```
Enter the number of points n:5
Enter the value of x:0
Enter the value of x:1
Enter the value of x:2
Enter the value of x:3
Enter the value of x:4
Enter the value of y:8
```

Cont'd

Output cont'd

```
Enter the value of y:12.5
Enter the value of y:18
Enter the value of y:24.5
Enter the value of y:32
The straight line is: Y=20.0X + -3.5X^2
```

ASSIGNMENT 14

Write a program in MATLAB to a curve of the form $y = ax + bx^2$ for the following:

(a)

x	1	2	3	4	7
y	2	4	8	6	128

(b)

x	0	1	2	3	5
$f(x)$	2	3	12	65	147

(c)

x	0.1	0.3	0.5	0.6	0.7
y	0.53	0.61	0.69	0.72	0.75

10.2 INTRODUCTION TO SCILAB

SCILAB is a programming language associated with a rich collection of numerical algorithms covering many aspects of scientific computing problems. From a software point of view, SCILAB is an interpreted language. This generally speeds up the development process, because the user directly accesses a high-level language, with a rich set of features provided by the library. SCILAB users can develop their own modules so they can solve their particular problems. The SCILAB language allows dynamic compiling and linking of other languages such as FORTRAN and C in such a way that the external libraries can be used as if they were a part of SCILAB built-in features.

From a scientific point of view, SCILAB comes with many features. The following is a short list of its capabilities:

- Linear algebra, sparse matrices
- Polynomials and rational functions
- Interpolation, approximation
- Linear, quadratic and non-linear optimisation
- Ordinary differential equation solver and differential algebraic equation solver
- Classic and robust control, linear matrix inequality optimisation
- Differentiable and non-differentiable optimisation
- Signal processing
- Statistics

SCILAB provides many graphics features, including a set of plotting functions, which allow the creation of 2D and 3D plots as well as user interfaces.

10.2.1 SCILAB Overview

The basic configuration of SCILAB is quite similar to that of MATLAB. In SCILAB, we use the same syntax as in MATLAB. If we start SCILAB, the command prompt '-->' appears in the main window.

10.2.2 Operating the SCILAB Package

The SCILAB package primarily involves the construction of matrices. When we start SCILAB, the command prompt '-->' will appear. A scalar in MATLAB is a $\mathbf{1}\times\mathbf{1}$ matrix and a vector is a $\mathbf{1}\times\mathbf{n}$ or $\mathbf{n}\times\mathbf{1}$ matrix. To construct a matrix in SCILAB, we use the matrix constructor operator **[]**.We will now discuss the procedure used to construct different types of matrices.

(i) One-dimensional row matrix: Create a row in the matrix constructor operator [] by entering elements separated by a **comma** or **space**. For example, if we want to create a one-row matrix of four elements such as A= (1, 2, 3, 4) in SCILAB, we must follow the procedure given below:

1. Type **A** = [1 2 3 4] after the command prompt '-->', i.e., --> A = [1 2 3 4]
2. Press the ENTER button. The screen will display

```
A =
    1.  2.  3.  4.
```

(ii) One-dimensional column matrix: This is quite similar to the previous type. The only difference is that in this case, each element is separated by a semi-colon.

For example, the column matrix $C = \begin{pmatrix} 1 \\ 2 \\ 3 \\ 4 \end{pmatrix}$ is stored in SCILAB by typing `C = [1;2;3;4]` after the command prompt `'-->'`, as shown below:

1. `--> C = [1;2;3;4]`
2. Press the ENTER button. The screen will display

```
C =
    1.
    2.
    3.
    4.
```

(iii) Multi-dimensional matrix: This is a combination of procedures (i) and (ii). In this case, each row is separated by a semi-colon. However, it should be noted that all rows must contain the same number of elements.

For example, consider the matrix $A = \begin{pmatrix} 1 & 2 & 3 & 4 \\ 5 & 6 & 7 & 8 \\ 9 & 10 & 11 & 12 \end{pmatrix}$. Clearly, the matrix has 3 rows and 4 columns and its representation in SCILAB can be done by typing `A = [1 2 3 4;5 6 7 8;9 10 11 12]` after the command prompt '`-->`', as shown below:

```
--> A = [1 2 3 4;5 6 7 8;9 10 11 12]
 A =
     1.    2.    3.    4
     5.    6.    7.    8
     9.   10.   11.   12
```

Another syntax used to write the above matrix A is shown below:

```
--> A = [1 2 3 4
--> 5 6 7 8
--> 9 10 11 12]
 A =
     1.    2.    3.    4
     5.    6.    7.    8
     9.   10.   11.   12
```

Note: If we are not interested in displaying the matrices, we must insert a **semi-colon** at the end of []. Thus, if we type `A = [1 2 3 4;5 6 7 8;9 10 11 12];` after the command prompt '-->', SCILAB will store the elements of the matrix but will not display it.

10.2.3 Operations on Matrices

Operations such as addition, subtraction and multiplication of matrices are the same as in MATLAB. Here, we provide some examples to illustrate these operations.

EXAMPLE 10.12 Evaluate C = A + B, D = 2A – B, where A = (2 4 3 1) and B = (1 –3 2 1).

Solution:

```
-->A = [2 4 3 1]
 A =
     2.  4.  3.  1
-->B = [1 -3 2 1]
 B =
     1.  -3.  2.  1
-->C = A+B
 C =
     3.  1.  5.  2
-->D = 2*A-B
 D =
     3.  5.  4.  1
```

EXAMPLE 10.13 Using SCILAB, evaluate C = AB and D = BA, where $A = \begin{pmatrix} 5 & 2 & 6 \\ 1 & 4 & 2 \end{pmatrix}$ and $B = \begin{pmatrix} 1 & 7 \\ 6 & -2 \\ 5 & 4 \end{pmatrix}$.

Solution:

```
-->A = [5 2 6;1 4 2]
```

```
A =
    5.  2.  6
    1.  4.  2
-->B = [1 7;6 -2;5 4]
 B =
    1.   7
    6.  -2
    5.   4
-->C = A*B
 C =
    47.  55
    35.   7
-->D = B*A
 D =
    12.    30.  20
    28.     4.  32
    29.  26.  38
```

EXAMPLE 10.14 Find the transpose and inverse of the matrix $A = \begin{pmatrix} 5 & 2 \\ 7 & 3 \end{pmatrix}$.

Solution:

```
-->A = [5 2;7 3]
 A =
    5.  2
    7.  3
-->B = A'
 B =
    5.  7
    2.  3
-->C = inv(A)
 C =
    3.   -2
   -7.       5
```

EXAMPLE 10.15 Find A^{-1} when $A = \begin{pmatrix} 3 & 4 \\ 8 & 11 \end{pmatrix}$ using MATLAB, and check the relation $AA^{-1} = I$, where I is the identity matrix of order 2.

Solution:

```
-->A = [3 4;8 11]
 A =
    3.    4
    8.   11
-->B = inv(A)
 B =
    11.   -4
    -8.    3
```

```
-->C = A*B
  C =
     1.  0
     0.  1
```

10.3 INTRODUCTION TO LabVIEW

LabVIEW is an abbreviation of **lab**oratory **v**irtual **i**nstrumentation **e**ngineering **w**orkbench. It is a graphical computing environment for instrumentation. It allows us to control the instruments used in labs. In addition, LabVIEW enables us to obtain the information recorded in the computer; we can then use a variety of math functions to alter it, fit it and analyse it using statistical functions, plot it on a graph, and store it in a file.

10.3.1 LabVIEW Overview

LabVIEW has a graphical environment where we can create files, called VIs, and this is quite similar to the creation of an instrument.

LabVIEW includes the following tools to help us analyse data:

- Analysis VIs for differential equations, optimisation, curve fitting, calculus, linear algebra and statistics
- Signal processing VIs for filtering, windowing, transforms, peak detection, harmonic analysis and spectrum analysis

10.3.2 Operating the LabVIEW Package

When we start LabVIEW on a computer, we type the commands given below:

```
New>>Blank VI
```

LabVIEW programs are called virtual instruments (VIs). Each VI has two windows:

- **Front panel**
  - User interface (UI)
    - Controls = Inputs
    - Indicators = Outputs
- **Block diagram**
  - Graphical code
    - Data travels on wires from controls through functions to indicators
    - Blocks execute by dataflow

With the help of these two windows, users can create VIs, debug them, find the errors and develop graphs.

10.4 INTRODUCTION TO MATHEMATICA

Mathematica is an immense computer software package for performing mathematical computation and exploration. It contains hundreds of mathematical functions, commands for producing graphics, and a complete programming language. By virtue of this software package, users can solve technical computing problems and analyse data graphically in various aspects such as **numerical analysis, time series, linear algebra and optimisation.**

10.4.1 Operating the Mathematica Package

At the most basic level, Mathematica may be used interactively, like a calculator. Commands to be evaluated are entered into **input cells,** which are displayed in bold. They are evaluated by pressing the ENTER key. (Note that **ENTER** is different from **RET** which is used to start a new line). Results are displayed in **output cells**. Here is a simple example:

```
2 + 2
4
```

You can enter longer expressions over several lines. You can also include comments in input cells by enclosing them in **(*.*)**

```
(5 * 7 - 32/6)/4
(* Here is a comment *)
```

$$\frac{89}{12}$$

Notice that we obtain a fraction. Mathematica is capable of exact computations, rather than decimal approximations. We can always obtain numerical approximations using the **N** function (for numerical). In the following line, **%** refers to the previous result and **// N** passes that result to the **N** function.

```
%//N
7.41667
```

- Mathematica's ability to perform exact computations is truly impressive.

```
2 ^ 1000
10 715 086 071 862 673 209 484 250 490 600 018 105 614 048 117 055
336 074 437 503 883 703 510 511 249 361 224 931 983 788 156 958
581 275 946 729 175 531 468 251 871 452 856 923 140 435 984 577
574 698 574 803 934 567 774 824 230 985 421 074 605 062 371 141
877 954 182 153 046 474 983 581 941 267 398 767 559 165 543 946
077 062 914 571 196 477 686 542 167 660 429 831 652 624 386 837
205 668 069 376
```

Approximations may be computed to any desired degree of accuracy. Here, we use the **N** function again, along with an optional second argument, to obtain 100 decimal digits of π.

```
N [Pi, 100]
3.141592653589793238462643383279502884197169399375105820974944592
307816406286208998628034825342117068
```

- Mathematica has extensive knowledge of mathematical functions.

```
Sin [Pi/3]
```

$$\frac{\sqrt{3}}{2}$$

Note: Built-in function names are always capitalised and the arguments are enclosed in square brackets.

- Mathematica can also be used for algebraic operations.

```
Expand [(x + 1)^10]
```

$1 + 10\ x + 45\ x^2 + 120\ x^3 + 210\ x^4 + 252\ x^5 + 210\ x^6 + 120\ x^7 + 45\ x^8 + 10\ x^9 + x^{10}$

```
Factor [x^4 - x^3 - 2 x - 4]
```

$(-2 + x)(1 + x)(2 + x^2)$

- Mathematica can perform calculus.

```
D [x^2, x]
2x
Integrate [Sin[x], {x, 0, Pi}]
2
```

- Mathematica can compute sums.

```
Sum [2/9^n, {n, 1, Infinity}]
```

$\frac{1}{4}$

Brackets [], braces {}, and parentheses (): Brackets, braces, and parentheses all have distinct purposes in Mathematica. Brackets `[]` are used to enclose the arguments of a function, such as:

```
N [Tan[1], 20]
1.5574077246549022305
```

Parentheses `()` are used for grouping mathematical expressions, such as:

```
(2 * 5 - 1)/3
3
```

Braces `{}` are used to form lists. Lists play a very important role in Mathematica as they are a fundamental data structure. Many functions automatically act on the individual elements of a list.

```
N [{Sqrt[2], E}, 20]
{81 4142135623730950408, 2.7182818284590452354}
```

Defining constants and functions: Constants may be defined in an intuitive way.

```
c = 7
7
```

The symbol **c** will now be treated as 7.

```
c + 3
10
```

This is how the function $f(x) = x + 2$ is defined:

```
f[x_]: = x + 2
```

We may now enter any value we want for **x**.

```
f[3]
5
f[y²]
2 + y²
```

Solving equations: Mathematica has powerful built-in techniques for solving equations algebraically and numerically. The basic command is `Solve`. Here is an example of how to use `Solve` to find the roots of a polynomial:

```
Solve [x³ - 2 x - 4 == 0, x]
{{x → -1 - i}, {x → -1 + i}, {x → 2}}
```

Note that the `NSolve` command is similar, but returns numerical approximations.

```
N Solve [x⁵ - 2 x - 3 == 0, x]
{{x→-0.958532 -0.498428 i}, {x→-0.958532 + 0.498428 i},{x→0.246729 -
1.32082}, {x→0.246729 + 1.32082 i},{x→1.42361}}
```

Appendix

Fourier Series and Fourier Transforms

INTRODUCTION

In many engineering fields such as conduction of heat and electrodynamics, one of the important aspects is the study of periodic phenomena. In such problems of study, it becomes necessary to express a function as a series of **sines** and **cosines**. The Fourier series is one such series.

PERIODIC FUNCTIONS

A real-valued function $f(x)$ is said to be periodic if there exists a least positive real number T such that $f(x+T)=f(x)$ for all values of x. For example, $f(x)=\sin x$ is a periodic function with period 2π because $\sin(2\pi+x)=\sin x$. Moreover, if $f(x)$ is a periodic function with period T, then $f(ax)$ is a periodic function with period $\frac{T}{a}$.

FOURIER SERIES OF A REAL-VALUED FUNCTION

Let $f(x)$ be defined and integrable in the interval $(-T, T)$. We can extend the function to a periodic function of period $2T$ by defining $f(x+2T)=f(x)$ for all values of x. Then, the Fourier series of $f(x)$ is given by

$$\frac{a_0}{2}+\sum_{n=1}^{\infty}\left(a_n\cos\frac{n\pi x}{T}+b_n\sin\frac{n\pi x}{T}\right)$$

where a_n, b_n are called Fourier coefficients and are determined by

$$a_0=\frac{1}{T}\int_{-T}^{T}f(x)dx$$

$$a_n=\frac{1}{T}\int_{-T}^{T}f(x)\cos\frac{n\pi x}{T}dx$$

$$b_n=\frac{1}{T}\int_{-T}^{T}f(x)\sin\frac{n\pi x}{T}dx$$

When $T = \pi$, i.e., $f(x)$ is defined and integrable in the interval $(-\pi, \pi)$, the Fourier series of $f(x)$ is given by

$$\frac{a_0}{2} + \sum_{n=1}^{\infty}(a_n \cos nx + b_n \sin nx)$$

where

$$a_0 = \frac{1}{\pi}\int_{-\pi}^{\pi} f(x)\,dx$$

$$a_n = \frac{1}{\pi}\int_{-\pi}^{\pi} f(x)\cos nx\,dx$$

$$b_n = \frac{1}{\pi}\int_{-\pi}^{\pi} f(x)\sin nx\,dx$$

ODD FUNCTION AND EVEN FUNCTION

- **Even function:** A real-valued function $f(x)$ is said to be even if $f(-x) = f(x)$. For example,
  (a) $\cos x$, $\sec x$, etc.
  (b) Any constant or **even** power of x such as x^2, x^4, etc.
- **Odd function:** A real-valued function $f(x)$ is said to be odd if $f(-x) = -f(x)$. For example,
  (a) $\sin x$, $\tan x$, etc.
  (b) Any **odd** power of x such as x, x^3, etc.

We use the following rule to obtain the Fourier series of an odd or even function:

$$\int_{-a}^{a} f(x)\,dx = \begin{cases} 0 \text{ if } f(x) \text{ is odd} \\ 2\int_{0}^{a} f(x)\,dx \text{ if } f(x) \text{ is even} \end{cases}$$

Fourier Series of an Odd Function If $f(x)$ is an odd function in the interval $(-\pi, \pi)$, then $a_0 = 0$, $a_n = 0$ and

$$b_n = \frac{2}{\pi}\int_{0}^{\pi} f(x)\sin nx\,dx$$

Then, its Fourier series is given by

$$\sum_{n=1}^{\infty} b_n \sin nx$$

Fourier Series of an Even Function If $f(x)$ is an even function in the interval $(-\pi, \pi)$, then $b_n = 0$ and

$$a_0 = \frac{2}{\pi}\int_{0}^{\pi} f(x)\,dx$$

$$a_n = \frac{2}{\pi}\int_{0}^{\pi} f(x)\cos nx\,dx$$

Then, its Fourier series is given by

$$\frac{a_0}{2}+\sum_{n=1}^{\infty} a_n \cos nx$$

EXAMPLE 1 Find the Fourier series of $f(x) = x, \ -\pi<x<\pi$.

Solution: We extend the given function $f(x) = x, \ -\pi<x<\pi$ to a periodic function with period 2π by defining $f(x + 2\pi) = f(x)$. Then, its Fourier series is given by

$$\frac{a_0}{2}+\sum_{n=1}^{\infty}\left(a_n \cos nx + b_n \sin nx\right)$$

where

$$a_0 = \frac{1}{\pi}\int_{-\pi}^{\pi} f(x)\,dx$$

$$a_n = \frac{1}{\pi}\int_{-\pi}^{\pi} f(x)\cos nx dx$$

$$b_n = \frac{1}{\pi}\int_{-\pi}^{\pi} f(x)\sin nx dx$$

As $f(x)$ is an odd function, $a_0 = 0$, $a_n = 0$ and

$$b_n = \frac{2}{\pi}\int_{0}^{\pi} f(x)\sin nx dx = \frac{2}{\pi}\int_{0}^{\pi} x\sin nx dx$$

$$= \frac{2(-1)^{n+1}}{n}$$

Therefore, the required Fourier series is

$$2\sum_{n=1}^{\infty}\frac{(-1)^{n+1}}{n}\sin nx$$

EXAMPLE 2 Find the Fourier series of $f(x) = x^2, \ -\pi<x<\pi$.

Solution: We extend the given function $f(x) = x^2, \ -\pi<x<\pi$ to a periodic function with period 2π by defining $f(x + 2\pi) = f(x)$. Then, its Fourier series is given by

$$\frac{a_0}{2}+\sum_{n=1}^{\infty}\left(a_n \cos nx + b_n \sin nx\right)$$

where

$$a_0 = \frac{1}{\pi}\int_{-\pi}^{\pi} f(x)\,dx$$

$$a_n = \frac{1}{\pi}\int_{-\pi}^{\pi} f(x)\cos nx dx$$

$$b_n = \frac{1}{\pi}\int_{-\pi}^{\pi} f(x)\sin nx dx$$

As $f(x)$ is an even function, $b_n = 0$ and

$$a_0 = \frac{2}{\pi}\int_{0}^{\pi} f(x)dx = \frac{2}{\pi}\int_{0}^{\pi} x^2 dx$$

$$= \frac{2\pi^2}{3}$$

$$a_n = \frac{2}{\pi}\int_{0}^{\pi} f(x)\cos nx dx = \frac{2}{\pi}\int_{0}^{\pi} x^2 \cos nx dx$$

$$= \frac{4(-1)^n}{n^2}$$

Therefore, the Fourier series is given by

$$\frac{\pi^2}{3} + 4\sum_{n=1}^{\infty}\frac{(-1)^n}{n^2}\cos nx$$

11.1 DIRICHLET'S CONDITIONS

Though a function defined in a finite interval can be expanded to its Fourier series, it may not be equal to its Fourier series. The conditions under which a function is equal to its Fourier series are given by Dirichlet.

Let a function $f(x)$ be defined and in the interval $[-T, T]$. Then, $f(x)$ will be equal to its Fourier series if it satisfies either of the following two conditions:

(i) $f(x)$ is bounded in $[-T, T]$ and the interval $[-T, T]$ can be broken into a finite number of subintervals such that $f(x)$ is monotonic on each subinterval.

(ii) $f(x)$ has a finite number of infinite discontinuity in $[-T, T]$ and when arbitrary neighbourhoods of these points are excluded, $f(x)$ becomes bounded in the remaining part. This remaining part can then be decomposed into a finite number of subintervals such that $f(x)$ is monotonic in each subinterval. Moreover, the improper integral $\int_{-T}^{T} f(x)dx$ is absolutely convergent.

The above two conditions are known as Dirichlet's conditions.

Note: Neighbourhood of a point c means $[c - \delta, c + \delta]$, where $\delta > 0$ is, however, small.

EXAMPLE 3 Find the Fourier series of $f(x) = x$, $-\pi<x<\pi$ and hence show that $1-\frac{1}{3}+\frac{1}{5}-\cdots=\frac{\pi}{4}$.

Solution: From Example 1, we have the Fourier series of $f(x) = x$, $-\pi<x<\pi$ as

$$2\sum_{n=1}^{\infty}\frac{(-1)^{n+1}}{n}\sin nx = 2\left(\sin x - \frac{\sin 2x}{2} + \frac{\sin 3x}{3} - \cdots\right)$$

We see that the given function is bounded, continuous and increasing in $-\pi<x<\pi$. So it satisfies Dirichlet's condition. Therefore, we have

$$f(x)=2\left(\sin x-\frac{\sin 2x}{2}+\frac{\sin 3x}{3}-\cdots\right), \text{ i.e., } x=2\left(\sin x-\frac{\sin 2x}{2}+\frac{\sin 3x}{3}-\cdots\right) \qquad (1)$$

Putting $x=\frac{\pi}{2}$ in (1), we get

$$\frac{\pi}{2}=2\left(\sin\frac{\pi}{2}-\frac{\sin\pi}{2}+\frac{\sin\frac{3\pi}{2}}{3}-\cdots\right), \text{ i.e., } 1-\frac{1}{3}+\frac{1}{5}-\cdots=\frac{\pi}{4}$$

HALF-RANGE SINE SERIES

If $f(x)$ is defined on $(0, \pi)$, we extend the domain of definition to $(-\pi, 0)$ by $f(-x) = -f(x)$ so that $f(x)$ becomes odd in the interval $(-\pi, \pi)$. Then, $a_0 = 0$, $a_n = 0$ and

$$b_n=\frac{2}{\pi}\int_0^{\pi} f(x)\sin nx\,dx$$

Therefore, the Fourier series will be

$$\sum_{n=1}^{\infty} b_n \sin nx$$

HALF-RANGE COSINE SERIES

If $f(x)$ is defined on $(0, \pi)$, we extend the domain of definition to $(-\pi, 0)$ by $f(-x) = f(x)$ so that $f(x)$ becomes even in the interval $(-\pi, \pi)$. Then $b_n = 0$ and

$$a_0=\frac{2}{\pi}\int_0^{\pi} f(x)\,dx$$

$$a_n=\frac{2}{\pi}\int_0^{\pi} f(x)\cos nx\,dx$$

Then, its Fourier series is given by

$$\frac{a_0}{2}+\sum_{n=1}^{\infty} a_n \cos nx$$

FOURIER TRANSFORMS

Definition 1: The Fourier transform of $f(x)$, $-\infty<x<\infty$ is defined as

$$\mathcal{F}\{f(x)\}=\int_{-\infty}^{\infty} f(x)e^{isx}\,dx=F(s)$$

Definition 2: The Fourier cosine transform of $f(x)$, $0<x<\infty$ is defined as

$$\mathcal{F}_c\{f(x)\} = \int_0^\infty f(x)\cos sx\,dx = F_c(s)$$

Definition 3: The Fourier sine transform of $f(x)$, $0<x<\infty$ is defined as

$$\mathcal{F}_s\{f(x)\} = \int_0^\infty f(x)\sin sx\,dx = F_s(s)$$

Theorem: If $F_c(s)$ and $F_s(s)$ be, respectively, Fourier cosine and sine transforms of a function $f(x)$, $-\infty<x<\infty$, then its Fourier transform $F(s)$ is given by

$$F(s) = \begin{cases} 2F_c(s), \text{if } f(x) \text{ is even} \\ 2iF_s(s), \text{ if } f(x) \text{ is odd} \end{cases}$$

Properties of Fourier Transforms

- **Change of scale property:** If $F(s) = \mathcal{F}\{f(x)\}$, then $\mathcal{F}\{f(ax)\} = \frac{1}{a}F\left(\frac{s}{a}\right)$
- **Shifting property:** If $F(s) = \mathcal{F}\{f(x)\}$, then $\mathcal{F}\{f(x-a)\} = e^{ias}F(s)$

Formulae on Fourier Transforms

- If $F(s) = \mathcal{F}\{f(x)\}$, then $\mathcal{F}\left\{f^{(n)}(x)\right\} = (-is)^n F(s)$
- If $F(s) = \mathcal{F}\{f(x)\}$, then $\mathcal{F}\{xf(x)\} = -i\frac{d}{ds}F(s)$

Problem Set I: Find the Fourier transforms of the following functions:

(i) $f(x) = e^{-x^2/2}$

(ii) $f(x) = e^{-x^2/3}$

(iii) $f(x) = xe^{-4x^2}$

(iv) $f(x) = e^{-2(x-3)^2}$

(v) $f(x) = e^{-4(x-3)^2}$

(vi) $f(x) = e^{-a|x|}$

(vii) $f(x) = 2e^{-|x|} - 3e^{-2|x+2|}$

Problem Set II: Find the Fourier cosine and sine transforms of the following functions:

(i) $f(x) = e^{-ax}$, a>0

(ii) $f(x) = e^{-3x}$

(iii) $f(x) = e^{-2x+6}$

Problem Set III

(i) Find the Fourier cosine transform of $f(x)=\dfrac{1}{1+x^2}$ and hence deduce the Fourier transform of $g(x)=\dfrac{x}{1+x^2}$.

(ii) Find the Fourier sine transform of $f(x)=\begin{cases}0, 0<x<a\\ x, a\le x\le b\\ 0, x>b\end{cases}$.

(iii) Find the Fourier cosine transform of $f(x)=e^{-x^2}$ and hence deduce the Fourier sine transform of $g(x)=xe^{-x^2}$.

(iv) Find the Fourier sine transform of $f(x)=\dfrac{1}{x\left(x^2+a^2\right)}$.

FOURIER INVERSE TRANSFORMS

Definition 1: If $F(s)=\mathcal{F}\{f(x)\}$, then

$$f(x)=\mathcal{F}^{-1}\{F(s)\}=\begin{cases}\dfrac{1}{2\pi}\displaystyle\int_{-\infty}^{\infty}F(s)e^{isx}\,ds, \text{ if } x \text{ is the point of continuity}\\ \dfrac{1}{2}\left[\lim\limits_{t\to x-}f(t)+\lim\limits_{t\to x+}f(t)\right], \text{ if } x \text{ is the point of discontinuity}\end{cases}$$

Definition 2: If $F_c(s)=\mathcal{F}_c\{f(x)\}$, then

$$f(x)=\mathcal{F}_c^{-1}\{F_c(s)\}=\begin{cases}\dfrac{1}{2\pi}\displaystyle\int_{0}^{\infty}F_c(s)\cos sx\,ds, \text{ if } x \text{ is the point of continuity}\\ \dfrac{1}{2}\left[\lim\limits_{t\to x-}f(t)+\lim\limits_{t\to x+}f(t)\right], \text{ if } x \text{ is the point of discontinuity}\end{cases}$$

Definition 3: If $F_s(s)=\mathcal{F}_s\{f(x)\}$, then

$$f(x)=\mathcal{F}_s^{-1}\{F_s(s)\}=\begin{cases}\dfrac{1}{2\pi}\displaystyle\int_{0}^{\infty}F_s(s)\sin sx\,ds, \text{ if } x \text{ is the point of continuity}\\ \dfrac{1}{2}\left[\lim\limits_{t\to x-}f(t)+\lim\limits_{t\to x+}f(t)\right], \text{ if } x \text{ is the point of discontinuity}\end{cases}$$

Problem Set IV: Find the functions whose Fourier transforms are given below:

(i) $\dfrac{s}{1+s^2}$

(ii) $\dfrac{e^{-as}}{s}$

(iii) $\dfrac{1}{s^2+4s+13}$

Problem Set V

(i) Find the Fourier transform of $f(x)=\begin{cases}1, |x| \le 1\\ 0, |x| > 1\end{cases}$.

Hence evaluate (a) $\int_0^\infty \frac{\sin x \cos \lambda x}{x} dx$ and (b) $\int_0^\infty \frac{\sin x}{x} dx$, λ being a constant.

(ii) Find the Fourier transform of $f(x)=\begin{cases}1-x^2, |x| \le 1\\ 0, |x| > 1\end{cases}$.

Hence evaluate $\int_0^\infty \frac{x\cos x - \sin x}{x^3} \cos\frac{x}{2} dx$.

(iii) Find the Fourier sine transform of $(x)=\begin{cases}1, 0 \le x \le \pi\\ 0, x > \pi\end{cases}$.

Hence evaluate $\int_0^\infty \frac{1-\cos \pi s}{s} \sin sx ds$.

Index